THE EARLY TYPE STARS

ASTROPHYSICS AND SPACE SCIENCE LIBRARY

A SERIES OF BOOKS ON THE RECENT DEVELOPMENTS
OF SPACE SCIENCE AND OF GENERAL GEOPHYSICS AND ASTROPHYSICS
PUBLISHED IN CONNECTION WITH THE JOURNAL
SPACE SCIENCE REVIEWS

VOLUME 6

ANNE B. UNDERHILL
Sterrewacht 'Sonnenborgh', Utrecht, The Netherlands

THE EARLY TYPE STARS

SPRINGER-SCIENCE+BUSINESS MEDIA, B.V.

ISBN 978-94-010-3558-3 ISBN 978-94-010-3556-9 (eBook)
DOI 10.1007/978-94-010-3556-9

Originally published by D. Reidel Publishing Company in 1966
Softcover reprint of the hardcover 1st edition 1966

To the memory of Otto Struve
whose energy and enthusiasm for observing and
interpreting the spectra of the early type stars
laid the basis for the work reported in this book

PREFACE

At this time when astronomers are being surprised by the discovery of objects which emit a fabulously large amount of energy, that is the quasi-stellar radio sources and the quasi-stellar galaxies, and when by the means of space vehicles X rays, gamma rays and cosmic rays are being observed to come from the depths of interstellar space, one may ask why write a book about stars. Stars seem to be almost incidental when one looks at the universe in terms of exceedingly great energies. Nevertheless, stars exist. They are accessible to study and they have not yet revealed all their secrets. This is enough to arouse interest and to cause one to try to find answers to the questions which arise.

The early type stars are particularly interesting because they are spendthrift stars pouring out their energy at a great rate. But their brilliance is also their undoing. They must evolve rather quickly, on an astrophysical scale. Thus by studying these stars we are studying a population in change. The implications from the local and from the cosmological viewpoint are important if one wishes to understand the details of stellar evolution and of galactic structure. Perhaps one of the simplest reasons for writing a book about the early type stars is to see if some of the conundrums presented by the spectra of these stars can be unravelled when all the available information is brought together.

There is much information available about the early type stars, for many astronomers have devoted much time to them. One who had a particular interest in the B type stars was Otto Struve. With his great energy he assembled many observations, and with his keen physical insight he isolated many of the spectroscopic phenomena which have proved key factors in understanding stellar spectra. It is with great pleasure that I dedicate this book to the memory of Otto Struve. His interest in the early type stars sparked my interest in them. I think he would have agreed with me that now is the time to look at what the observations of over sixty years have revealed. I hope he would have been pleased at the progress; I am sure he would have unerringly placed his finger on some points and said there we must observe further in order to understand.

I am most grateful for the continuing accurate and enthusiastic help of Dr. J. R. W. Heintze who assisted me with preparing the manuscript for the printer. My thanks go to Mrs. T. Crouwel, Mrs. I. de Haan and Mrs. J. Odijk for typing the manuscript; to Miss G. Dorresteijn who assisted with preparing some of the tables and diagrams, and to Mr. R. Staleman who drew some of the diagrams. I am indebted to the Dominion Astrophysical Observatory for permission to reproduce Figures 2, 3, 13, 25, 26, 31,

41, 42, 43, 44, 45, 46, 47, 54, 55 and 56; to the Springer Verlag for Figure 33; and to the Editors of the Monthly Notices for Figures 17 and 18. The remainder of the diagrams have been prepared at the Utrecht Observatory.

I hope I have given a reasonably complete and fair summary of the observations that have been made of early type stars. Undoubtely I have missed mentioning some papers, possibly some of great significance. I do hope that the authors of such works will excuse me. This book is one person's view of the early type stars. Consequently the viewpoint is more or less unified; unfortunately, however, the vision may not be sharp in all directions. I resign myself to inevitable omissions and errors, with the thought that imperfections, when not too large, lend piquancy to a creative work.

Utrecht, The Netherlands, October 1965 ANNE B. UNDERHILL

TABLE OF CONTENTS

CHAPTER I

AN INTRODUCTION TO THE EARLY TYPE STARS AND TO THE MEANS OF OBSERVING THEM

1. The early type stars

The early type stars are relatively few in number and few of them are apparently bright, but because of their importance for studies of galactic structure and for studies of stellar structure, as well as because of the interesting problems presented by their spectra, numerous research papers have been devoted to them. Here an attempt is made to assemble much of the information that has been found so that the whole picture may be viewed critically. Two types of information exist: astrophysical information about the structure and constitution of early type stars, and kinematical information about the motions and distribution of these stars.

This information is found chiefly by studying the spectra of the early type stars, their colours, and their magnitudes. Some knowledge of the dust and gas in the space between the earth and the stars under study results as a by-product. The technical difficulties of observing the early type stars have imposed very definite limits on our knowledge and it is well to recognize these limitations. Most work done before 1940 was limited to the spectral range 3800 Å to 5000 Å (glass optics, prism spectrographs and blue-sensitive photographic plates). The spectral range which may be investigated effectively from ground-based observatories is from 3050 Å, the ultraviolet cut-off imposed by ozone absorption in the earth's atmosphere, to about 7000 Å, the longward range of fast red-sensitive photographic emulsions.

The early type stars comprise the following spectral classes: B type stars, including Be stars and B type stars with a shell; O type stars, including the Of stars; Wolf-Rayet stars; and subluminous stars containing absorption lines of He I and He II in their spectra.

A typical stellar spectrum of early type is dominated by absorption lines, but the Balmer lines of hydrogen and a few other lines are sometimes seen in emission superposed on the normal absorption lines. Spectroscopically forbidden lines are not usually seen in emission in the spectra of early type stars. The spectra of the Wolf-Rayet stars are dominated by broad, strong emission features. In this group alone of the spectral classes usually collected under the heading "the early type stars" are the absorption lines not the dominant features of the spectrum.

2. Techniques by which information is obtained

Information may be obtained from measurements on stellar spectrograms, from photometric measures, making use of magnitudes and colours measured through broad-band and through narrow-band filters, from astrometric measures of parallax and proper motions, and from the apparent distribution in the sky of the early type stars.

Much information can be obtained from stellar spectrograms. Because the light from the star is usually highly dispersed by prisms or gratings, an efficient spectrograph and recording device is required at the focus of a telescope of great light gathering power if any but the apparently brightest stars are to be studied. The spectra of only a few of the apparently bright early type stars have been studied in detail using high-dispersion spectra. Moderate and low dispersion spectra, having linear dispersions between 30 Å/mm and about 200 Å/mm have been obtained for many early type stars brighter than the 10th magnitude. Very few slit spectra of O and B type stars fainter than the 10th magnitude have been studied. Photometric techniques can be used efficiently with apparently faint stars, and attempts are being made to use the information secured through narrow-band filters to extend our astrophysical knowledge of stars to studies of faint objects.

Astrometric measures are useful only for the relatively nearby stars. Since in the vicinity of the sun, O and B type stars are rare per unit volume, a complete sampling of typical objects cannot be made within the region where astrometric measures have their greatest value. Consequently, astrometric measures of the motions of early type stars are not such a satisfactory source of information for the early type stars as they are for the more abundant, less luminous later type stars.

The apparent distribution of the early type stars is far from uniform. The first spectral type surveys clearly showed that the O and B type stars are confined closely to the plane of the galaxy where the interstellar dust and gas is concentrated. The stars of type B2 and earlier tend to occur in clusters or in the somewhat looser and larger aggregations known as associations. The more numerous, less luminous stars of types B7, B8, and B9 appear to be distributed more uniformly in the volume of space which has so far been examined.

3. Stellar spectrographs

If one is to appreciate fully the value and the limitations of the available observational material it is necessary to know something of the equipment with which the material was obtained. It is recommended that the reader consult the original research papers that will be mentioned to find the exact details of each spectrograph and the quality of the spectrograms which are used for each piece of research that is discussed. Most of the information about the physics of the early type stars has been obtained from the study of spectra. It is essential to have a grasp of the quality of the different spectrograms that have been used because what is known is quite definitely limited by the quality of the spectra that have been available for study.

High linear dispersion, that is 10 Å/mm or better, coupled with high spectral purity is required for astrophysical studies of early type stars. It is also most advantageous to observe a wide wave-length range, preferably from the ozone cut-off in the ultraviolet to the long wave-length end of fast, red-sensitive emulsions near 7000 Å. A few observations of early type stars have been made at wave lengths longer than 7000 Å, but on the whole the infra-red recording devices (photographic emulsions or photoconductive cells) have been too insensitive to permit one to obtain spectra except at very low dispersion. The most efficient tool for astrophysical studies is a high resolution grating spectrograph placed at the coudé focus of a large reflecting telescope. DUNHAM (1956) and BOWEN (1962) have described some of the most efficient designs. Much useful work on early type stars has also been carried out with prism spectrographs of moderate dispersion. The system of spectral classification is based on spectra covering the range 3800–5000 Å. Almost all radial-velocity results are obtained from moderate and low dispersion prismatic spectra covering the same wave-length range.

A schematic diagram of a spectrograph is shown in Figure 1. The slit is illuminated

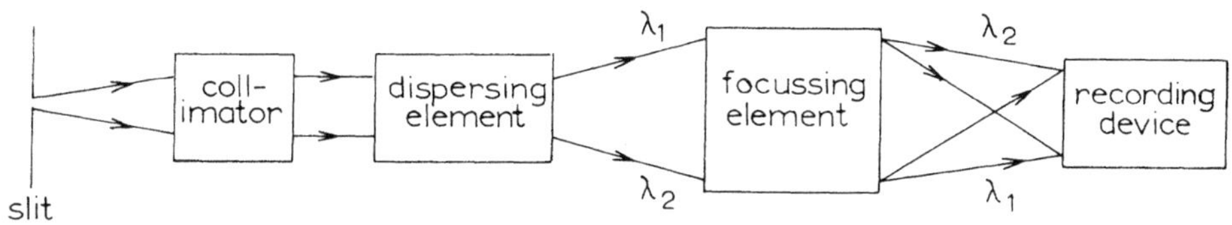

Fig. 1. Schematic diagram of a spectrograph.

by the image of a star. The spectrograph will be more efficient if the slit is opened sufficiently wide to accommodate the whole image. However, a wide slit will reduce the spectral purity. The purity is measured by the width of the projected image of the slit at the recording device. If w is the width of the slit in millimeters, then the projected slit width, p, in angstroms at the recording device is given by

$$p = w\Delta f_{\mathrm{cam}}/f_{\mathrm{coll}}, \tag{I-1}$$

where Δ is the dispersion in angstroms per millimeter at the recording device, f_{cam} is the effective focal length of the camera, and f_{coll} is the effective focal length of the collimator. For astrophysical studies p should be of the order of 0.5 Å or less. Much radial-velocity work and spectral classification work is done with p greater than 1.5 Å. The resolving power of stellar spectrographs is usually limited by the slit width and not by the theoretical resolving power of the dispersing element. A comparison spectrum, usually the iron arc burning in air, is exposed on either side of the stellar spectrum. Since stellar spectrograms are normally used for radial-velocity work as well as for spectrophotometric studies, it is not desirable to have the comparison spectrum and the stellar spectrum superposed. It is most important that the light path of the comparison spectrum and of the light from the star be closely the same, particularly if the spectrograph is attached to the telescope at the Cassegrain focus, the Newtonian focus, or the Prime focus, for the optical axis of the spectrograph may shift slightly as the telescope adopts different attitudes. The great advantage of putting the spectrograph at the coudé focus is that the spectrograph is stationary. However, at the coudé focus the direction of the slit rotates with respect to the sky as the exposure proceeds, and if this rotation is undesirable (for instance when one component of a binary star is being observed) an auxiliary apparatus called an image rotator must be used to counteract this motion.

With modern spectrographs, the collimator and the camera are often mirrors rather than lenses. When mirrors are used there is no problem with chromatic aberration or with the transmission being good in only a restricted wave-length region. However, with a mirror camera, the plate holder assembly requires special design so as to be readily accessible yet not to occult too much light from the beam. With aluminised mirrors it is possible to observe to the ultraviolet limit imposed by the ozone absorption of the earth's atmosphere. A further advantage of mirror optics is that mirrors can be made as large as desired, thus one can use a large beam. BOWEN (1963) has shown that the effective speed of a stellar spectrograph is related to the size of the telescope by the ratio d/D where d is the diameter of the collimator and D is the diameter of the telescope objective. Furthermore, the speed of a spectrograph depends upon the ratio $f_{\mathrm{coll}}/f_{\mathrm{cam}}$. Thus to obtain a highly efficient, fast spectrograph when the telescope objective has a diameter D, one should use a collimator with large aperture and long focal length together with a camera of as short a focal length as possible. To obtain high linear dispersion one requires large angular dispersion from the dispersing element. In practice, the size of the beam that can be used is limited by the size of the available dispersing elements.

The first successful stellar spectrographs used prisms as the dispersing element. Since the largest available aperture was about 3 inches, the speed of the spectrograph was limited by this beam size. Prisms have advantages for work at moderate and low dispersion. First there is the fact that there are no overlapping orders which have to be removed by suitable filter systems; secondly the reduced dispersion in the yellow-red spectral region is an advantage when you wish to scan the full spectrum of an early type star in one exposure because the reduced dispersion compensates for the lower intensity of red and yellow light with respect to blue and violet light in the spectra of the hot stars. Finally, there are not so many stellar absorption lines in the red and yellow parts of the spectra of early type stars as there are in the blue and the violet, thus high dispersion is not necessary in order to obtain a clear impression of what is present in that part of the spectrum.

For high resolution, high dispersion studies, gratings are invaluable as the dispersing element. Excellent gratings are now made in large sizes and they may be blazed very effectively so that the light is strongly concentrated in one order for a given wave-length range. The most efficient stellar spectrographs use a mirror collimator, a reflection grating, and a mirror camera (with Schmidt correcting plate when necessary). Several cameras and gratings are usually available for each spectrograph in order to give the desired dispersion in each wave-length range.

The recording device for most stellar spectrographs is a photographic plate or film. The advantages of photographic recording are efficiency and compactness. One may quite easily record the information in 2000 Å or more at one time, the resolution being set by the projected slit-width. The projected slit-width should be greater than the size of the photographic grain, about five microns. A comparison spectrum is recorded on either side of the spectrum for reference when determining wave lengths and radial velocities. One must have a wave-length reference external to the stellar spectrum in order to find the wave-length displacements caused by the motion of the earth, by the motion of the star, and possibly by sections of the stellar atmosphere. The disadvantages of photographic plates are that they do not respond linearly to varying light intensity and that the resolution is limited by the size of the photographic grain. An intensity calibration spectrum must be exposed with every stellar spectrogram and developed together with the stellar spectrogram. The spectrograms are traced with a microphotometer in order to derive information about the variation of the intensity of light in the star spectrum with wave length. The principle methods used for spectrophotometry are reviewed in a later chapter.

Photoelectric recording of stellar spectra has been attempted at several observatories. The advantage of this technique is that the response is linear over a large intensity range and absolute intensities may be derived more easily than from photographs. However, direct recording of high-dispersion spectra over a wide wave-length range with high resolution has not generally been attempted, for it is very wasteful of telescope time. WHITFORD (1962) and CODE and LILLER (1962) have reviewed the techniques which have been used. The chief difficulties that occur are (1) compensating for fluctuations due to seeing changes in the intensity of the star light entering the spectrograph and to motion of the star image across the slit, (2) handling electronically the very low light level received in the selected range $\Delta\lambda$ when high resolution is required, and (3) establishing precise external wave-length standards in the recorded spectrum.

It would be technically possible to record spectra on magnetic tape by receiving the spectrum on a sensitive surface that might be scanned by television techniques and storing the information. One would have to devise a method for coding sufficiently accurately the required material, that is intensity of light in each range $\Delta\lambda$, and one would have to introduce a method for recognising random errors due to uncontrolable imperfections in the image.

In the case of astrophysics the question to be decided is which method of recording the spectrum collects the most bits of information in the least telescope time, for the limiting factor in studying stellar spectra is the amount of observing time that is available with equipment that will give the desired spectral range and resolution. From this point of view, photographic plates are a very advantageous method of recording and storing information. Furthermore the stellar spectrum with its comparison spectrum can be readily examined visually at any later date and scanned for any desired piece of information. Also it is easy to recognize and reject false information due to flaws in the record. In addition spectrograms may be stored quite compactly. Rolls of paper tracings such as those produced by the usual pen and ink recorder attached to a photometer are much more difficult to store and to search quickly for relevant information. On a pen and ink recording of a stellar spectrum it is not easy to distinguish false information resulting from imperfect compensation for seeing variations or from imperfections of the recording apparatus. Magnetic tapes can store a great density of information, but they can only be searched by means of the appropriate decoding devices. It would appear that the photographic recording of stellar spectra is still of great value. The greatest need is to develop greater sensitivity to low light levels, or alternatively to use some method of intensifying the stellar spectrum before it is recorded by the photographic plate. Experiments are being made in this direction at several observatories. Some fine spectra using a Lallemand tube have been obtained at the Lick Observatory (see for instance LALLEMAND, DUCHESNE, and WALKER, 1960) to give only one example. Practically all the detailed spectral information about early type stars that is discussed in this book has been obtained from photographically recorded stellar spectra.

4. The classification and description of early type spectra

Since there are very many stars, obviously each star cannot be studied in detail. One purpose of spectral classification is to find representative stars which are characteristic of a group. Then one or two members of the group may be studied carefully to find the characteristic temperatures, pressures and chemical composition of these stars. Another purpose is to study the distribution in space of the various types of star. The spectral criteria selected for classification purposes are chosen first to arrange the stars in a series of increasing surface temperature. The spectral types are so arranged that the temperature in the stellar atmosphere increases as one progresses from spectral type B8 through B5 to B0 and on to the O stars. When this ordering has been performed, other criteria are used to arrange the stars according to the gas pressure in their atmospheres. The stars with the lowest gas pressure in their atmospheres have generally the greatest luminosity. Changing temperatures and pressures will account qualitatively for the appearance of stellar spectra of different types. However, when one begins to examine stellar spectra closely, further systematic differences are found. It seems that the stars can be ordered according to a third parameter. One of the present problems of spectral classification is to isolate the most meaningful third parameter and to establish criteria for recognising it. A possible third parameter is the age of the star, which is presumably shown by the changing composition of the stellar atmosphere. Other physical causes that may affect strongly the appearance of the stellar spectrum are magnetic fields and the fraction of the energy stream which is transported by mechanical means.

The early history of spectral classification has been summarised by R. H. CURTIS (1932). The first observations of stellar spectra were made by direct vision spectroscopes, some of the most systematic work being done by Secchi. Huggins and Lockyer in the last years of the 19th century studied the spectra of the apparently bright stars and attempted to set up sequences illustrating the evolution of stars. Because the stars which we now classify as types O and B have relatively simple appearing spectra containing only a few strong lines of the light elements, it was thought (McLean, 1898) that they were stars in an early stage of development. Hence the name "the early type stars".

The classification systems suggested by Huggins and by Lockyer are cumbersome and were never much used. A simpler system was established at the Harvard College Observatory under the direction of Pickering and it forms the basis of the present systems of spectral classification. It was found possible to arrange the stars in a sequence of temperature by means of visual inspection of objective prism spectra which had a dispersion of about 200 Å/mm at Hγ. Only the major features of the spectrum could be used; the fine detail was not visible. The *Henry Draper Catalogue* and its *Extensions* (*Harvard Annals* **91–100** and **112**) contain the results of Pickering's enormous project to obtain the spectral type of every star, brighter than 8.5 magnitude, and to give the spectral types to much fainter magnitudes in the neighbourhood of the Magellanic Clouds.

The classification system which was adopted for the *Henry Draper Catalogue* was one-dimensional. A two-dimensional system which made use of spectra of higher dispersion was also developed at the Harvard Observatory in the early years of this century chiefly by Miss Maury. This system, which recognized that the pressure in the atmosphere as well as the temperature was an important factor in determining the appearance of the stellar spectrum, is the basis for the presently adopted two-dimensional classification schemes. In general the rather complicated symbolism of Miss Maury has not been adopted; rather the letters and numbers denoting spectral classes which were used for the *Henry Draper Catalogue* have been carried on with a few modifications and additions.

The stars with which we will be concerned are classified as type B and type O in the *Henry Draper Catalogue*. The dominant absorption lines appearing at type B are the Balmer lines of hydrogen and the He I lines. Most stars are classified as types B8, B5, B3, B2, B1, or B0; fewer stars are given the types B9, B7, B6, B4, B2.5, B1.5, and B0.5. In the O type stars the predominant absorption lines are the Balmer lines of hydrogen, the lines of He I and the Pickering series (4–*n*) of He II. The *Henry Draper Catalogue* places the pure absorption-line O type spectra in classes Od and Oe5. (The *e* of Oe5 does not indicate emission features.) A few stars are listed as types Oa, Ob, and Oc. These are Wolf-Rayet stars which are now given the types WC or WN.

In order to convey further information, the letters and types of the *Henry Draper Catalogue* are sometimes modified in later work by prefix or suffix letters. The following letters are frequently seen:

prefix letter	meaning
c	supergiant, or extremely luminous star
g	giant; a star definitely above the main-sequence, but not extremely luminous
d	dwarf or main-sequence star

suffix letter	meaning
n	nebulous or washed out lines, usually indicating high rotation
s	sharp lines
k	interstellar line due to Ca II is present
e	hydrogen lines appear in emission; sometimes e_α, e_β, etc., the Greek letter indicating the last line of the Balmer series seen in emission
p	peculiar; some lines unusually strong or unusually weak
f	emission at He II λ4686 and at the N III multiplet λ4634-41

The great majority of B type stars have a pure absorption-line spectrum. Rather few of these stars have sharp lines, most having greatly broadened lines due to rapid rotation. A considerable number of stars have been observed to change from B to Be with a shell in a matter of 20 to 30 years. Many of these changes are recorded in the *Mount Wilson Catalogue of* Be *and* Ae *stars* (MERRILL and BURWELL, 1933, 1943, 1949, 1950). A shell spectrum is an additional set of rather sharp absorption lines which mimic, more or less, an A type supergiant. This set of lines appears superposed on the normal B type spectrum. It is due to a temporary, extended atmosphere surrounding the B star. All shell stars of type B also produce hydrogen emission lines, but all Be stars do not show a shell-absorption spectrum.

The counterpart of shell stars among the O stars appears to be the stars classified as Of. The suffix f is used only with type O; the suffix e is not generally used at present with the O stars. The Balmer lines in the blue-violet region of the spectrum do not appear strongly in emission, probably owing to the high state of ionization of hydrogen which results from the high temperatures in O type atmospheres.

The *Henry Draper* types Oa, Ob, and Oc contain a great variety of objects. The spectra of these stars are characterised by broad emission lines, some of which are very strong. These are the Wolf-Rayet stars which are now given types WC or WN plus a number to indicate a subdivision. The present classification system for Wolf-Rayet stars is not entirely satisfactory. The problem of these stars will be discussed further in a later chapter. The letter C indicates that lines from the carbon ions are particularly strong, while the letter N indicates that lines from the nitrogen ions are particularly strong. It is relatively simple to differentiate between the two groups by means of low dispersion spectrograms which cover the spectral region 4600 to 5000 Å, because the strong, broad emission features at λ4640 (N III), λ4650 (C III and C IV), and at λ4686 (He II) give a distinctive pattern.

CHAPTER II

IMPROVED SYSTEMS OF SPECTRAL CLASSIFICATION

1. Systems depending upon the visual inspection of spectrograms

The systems of spectral classification that are described here have been developed for use with spectra which have a linear dispersion of at least 120 Å/mm at Hγ. Other systems of spectral classification have been developed for use with spectra of very low dispersion, generally objective prism spectra, the purpose being to further our understanding of galactic structure and of the distribution of the stars. They are reviewed, for instance, by Fehrenbach (1958).

1.1 *The Victoria Revised system*

Plaskett and Pearce (1930a) published a radial-velocity catalogue of all O and B stars north of $-11°$, brighter than magnitude 7.5, and having *Henry Draper* types B5 to B0. Revised spectral types are given in this catalogue. The spectral types were assigned chiefly by Pearce from visual inspection of the spectrograms which were obtained for the radial-velocity work. The dispersion is 30 Å/mm or 50 Å/mm at Hγ and the wave-length range covered is from 5000 Å to 3900 Å. Some idea of the character of the spectral lines is given by the letters s and n. The presence of an interstellar Ca line is indicated by the letter k. Typical B and O type spectra are shown in Figure 2 and their *Victoria Revised* types are given in the caption.

Throughout type B the Balmer lines of hydrogen are prominent. The He I lines are first seen at type B9 on moderate dispersion spectrograms and they increase in strength towards type B2. The He I lines reach maximum strength at about type B2 and decline somewhat in strength through the early B types and type O. It is generally stated that the He I lines are first visible at type B9. This is so for moderate dispersion spectrograms. Very weak lines due to He I may be found on high-dispersion spectra of early A type stars. The types between B9 and B5 are called the late B type stars. At type B8 the hydrogen lines are strong, and there are weak stellar lines of Ca II present at λ3933 (K), and at λ3968 (H). At type B5 the hydrogen lines are not so strong, but the He I lines are stronger. The lines of Si II and Mg II are weakening, and the Si III lines λ4552, λ4568, and λ4574 are just beginning to be visible on high-dispersion spectrograms. Some lines of C II and N II are visible on moderate dispersion spectrograms. The types B4, B6, and B7 are not often assigned, most stars being given the types B5, B8, or B9.

The types recognized among the early B type stars are B3, B2.5, B2, B1.5, B1, B0.5, and B0. At type B3 the He I lines are strong and the hydrogen lines have weak-

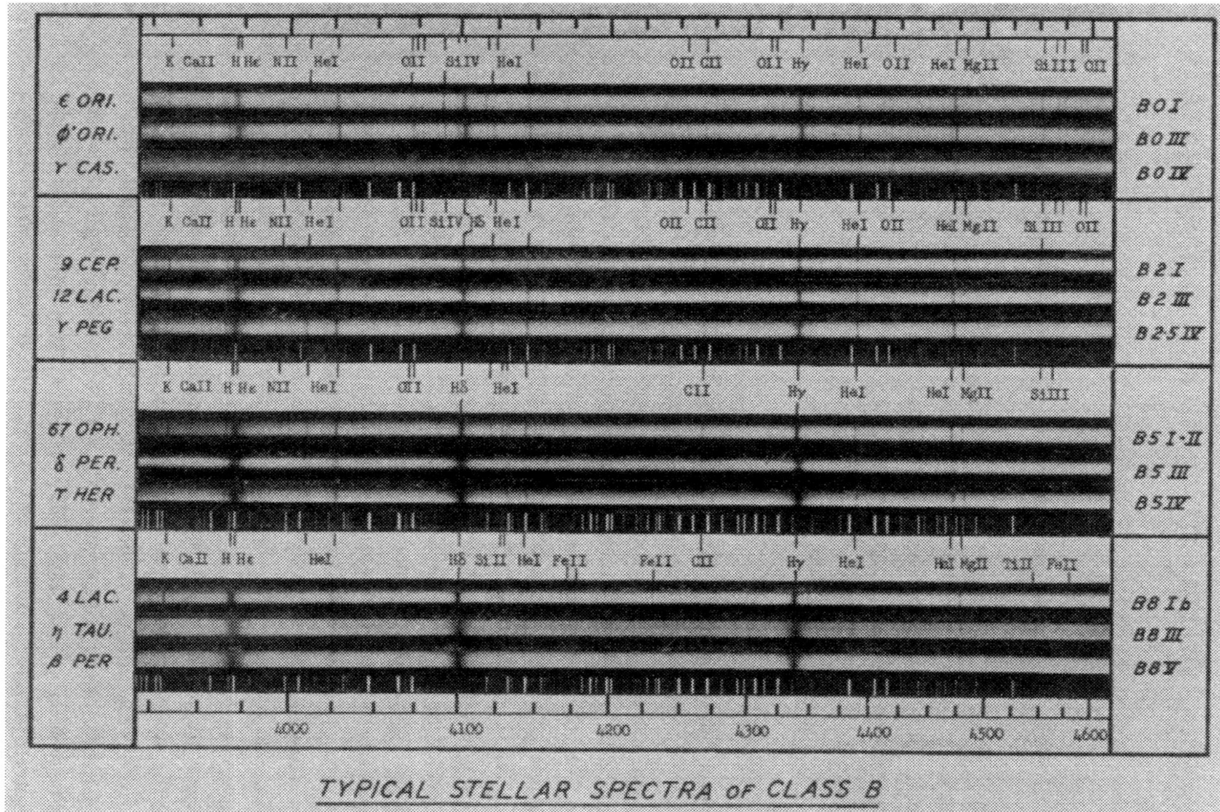

Fig. 2. Typical B type spectra. Note how the hydrogen lines weaken towards earlier types and the He I lines increase in strength. Luminosity effects are visible in the hydrogen lines. The star γ Cas is rapidly rotating. The Victoria revised types of these stars are ε Ori — B0k, φ^1 Ori — B0ssk, γ Cass — B0nne, 9 Cep — B2sk, 12 Lac — B12k, γ Peg — B2ss, 67 Oph — B8s, δ Per — B8n, τ Her — B7s, η Tau — B5n. Victoria types do not appear to exist for 4 Lac and β Per.

ened further from B5. Lines of Ne I, Si II, N II, C II, and Si III are readily visible on high-dispersion spectrograms. At type B2 the He I lines are at maximum strength; the H lines are still fairly strong, but weakening, and many lines of S II, O II, N II, C II, Si II, Si III, Si IV, and Fe III can be seen on high-dispersion spectrograms. At type B1 the He I lines are still strong; the hydrogen lines have weakened further. The Si II lines are very weak, and the Si III lines are near maximum intensity. The Si IV lines λ4088 and λ4116 and the O II lines are becoming quite prominent. A group of Al III lines is visible at λ4479; the Mg II line λ4481 has become very weak. At type B0 He II λ4686 may be seen as a weak line and possibly He II λ4541 on high-dispersion spectrograms. Lines of C III, O III, and Ne II appear, the lines of C II, O II, and Ne I being very weak; Si II is very weak; Si III is past maximum intensity and Si IV is strong. The hydrogen lines are present, but not very strong; the He I lines are strong, but weakened somewhat from B2. In spectral types earlier than B2 the He I lines from triplet levels strengthen with respect to those from singlet levels.

The B type stars may readily be separated into dwarf or main-sequence stars, giants, and supergiants by noting the shape of the wings of the hydrogen lines and the appearance of forbidden He I lines. Typical luminosity effects at Hγ and at He I λ4471 are shown in Figure 3. The Stark broadening of the Balmer lines is the chief characteristic that is used to make a separation according to luminosity. The Stark broadening of the He I lines and the appearance of [He I] λ4469 is also used on moderate dispersion spectra. In the main-sequence stars the wings of the hydrogen lines are very broad. Supergiants can be separated immediately by the bell-shaped appearance of their spectral lines, there being essentially no extensive wings for the Balmer lines and [He I] λ4469 is not visible. With spectra which extend to λ3600 one may also use the number of hydrogen lines visible to estimate the luminosity of the star (UNSÖLD and STRUVE, 1940). In main-sequence stars the Balmer series breaks off at about $n_m = 14$; in giants the series ends near $n_m = 19$; in supergiants the Balmer series may be followed further, the point of break-off depending to some extent upon the dispersion which is used and upon the amount of broadening of the hydrogen lines due to large scale motions in the stellar atmosphere. Here n_m is the upper quantum number of the last visible Balmer line.

Empirically it has been known for many years that in the A and B type stars the equivalent width of Hγ or of Hδ may be correlated uniquely with absolute magnitude. R. M. PETRIE (1949, 1952) has established quantitative relations between M_V, the visual absolute magnitude, and the equivalent width of Hγ which are valid for stars of spectral types A5 to B0. The manner in which any empirical measurement in a stellar spectrum is calibrated in terms of absolute magnitude is a critical step. This point will be discussed more fully in Chapter III. The purpose of such work is to be able by observations of stellar spectra to determine the intrinsic brightness of stars, and then, from a knowledge of their apparent brightness and of the amount of extinction due to interstellar dust and gas, to estimate the distances of the stars. It is a relatively simple matter to separate stars into classes of differing luminosity by considering the appearance of details in their spectra. It is quite another problem to

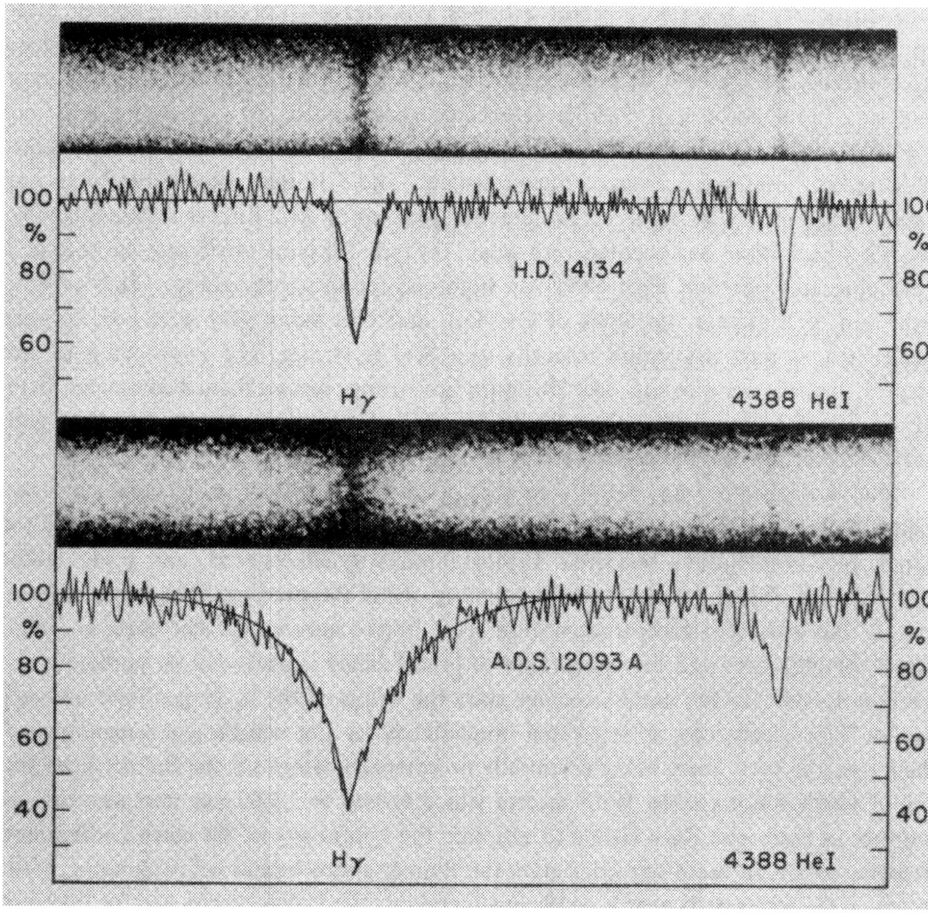

Fig. 3. Luminosity effects in B type spectra. The great change in the hydrogen line profiles due to Stark effect is illustrated here. The star HD 14134 is a supergiant; ADS 12093A is a main-sequence star

relate these classes, or the value of a parameter such as the equivalent width of Hγ, to intrinsic luminosity.

The Victoria classification system for the early type stars, that is the system developed and applied at the Dominion Astrophysical Observatory, consists of two parts: (1) the spectral type is estimated by means of visual inspection of spectrograms which have a dispersion of 50 Å/mm at Hγ, and (2) the luminosity is estimated by measuring the equivalent width of Hγ on these spectrograms by spectrophotometric techniques and using the luminosity calibration established by Petrie. The Victoria classifications for B stars are listed in the radial-velocity catalogues prepared at the Dominion Astrophysical Observatory. A comparable system for O stars was also initiated, chiefly based upon the pioneering studies of H. H. Plaskett (1922). The O stars are arranged essentially in order of increasing strength of the He II lines with respect to He I and H. The letter f was introduced to indicate those stars in which emission occurred at He II λ4686 and at the N III λ4634-41 multiplet. In the O stars emission is rarely seen at the hydrogen lines which are visible on blue sensitive plates, i.e. at Hβ, Hγ, or Hδ. The O stars are assigned types between O9.5 and O5 or O4.5. No star has yet been discovered for which the types O4, O3, O2, or O1 have been suggested. The stars of type O5 or O4.5 have absorption-line spectra of the highest known level of excitation and ionization.

1.2 *The MKK system*

In 1943 MORGAN, KEENAN, and KELLMAN (1943) published *An Atlas of Stellar Spectra* which defined a two-dimensional classification system for stars covering spectral types O to M. In this work (which establishes the MKK system of spectral classification) it was emphasized that spectral classes should be defined by means of standard stars, and that classification should be performed by intercomparing spectrograms of the same dispersion and width which had been taken on the same emulsion and developed in exactly the same way. Photographs are given of many stellar spectra in order to facilitate this comparison, and significant criteria for spectral type and luminosity are noted. These criteria are valid only for spectra of the same quality as those of the *Atlas*. The dispersion, resolution, and width of spectrum were carefully selected to make accurate classification possible on the basis of temperature variation (*Henry Draper* spectral type O, B, A, F, G, K, M, N, R, S) and of luminosity. The concept of *luminosity classes* was introduced to differentiate between stars of about the same surface temperature or excitation conditions, but differing luminosity. These classes are designated by roman numerals and they are related in the following way to the descriptive terminology used before this time to indicate the relative sizes (thus luminosities) of the stars:

Luminosity Class	Descriptive terminology
I	supergiant
II	bright giant
III	giant
IV	subgiant
V	dwarf or main-sequence star

Later it was found possible to differentiate further between the supergiants, the symbol Ia indicating the very luminous supergiants and Ib indicating the less luminous supergiants. Other modifications such as Iab and Ia-O (to indicate extremely luminous stars) are sometimes encountered. The subdwarfs may be indicated by the luminosity class VI.

The MKK system provides a means for placing stars in boxes where the overall properties of a star are about the same. The available divisions are defined by the sensitivity of the available spectral criteria at any spectral type. There is no reason to believe that recognizable differences in spectral type indicate regular differences in real physical variables such as the surface temperature or the luminosity. At some spectral types it is not possible to separate all the luminosity classes, nor are all subdivisions of each spectral class (O, B, etc.) from 0 to 9 available. This system is essentially an extention and improvement of the *Henry Draper* system. The calibration of the luminosity classes in terms of absolute magnitude is a problem distinct from the assignation of spectral types. It is discussed in Chapter III both for the MKK system and for the MK system, which is a revision of the MKK system.

1.3 *The MK system*

After a few years' use, it was found that the MKK system could be improved by providing more standard stars at certain spectral types and by defining better criteria in some cases. The revised system is known as the *MK system*. It is defined by the standard stars and spectral types listed by MORGAN and KEENAN (1951) and by MORGAN and JOHNSON (1953). In the latter paper, magnitudes and colour indices on the UBV system of broad band photometry are given for some of the standard stars as well as for a large number of other stars distributed over the Hertzsprung-Russell diagram. For many O and B stars spectral types on the MK system and magnitude and colour observations on the UBV system may be found in comprehensive papers by MORGAN, CODE, and WHITFORD (1955) and by HILTNER (1956).

The MKK and the MK systems of spectral classification make use of spectra having a dispersion of 120 Å/mm at Hγ. The spectral range covered is from about 3900 Å to about 4700 Å. The adopted spectral purity was carefully selected so that the broadening of the spectral lines of some stars due to the rapid rotation of the star would not give a noticeably different appearance to the spectrum than that recorded for non-rotating stars. Thus this system of spectral classification is not seriously affected by the geometric factor of rapid rotation. When it is desired to transfer the system to spectra of other dispersions and other purity, one must first investigate whether the classification criteria selected for the MK system remain valid on the new spectrograms. Fine detail is not generally well resolved on the MK classification spectra. Therefore, when a group of B2 V stars, for instance, is observed at high dispersion, there is no reason to expect that they will all show the same detail.

One of the original purposes of spectral classification was to permit one to select typical stars that were representative of a group and which could be used for detailed astrophysical investigations. Modern work suggests that among the O and B stars it is very difficult to find truly representative objects. Each star, when it is observed in detail, differs from others of approximately the same type. It is therefore not easy to determine representative values of physical parameters such as temperature and pressure at some well-defined level in the atmosphere for each recognizable spectral type, particularly among the O and early B stars. It must be recognized that although an assigned spectral type and luminosity class or absolute magnitude imply an estimate of the physical conditions in the atmosphere, this estimate may not be exact for any star of the group. Fairly accurate and reproducible systems for estimating spectral types and luminosities of the O and B stars have been established. Much work remains to be done in establishing the physical characteristics which these spectral types represent.

2. Systems depending on measured quantities

The assignment of spectral types by visual inspection of spectrograms can be done accurately and quickly and consistently, but there is no denying that the results are, to a certain extent, subjective. By making quantitative measurements of the strengths of the various spectral features one could hope to set up a more impersonal method of classification. Furthermore, if the measurements can be made with sufficient accuracy, they might be used with the appropriate theory of the formation of stellar spectra to establish the values of physical parameters such as temperature, pressure, and abundances of the elements in a stellar atmosphere. A few programs aimed at obtaining sufficient quantitative measurements in O and B type spectra for classification purposes and for astrophysical studies have been carried out. They will be reviewed in the following paragraphs.

2.1 *The work of E. G. Williams*

This work is reported in two papers (WILLIAMS, 1936a and 1936b). It forms a landmark in the study of O and B type spectra. Williams studied microphotometer tracings of prismatic spectra which were obtained at the Mount Wilson Observatory of 84 O and B type stars. The dispersions at Hγ were 27 Å/mm, 39 Å/mm, and 65 Å/mm. He measured the equivalent widths of all possible lines of H, He I, He II, and of the stronger lines of C II, C III, N II, N III, O II, O III, Mg II, Si II, Si III, and Si IV, and he gave the apparent central depths of the stronger lines. These measurements of equivalent width are very consistent and they agree with the best modern measures from spectrograms of the same dispersion. These equivalent widths are good, but they are not comparable to what can be obtained from high dispersion spectrograms. Unfortunately, it is now clear (WRIGHT, LEE, JACOBSEN and GREENSTEIN, 1963) that measures of equivalent width made on spectrograms of moderate or low dispersion which have rather modest spectral purity (p being of the order of 1 Å) usually are larger than the values found from high dispersion, high resolution spectrograms. With the stronger lines, there is difficulty in tracing the wings accurately; with weak lines the profile is purely instrumental; the very weak lines are entirely lost in the photographic grain.

From his data Williams set up line ratios which varied with spectral type and he reclassified his stars, noting as was also apparent from the first Victoria catalogues, that many stars which had the spectral type B8 or B5 in the *Henry Draper Catalogue* were really of earlier type. Williams also used the appearance of the spectrum, chiefly the amount of Stark broadening of H and He I lines, to give a luminosity classification into giants, intermediates, and dwarfs. The strength of the interstellar K line of calcium may be correlated roughly with the distance of the star. Williams used such a correlation to estimate absolute magnitudes for his stars. His work culminated in a table giving revised spectral types and absolute magnitudes for the 84 O and B type stars which he studied. An attempt was made to correct the absolute magnitudes for interstellar absorption. The results do not differ greatly from modern values.

Williams' material gives a clear quantitative impression of how the strengths of the major absorption lines in O and B type spectra vary with spectral type. Some of his material is reproduced here. Figure 4 illustrates the variation of the Balmer line

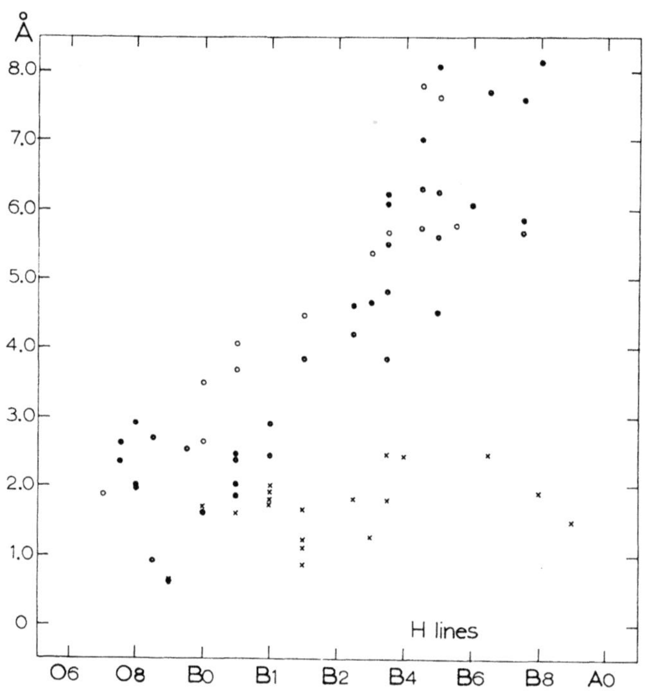

Fig. 4. Hydrogen line strengths *vs.* spectral type according to Williams. Supergiants are indicated by crosses, stars of intermediate luminosity by dots, and dwarfs by circles.

intensity with spectral type and with luminosity. The stars classified by Williams as giants are indicated by crosses, the intermediates by dots, and the dwarfs by open circles. It should be noted that no supergiants (giants in Williams' notation) are recognized earlier than type O9. The separation in hydrogen line intensity with luminosity becomes greater as later spectral types are approached. This is the well-known effect which Petrie (1949, 1952) has calibrated for use as a luminosity criterion for the B type stars. Figure 5 illustrates the variation of the intensities of lines of the 2^3P-n^3D series of He I with spectral type. A luminosity effect, due to Stark broadening, is clearly evident between B0 and B6. The lines reach maximum intensity near type B3.

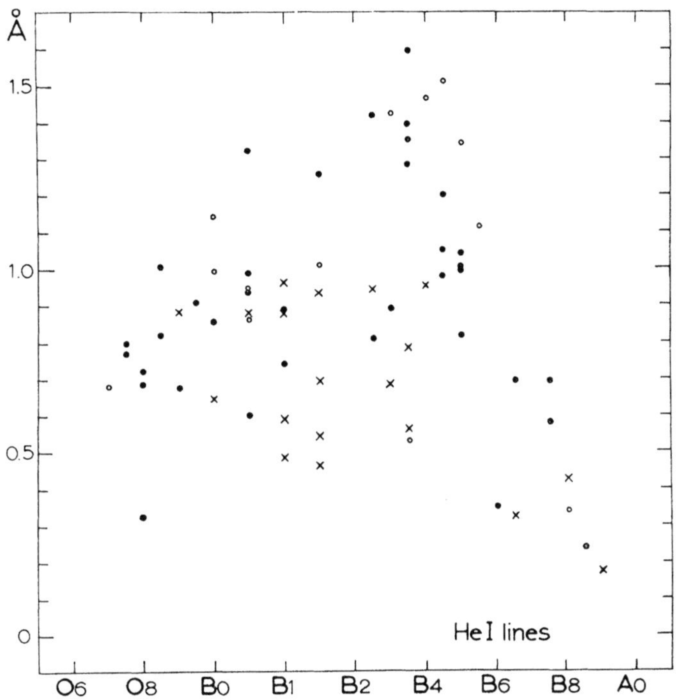

Fig. 5. The strength of the He I diffuse triplets *vs.* spectral type according to Williams. Supergiants are indicated by crosses, stars of intermediate luminosity by dots, and dwarfs by circles. The ordinate is (4471 + 4026 + 3820)/3.

Figures 6 and 7 show how the strengths of lines from the ions of C, N, O, and Si vary in O and B type spectra. The spectra which have the highest ionization potential reach maximum intensity at the earliest spectral type.

Williams reached the following conclusions:

1. In stars belonging to the main sequence, the lines of hydrogen show a steady five-fold increase in total absorption in passing from the earliest types to class A.

2. Very luminous B stars retain roughly the same low hydrogen intensity at all subtypes.

3. Main-sequence and very luminous stars cannot be distinguished at types earlier than O9. (It is not clear that at O7, say, supergiants exist at all. One is not surely known among the stars that have been observed.)

4. The central intensities in sharp-line stars decrease regularly from 60 per cent at O8 to 20 per cent at A0.

5. An absolute magnitude effect similar to that of hydrogen is found for the He I lines.

6. A two-fold increase occurs between O8 and B1 in the singlet-triplet ratio for the diffuse series of He I, but no luminosity effect on this ratio is detected. (Williams

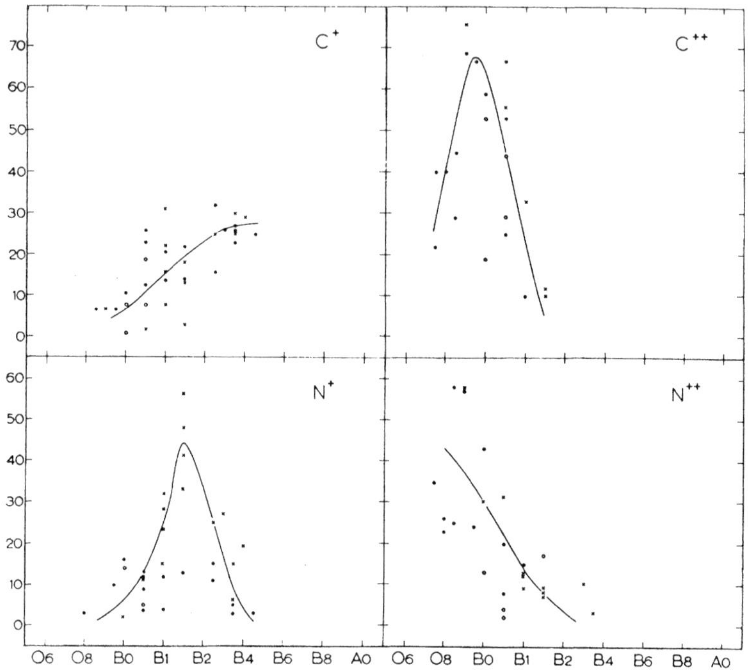

Fig. 6. The variation of the strength of lines of C and N ions with spectral type according to Williams. The symbols are as in Figures 4 and 5. The ordinate measures the sum of the equivalent widths of the available lines in each case.

also suspects that there is a fading of the sharp series singlets with respect to the triplets of He I as temperature increases. The He I spectrum dominates B type spectra and the relative changes in intensity of certain lines with varying temperatures and pressures give much information about the physical conditions in B and O type atmospheres. Because of the importance of this spectrum for the analysis of early type spectra, the spectral behaviour of He I is reviewed in Chapter VII. Particular attention is paid to the significance of the so-called "dilution" effects, which are the phenomena referred to in Williams' conclusion 6.)

7. The curve of growth for N II and for O II is steeper for the more luminous stars than it is for main-sequence stars. No appreciable difference is found between stars with sharp lines or stars with nebulous lines. (In the terminology of 1936 this phenomenon, which at present is interpreted to indicate the presence of increased micro- and macro-turbulence in the atmospheres of luminous stars, was called a gradient effect. It was noted that when one considered the observed relative intensities of intrinsically weak, intermediate, and strong lines in a multiplet, the gradient or slope of the straight line, more or less, relating observed strengths to f-values was steeper for the supergiants than it was for the less luminous stars.)

The trends in O and B type spectra which were evaluated quantitatively by

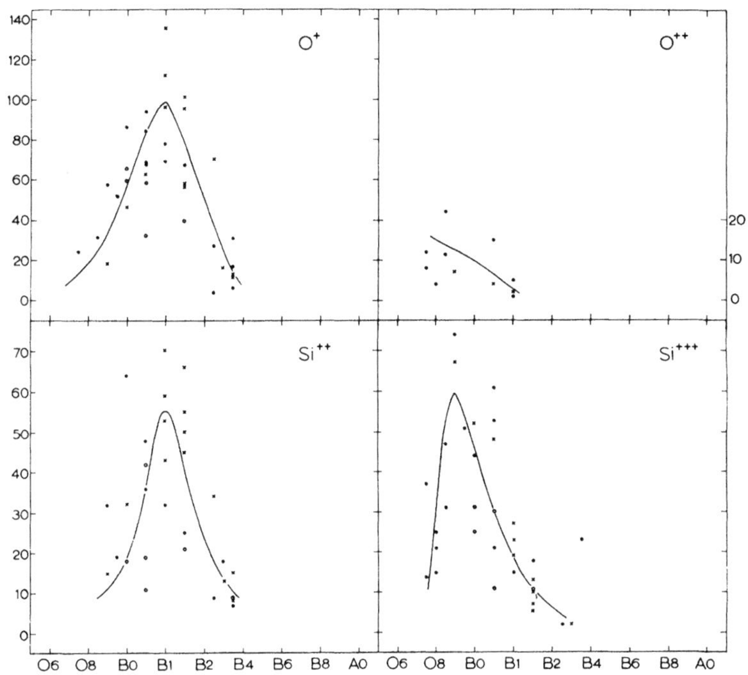

Fig. 7. The variation of the strengths of lines of O and Si ions with spectral type according to Williams. The symbols are as in Figures 4 and 5. The ordinate measures the sum of the equivalent widths of the available lines in each case.

Williams have been confirmed by all later work. Any series of models that is aimed at representing the atmospheres of the early type stars must reproduce these trends.

2.2 *The work of P. Rudnick*

In 1936 another paper on the intensities of lines in B type spectra appeared. RUDNICK (1936) measured the total absorption of the stronger lines of 70 O and B type stars on single-prism Yerkes spectrograms which have a dispersion of 26 Å/mm at Hγ. The total absorptions were measured using a photoelectric microphotometer and a square diaphragm. The diaphragm was large enough to include the full profile of the line being measured except in the case of very broad lines. The measurements were reduced to equivalent angstroms. The values given for the hydrogen lines are not total absorptions, but consistent fractions of the total absorptions. The method works reasonably well for those weak shallow lines of B type spectra which are not so close together that more than one line falls in the diaphragm at one time. Rudnick prepared diagrams like those of Williams showing the variation of line strength with spectral type. His chief conclusions are

1. that the triplets strengthen with respect to the singlets of the He I spectrum in stars of spectral type earlier than B2, and

2. that the metallic lines are stronger in giants than in dwarfs (the gradient effect).

2.3 *The work at the Dominion Astrophysical Observatory*

Following the appearance of the papers by E. G. Williams and by Rudnick, a program was instituted at the Dominion Astrophysical Observatory under the direction of R. M. Petrie to establish similar quantitative criteria for classifying O and B type spectra from the spectrograms obtained at Victoria for radial-velocity work and for assigning absolute magnitudes. PETRIE (1947a) noted that the early O type stars are poorly represented in the lists by Williams and by Rudnick, and he provided further material for twenty O type stars from prismatic spectrograms having a dispersion of 30 Å/mm at Hγ and extending from 5000 Å to 3900 Å.

In the O type stars, lines of the He II spectrum appear and they increase in strength as one proceeds to earlier types. Since alternate members of the 4–n series of He II (the Pickering series) fall close to the Balmer lines of hydrogen, the observed features at Hγ and Hδ are a blend of He II and H. To obtain an estimate of the strength of the hydrogen lines alone one must estimate how the blend should be resolved into a part representative of the hydrogen absorption and into another part representative of the He II absorption. In principle the problem is straight-forward; in practice it is difficult to solve exactly because it involves being able to predict quantitatively the formation in an extended atmosphere of lines arising from levels of very different excitation potential, the one from a neutral atom, the other from an ion of high ionization potential.

The simplest solution is to assume that the observed equivalent width may be divided into two parts:

$$W(\mathrm{H} + \mathrm{He\,II}) = W(\mathrm{H}) + W(\mathrm{He\,II}), \qquad \text{(II-1)}$$

and to estimate $W(\mathrm{He\,II})$ from the measured equivalent widths of neighbouring, unblended lines in the He II series. Subtraction yields the strength of the absorption due to hydrogen. This procedure is valid for a simple atmosphere consisting of one layer in which the absorption coefficient at each wave length, λ, in the line is so weak that the absorption at λ is proportional to the value of the line absorption coefficient. The value of the absorption coefficient at λ is the sum of the contributions from the two blending lines.

In general this simplified picture of line formation is not valid. There is quite conclusive evidence (UNDERHILL, 1962b, 1963a) that lines from levels of high excitation (for example the He II lines of the 4–n series) are formed in much deeper layers, where the temperatures and pressures are high, than lines from levels of low excitation (for example the Balmer lines of hydrogen) are formed. In order to derive precise information from the observed equivalent widths, the spectrum should be investigated by model atmosphere techniques such as those described in Chapter XI. The result of stratification and of taking account properly of the transfer of radiation through the atmosphere is to make the equivalent width of a feature due to blended lines less than the sum of the equivalent widths which would result from each line separately. Resolving the blend, therefore, by the simple method of subtracting an estimate of the equivalent width of one of the blended features, were it formed separately, will result in too low an equivalent width for the remaining feature.

In order to obtain an estimate of the strengths of the hydrogen lines in O type stars, Petrie used equation (II-1). He measured the intensities of the blended features at Hγ and at Hδ and subtracted from each an estimate of the strength of the component due to He II. This estimate was put equal to the average strength of He II λ4541 and He II λ4200. The resulting strengths of Hγ and Hδ of hydrogen are minimum values. A better separation cannot be performed without a detailed analysis of each O type atmosphere. The purpose of this procedure was to demonstrate that the strengths of the Balmer lines continued to decrease through the O stars, rather than remaining constant as is suggested by the data of Williams which is given in Figure 4. Petrie's material illustrating this point is reproduced in Figure 8. It strengthens our

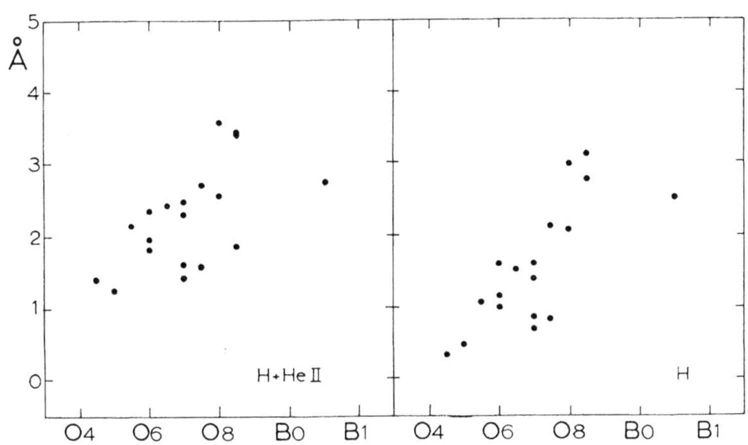

Fig. 8. The variation of the blend of H and He II in the O types according to R. M. Petrie.

impression that the absorption-line O type spectra form a continuation of B type atmospheres toward higher temperatures. Petrie has suggested that the hydrogen lines would disappear in absorption at type O4. In view of the theoretical difficulties of resolving the blend accurately, it seems premature to say that absorption in the hydrogen lines would no longer be observed at type O4. The expected hydrogen absorption, however, would be weak. In a star of type O4, one would expect to see the intensities of the 4–n He II series varying rather smoothly with n, only small increases of intensity occurring at even values of n where blends occur with the Balmer lines of hydrogen. In the late O type stars, hydrogen is the dominant component of the blend.

Petrie lists equivalent widths of the blended features at Hγ and Hδ; of He II λ4541 and λ4200; of He I $\lambda\lambda$4009, 4144, 4388, 4026, 4471; of Si IV λ4089; and of a line which he calls Mg II λ4481. It is rather surprising that an absorption line of measurable strength on Petrie's spectrograms should appear at the position of Mg II λ4481 in O type spectra, because the general level of ionization and excitation is such in O type spectra that one would expect most of the magnesium atoms to be at least doubly ionised. The persistence of ions in the singly ionised state may be an indication of a very extended atmosphere for an O star, the higher level of excitation in the O type shell being sufficient to keep the lower level of λ4481 (E.P. 8.86 volts) populated inspite of the relatively low number of magnesium ions expected, in contrast to the situation with B type shells where the Mg II λ4481 line is usually exceptionally weak. There is no doubt that a line of moderate strength does occur at λ4481.14 in the spectrum of 10 Lacertae (DE GROOT and UNDERHILL, 1963), and that it should be attributed to Mg II. The moderately strong Mg II line which is observed in O type spectra, see also WILSON (1957), can only come from an extended atmosphere. Petrie's observations of the strength of the singlet-triplet ratio in the He I spectrum in O type spectra give additional evidence that all O type stars are surrounded by extensive, low density atmospheres or shells. The significance of the He I observations is discussed more fully in Chapter VII. The variation with spectral type of the He I lines is shown in Figure 9.

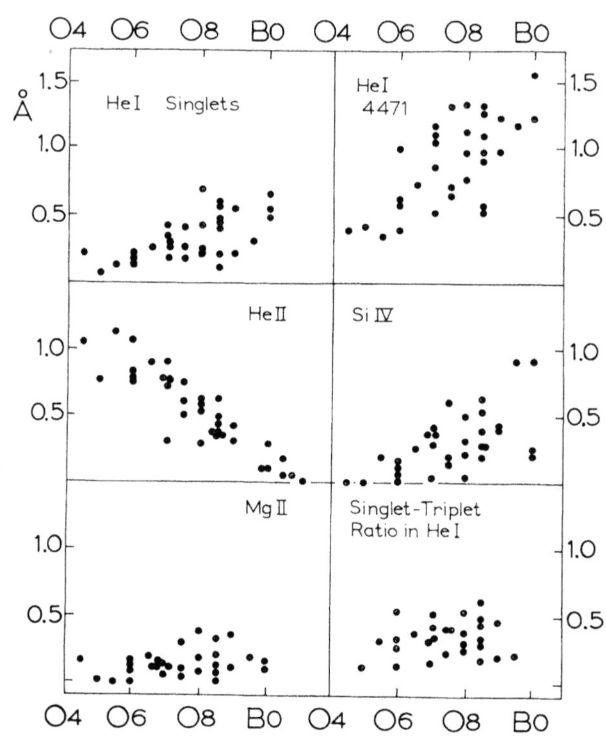

Fig. 9. The variation of line strengths in the O stars according to R. M. Petrie. The ordinates are in angstroms.

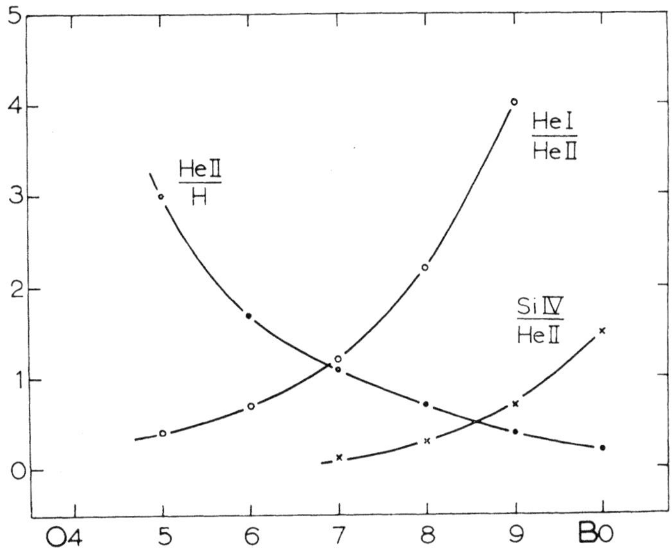

Fig. 10. The line intensity ratios for the classification of O stars.

From his material Petrie sets up three line-intensity ratios He I/He II, He II/H, and Si IV/He II useful for classifying O type spectra and he gives a revised classification for 20 stars. The variation with spectral type of the adopted ratios is shown in Figure 10. With these curves one may classify O type spectra accurately and consistently.

The line He II λ4686 is not useful for classifying O type spectra because it frequently appears in emission in O type spectra due to a cycle involving the absorption of the Lyman β radiation of hydrogen which is produced in the outer extended atmosphere of the star. It is difficult to decide how much the assigned spectral types for O stars represent the photosphere of the star, and how much they represent the extended atmosphere, for all the suggested classification ratios depend to some extent upon the strength of lines such as He I λ4471 which are largely formed in the extended atmosphere.

Petrie's paper on the O stars concludes with estimates of the absolute magnitudes of the O stars making use of measured strengths of the interstellar K line and an adopted relationship between K-line strength and distance. A general absorption of 0.8 magnitudes per kiloparsec is assumed. Petrie also attempts to estimate the ionization temperature in O type atmospheres from the relative strengths of He I λ4471 and He II λ4541. Since it is now clear that He I λ4471 is largely formed in an extended shell around the O star (see Chapter VII) whereas the He II lines are formed in the deeper layers of the photosphere where the temperatures and pressures are distinctly higher, the numerical values determined by Petrie have little real meaning. It is difficult to find suitable lines for estimating temperatures and pressures in O type atmospheres by simple methods of analysis, for the atmospheres appear to be very extended and they produce lines from atomic levels of greatly different excitation energies. If a simple method of analysis is to have any validity it must deal with spectral features that are formed, more or less, in the same parts of the atmosphere. The He I and He II lines are not formed in the same part of the atmosphere. The problem of estimating the temperatures and pressures in O type atmospheres is discussed more fully in Chapter XI.

Petrie notes, from his measurements, that the strengths of the readily measurable lines in O type spectra do not show any clear luminosity effects. He was unable to demonstrate whether this observed state of affairs, which is in contrast to that for the B stars, is due to no suitable lines being available for measurement, or whether it occurs because his limited observing list covered stars of only a small range in absolute magnitude. The absolute magnitudes of the O stars are discussed further in Chapter III.

Further measurements of the strengths of the strong lines in O type spectra made from spectrograms of moderate and low dispersion obtained at the Dominion Astrophysical Observatory have been published by UNDERHILL (1955). These observations confirm and extend the results of Petrie. Work is underway at the Dominion Astrophysical Observatory to provide similar material for B type stars from the moderate dispersion spectrograms used for the radial-velocity programs.

2.4 *The work at the Royal Observatory, Edinburgh*

Under the direction of Greaves and Baker a program was commenced at the Royal Observatory, Edinburgh to obtain accurate spectrophotometric measurements in the spectra of the early type stars brighter than magnitude 5.4 which were accessible from Edinburgh. About 100 stars having *Henry Draper* types between Oe5 and B3 were selected for study. The results were obtained by photographic spectrophotometry of prismatic spectrograms which have a dispersion of 134 Å/mm at Hα, 46 Å/mm at Hβ, and 29 Å/mm at Hγ. Care was taken to keep the exposure and development techniques uniform and a particular effort was made to eliminate so far as possible the uncertainties due to plate grain by measuring the transmission of the spectrograms at a fixed grid of wave lengths, reducing the readings to relative intensities with respect to a well defined reference continuum, and then taking means. Many spectrograms were measured for each star. In this way the random fluctuations due to photographic grain should cancel out, and weak diffuse absorption or emission features become visible. The techniques which were used are described by BAKER (1949), and the mean continuum for Oe5 stars, which is used as the reference intensity distribution is derived. Equivalent widths for some lines in the spectra of 7 O stars are given in this paper. The Edinburgh observers list the equivalent width between wave-length limits λ_1 and λ_2 which they specify for each line or group of lines, and they give the most probable identifications for the absorption (or emission) occurring in the listed range. The lines measured for equivalent width fall in the range λ4100 to λ6700. The second paper of the series (GREAVES, BAKER, and WILSON, 1955) gives some results for 7 stars of type B1; the third paper (WILSON, 1956) gives further results for another 7 stars of type B1; the fourth paper (WILSON, 1957) gives results for 16 stars of types O6 to B0;

the fifth paper (BUTLER and SEDDON, 1957) gives results for 20 stars of type B2; the sixth paper (BUTLER and SEDDON, 1959) gives results for 25 stars of type B3; and the seventh paper (BUTLER and THOMPSON, 1960) gives results for 10 stars of type B5 and for 7 stars of miscellaneous types. This series of papers gives one of the most detailed discussions available of the equivalent widths of the stronger lines in B type spectra and it gives the only available comprehensive lists of the equivalent widths of lines in the green-yellow-red spectral region. It is unfortunate that the resolution and spectral purity is so low (projected slit width between 12 and 60 microns); for instance the Si II lines λ4128 and λ4130 are not resolved. Each line is discussed separately and estimates are given of the contributions due to the various blends that may exist.

The spectrophotometric methods adopted by Greaves *et al.* are particularly valuable for revealing broad shallow spectral features. These authors note that the diffuse interstellar feature λ4430 extends from λ4368 to λ4498 and they give estimates of the equivalent width of this feature.

Extensive diagrams are given showing the detailed distribution of intensity with wave length in the spectra of stars of type O, B1, B2, and B3. Although these spectral plots do not have very high resolution, they give a useful impression of the distribution of light in B type spectra both for main-sequence and for supergiant stars.

The authors establish a luminosity ratio, ϕ_L, which is essentially the sum of the equivalent widths of 18 lines in the spectrum not due to H, He I, or He II and they show that ϕ_L increases with luminosity. This conclusion is another representation of the "gradient effect" discussed by Williams and by Rudnick. The stronger absorption lines generally have greater equivalent width in luminous stars than in main-sequence stars, if one excludes lines subject to Stark effect. WILSON (1956) notes that Mg II λ4481 at type B1 appears to be insensitive to luminosity, a conclusion reached earlier by MORGAN (1933) from visual inspection of spectrograms. This behaviour is related to the problem discussed in section 2.3 regarding the presence of Mg II λ4481 in O type spectra. Butler gives further information about this line.

WILSON's (1957) careful discussion of the spectra of O stars led to the discovery of the 5–*n* series of He II in absorption (near λ6500) and of very broad (±1500 km/sec), low intensity emission wings for the emission lines He II λ4686, N III λλ4634, 40, 41, and C III λ5696 in Of stars. The latter observation demonstrated for the first time the relation of the Of stars to the Wolf-Rayet stars and to the Be shell stars. Wilson also discussed the problem of finding luminosity criteria in the O stars.

Attempts are made by Butler and his associates to detect and interpret peculiarities in the strengths of the He I lines and of Mg II λ4481 in some of the stars of type B2 and later, but the available observation material is barely adequate for such a study. They indicate that Fe II is weakly present in absorption at type B5, an observation that is to be expected according to model atmosphere studies.

2.5 *The work of Barbier, Chalonge, and Divan*

In France a classification system for stars of spectral types O, B, A, and F was established by Barbier, Chalonge, and Divan making use of low-dispersion prismatic spectrograms which were obtained at the Jungfraujoch observing station and which extended to the ozone cut-off near 3100 Å. The dispersion is 250 Å/mm at Hγ. Spectrophotometric measurements are made of the shape of the continuous spectrum between λ3100 and about λ4800, careful corrections being applied for the extinction in the ultraviolet caused by the earth's atmosphere. The appearance of the spectrum in the neighbourhood of the Balmer jump is used to define spectral type and luminosity class.

Preliminary results were given by BARBIER and CHALONGE (1941); definitive results are given by CHALONGE and DIVAN (1952). The method has been developed from suggestions made by BARBIER and CHALONGE (1939) and by BARBIER, CHALONGE, and CANAVAGGIA (1947). Stars brighter than the sixth magnitude having spectral types between O6 and F5 are studied.

The parameters which are used to specify the spectral type are D and λ_{I}. The definition of these quantities is illustrated in Figure 11. Here a line is drawn to

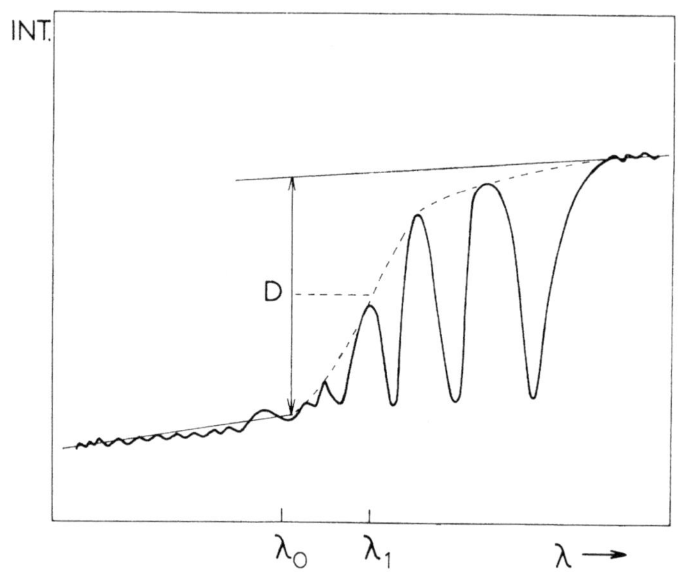

Fig. 11. The definition of the parameters D and λ_1.

represent a microphotometer tracing of a spectrogram in the region of confluence of the Balmer lines of hydrogen. The continuum above and below the Balmer discontinuity is approximated by straight lines, allowance being made for the overlapping of the wings of the hydrogen lines. The wave length at which the Balmer discontinuity begins, as one approaches from short wave lengths, is called λ_0. Let F_R be the monochromatic flux a this wave length on the projected line which represents the continuous spectrum at longer wave lengths and let F_V be the flux on the line representing the continuous spectrum at short wave lengths. Then the quantity D is

$$D = \log F_R - \log F_V, \tag{II-2}$$

The wave length λ_1, which represents the effective position of the Balmer jump is found by drawing a line parallel to the continuum of longer wave length at a height $\frac{1}{2}D$ and noting at which wave length this line intersects the microphotometer tracing. The parameter D is a measure of the spectral type (temperature in the outer layers of the stellar atmosphere) and λ_1 is a measure of the luminosity (pressure in the outer layers of the stellar atmosphere).

Barbier, Chalonge, and Divan also measure the slopes of the straight lines representing the continuous spectrum and give these as gradients φ_{uv}, φ_b, and φ_r for spectra which extend to long wave lengths. If it is assumed that the shape of the continuous spectrum may be represented by a Planck curve over the range of wave length in which each gradient is measured, colour temperatures may be deduced for the ultraviolet, blue, and red regions of the spectrum. The blue gradient, φ_b, is measured between λ4139 and λ4509, with a mean wave length of 4250 Å. This quantity is used as a third parameter by Chalonge and Divan to differentiate between F type and B type stars having the same values of D and λ_1.

The parameter D is useful for estimating the equivalent spectral types of B type model atmospheres, for the continuous spectrum is one of the first things that is computed. From the results of CHALONGE and DIVAN (1952) one may compile a table showing how D varies with spectral type. This is given in Table 1. The observed

TABLE 1

The observed variation of D with spectral type

Spectral type	D	Spectral type	D	Spectral type	D
O9	0.047, 0.049	B1	0.092, 0.095	B5	0.260, 0.319
O9.5	0.07	B2	0.143	B8	0.350, 0.405
B0	0.089	B2.5	0.155, 0.156	B9	0.431, 0.517
B0.5	0.087	B3	0.216, 0.217	A0	0.494, 0.504

extreme range of values of D for each spectral type is listed. The data are taken from Table II of CHALONGE and DIVAN (1952). The precise shape of the observed relation, D-spectral type, is to some extent dependent upon the adopted spectral types. When one consults different lists of spectral type for the bright O and B stars, one may find different types assigned to the same star. For instance, the star 10 Lacertae is designated Oe5 in the *Henry Draper Catalogue*, O9sk in the *Victoria Revised* system (PLASKETT and PEARCE, 1930a), O8.5ss, intermediate by WILLIAMS (1936b), O8.5 by PETRIE (1947a), and O9V in the MK system (MORGAN and JOHNSON, 1953). These differences are of some significance when one wishes to estimate temperatures and pressures accurately and to establish unique relationships between temperature and spectral type. Another example is offered by the star ζ Cassiopeiae. The type is B3 in the *Henry Draper Catalogue*, B2sk on the Victoria Revised system, B2.5*s*, intermediate according to Williams, and B2V on the MK system. Variations such as these occur in the estimated types because somewhat different classification criteria are used in each case. They serve to remind us that in reality the stars of any spectral type, whatever the system, are not a completely homogeneous group. It is well to be aware that these inhomogeneities exist, and to investigate the range of the inhomogeneity before concluding that stars of one spectral type are similar objects suitable for investigating the variation of some parameter such as interstellar reddening by means of measurements of the varying intensity distribution of light with wave length. Some of the observed variation may be due to the differences between the stars themselves. The effects of such inhomogeneities in the various systems of spectral classification would tend to cancel out if one could observe a large number of objects of the same spectral type. One of the difficulties of drawing general conclusions about early type stars is that the O and B stars are very scarce and one can rarely study a large enough sample that the statistical fluctuations do in fact cancel out.

Chalonge and Divan have correlated their parameters D, λ_1, and φ_b with the spectral types and luminosity classes of the MK system and, where available, with n_m the upper quantum number of the last visible Balmer line, and with g, the acceleration of gravity in the atmosphere. Figure 12 shows the correlation between D

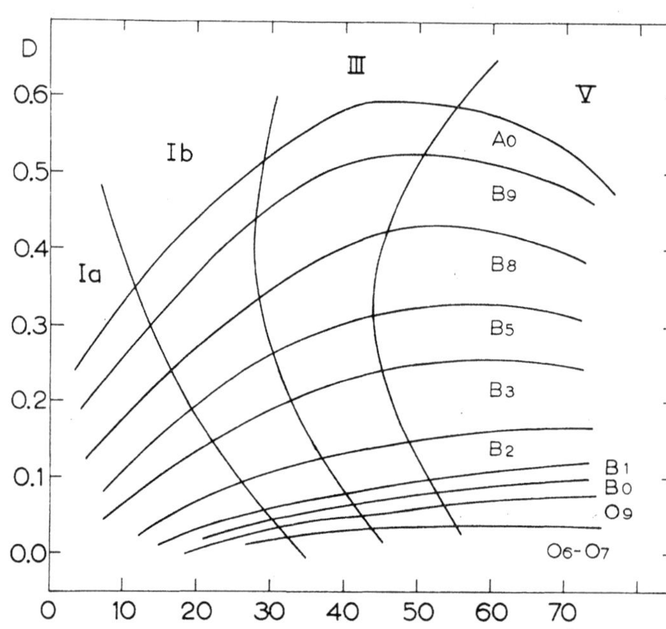

Fig. 12. The correlation between (D, λ_1) and MK spectral type.

and λ_1 and MK spectral type for stars of types A0 to O6. No independent attempt has been made to calibrate the parameters D and λ_1 in terms of absolute magnitude.

2.6 *The work at the Crimea Observatory*

KOPYLOV (1958a) has measured the equivalent widths of many of the stronger lines in 109 O and B stars. The spectrograms have a dispersion of 75 Å/mm at Hγ. Kopylov has compared his results with the material discussed above and with other less extensive series of measurements, and he has given linear transformations of the other series of measures to his own system. Thereby he is able to assemble lists of equivalent widths in the spectra of 238 O5–B7 stars.

In a following paper (KOPYLOV, 1958b), Kopylov has investigated the variation of line strength with spectral type such as Williams and all later observers have done and he has derived quantitative criteria which are about equivalent to the MK spectral type criteria. Kopylov has also attempted to set up absolute magnitude criteria in addition to the well known variation of the strength of the hydrogen lines with absolute magnitude. Kopylov's adopted absolute magnitudes are chiefly based on membership in various groupings of stars for which Kopylov or others have derived the distance. With the early B stars and the O stars it becomes difficult to find spectral criteria which distinguish between shell stars and true supergiants. Kopylov does not discuss this question.

2.7 *The work at the Stockholm Observatory*

SINNERSTAD (1961a, 1961b) has measured the equivalent widths and apparent central depths of the most prominent lines in 179 stars of MK types between O6 and F8 of all luminosity classes, and he has demonstrated how the various lines vary in strength with spectral type and with luminosity. His spectrograms have a dispersion of 75 Å/mm at Hγ. SINNERSTAD (1961b) has made a particular study of how the Balmer lines vary with luminosity and he has shown that the line depth measured on low dispersion, low resolution spectrograms is a useful parameter for classification in addition to the equivalent width. The line depth is useful in separating supergiants from main-sequence stars and for confirming spectral type.

CHAPTER III

THE LUMINOSITIES OF THE O AND B TYPE STARS

1. General principles of calibrating luminosity criteria

When the B type stars are classified, they are arranged into groups of differing luminosity at each spectral type by considering the strengths and shapes of lines which are sensitive to changes in pressure and to the extent of atmosphere. In the atmospheres of the stars of high luminosity, the pressure is lower than in the atmospheres of the stars of lower luminosity and the atmosphere is more extended. The most striking spectral features demonstrating this in B type spectra are the Balmer lines of hydrogen and the He I lines. In main-sequence stars the Balmer lines have wide wings of a typical shape due to Stark effect. For a given spectral type, as the luminosity increases and the electron pressure in the atmosphere decreases, the Stark effect becomes less and the Balmer lines of hydrogen become narrower. In the most luminous stars, the supergiants, large-scale motions in the atmosphere dominate over Stark effect as a broadening agent, and the absorption lines have a bell shape rather than a sharp, narrow core with extended wings. These changes in line profile produce readily measurable changes in the equivalent width of each Balmer line.

The Victoria classification system uses the changing equivalent width of Hγ, $W(\mathrm{H}\gamma)$, to measure the luminosities of the B stars. The MK system assesses the change in appearance of the spectrum, noting also the relative change in intensity of some of the lines from the silicon ions, O II and so forth, and assigns luminosity classes. The problem is to relate the value of the luminosity parameter, $W(\mathrm{H}\gamma)$ or luminosity class, uniquely to visual absolute magnitude, M_V, for all spectral types among the B stars. Furthermore, one would like to find similar parameters that would be useful for the O stars. The equivalent width of Hγ is a useful parameter from type B0 to at least spectral type A2. In the O stars the feature at Hγ is a blend of H and He II, thus the equivalent width of the gross feature cannot be used as a criterion of luminosity for the O stars.

In principle it is simple to calibrate a luminosity criterion in terms of absolute magnitude. Merely measure the criterion in the spectra of stars of known absolute magnitude and derive a correlation between the strength of the criterion and the absolute magnitude. For the O and B type stars the problem becomes difficult because the O and B stars are few in number and not many are close to the sun. Since it is difficult to obtain accurate distances for distant objects, it is difficult to find standard B type stars of known luminosity.

From the definition of visual absolute magnitude one may readily show that the error in visual absolute magnitude, M_V, which will be caused by an error Δr in the estimate of the distance of the object is

$$\Delta M_V = 2.17\,\Delta r/r\,. \tag{III-1}$$

If it is desired that the uncertainty ΔM_V be less than 0.2 magnitude, the distance must be known to within ten per cent. Conversely if one wishes to discuss the distribution in space of the O and B stars to an accuracy of ten per cent, one must be able to assign absolute magnitudes with an accuracy of ± 0.2 magnitude. Many of the interesting groups of O and B stars appear to have distances of 1000 to 2000 parsecs. Since, on the average, it is impossible to assign absolute magnitudes with an accuracy better than ± 0.2 magnitude, the positions in space of these stars are not known within a matter of some 100 to 200 parsecs. Such uncertainties in distance are not small when one thinks in terms of the diameters of aggregates such as open clusters or even associations.

One major purpose for desiring to know accurately the absolute magnitudes of the O and B stars is to be able to use these stars as distance indicators for studying galactic structure. Therefore it is not permissible to use assumptions regarding galactic structure for determining the distances of the standard stars which are to be used to calibrate the luminosity criteria. This means that the radial-velocity displacements due to galactic rotation and the strength of the interstellar calcium or sodium lines may not be used to establish the distances of the standard stars. On these grounds estimates of the absolute magnitude of the O and B stars such as those made by PETRIE (1947a) and by WILLIAMS (1936b) may not be used for calibrating the luminosity criteria.

The procedures that are valid are to make use of trigonometric parallaxes, after correcting the apparent visual magnitude of the star for interstellar absorption, and to select as standard stars those O and B type stars which are members of groups of stars (binaries or clusters), the distance of which have been determined independently of any assumptions about galactic structure or of the luminosities of the O and B stars. The relationship between parallax and visual absolute magnitude is the following:

$$M_V = m_0 + 5 + 5\log\pi\,, \tag{III-2}$$

where m_0 is the apparent magnitude of the star corrected for interstellar absorption. Consider a fifth magnitude B type star which has an absolute magnitude of -1.5 (which is an average value for a middle B type star). Substitution of these values in equation (III-2) shows that the parallax of the B star will be 0.″005. Since the average probable error of a trigonometric parallax is of the order of ±0.″006, it is clear that most if not all B stars are too distant for trigonometric parallaxes to yield useful information about their distances. Most of the standard stars which are available for calibrating the luminosity criteria must be selected from binaries and clusters whose distances are known from other sources.

The first thing to do is to demonstrate that the standard star is a physical member of the binary system or cluster. This is best done by showing that the motion of each star in the group is the same. For most secure results, both radial velocities and proper motions are required and one must show that the peculiar motions of the stars, that is the residual motions which remain after the reflex solar motion has been subtracted, are identical. It is very difficult to do this, particularly for binaries or clusters at distances greater than a few hundred parsecs. Other criteria for membership in the group are apparent propinquinity in the sky, and apparent position in the colour-magnitude diagram for the group. The latter condition presumes that we know approximately what the absolute magnitudes are for each spectral type and precisely where the stars should lie in the colour-magnitude diagram. Many early type stars are close binaries, unresolved when magnitude and colour measurements are made. This fact introduces an unavoidable scatter in the colour-magnitude diagram of any cluster or group of early type stars.

Another method of estimating the absolute magnitudes of stars is from their effective temperatures and radii. This method yields quite consistent results for those stars which are not such close binaries that the shapes of the stars are distorted by tidal effects and when not much light is contributed to the system by gas streams. For a simple, spherical star the luminosity is given by

$$L_* = \pi R_*^2 T_{\text{eff}}^4 \tag{III-3}$$

where T_{eff} is the effective temperature of the star. It follows that the difference in bolometric absolute magnitude between a star and the sun is

$$\Delta M_B(\odot - *) = 2.5 \log \frac{L_*}{L_\odot} = 5 \log \frac{R_*}{R_\odot} + 10 \log \frac{T_{\text{eff}}(*)}{T_{\text{eff}}(\odot)}. \tag{III-4}$$

Knowing the bolometric absolute magnitude of the sun, its radius and effective temperature, one may estimate the bolometric absolute magnitude of the star, once one has R_* and $T_{\text{eff}}(*)$. Then, if one can estimate the bolometric correction, one may derive the visual absolute magnitude of the star. This method is used to derive standard absolute magnitudes for B type stars from the radii estimated from the light curves and radial-velocity orbits of eclipsing binaries, the effective temperatures estimated from the spectral types, and estimates of the bolometric correction from theoretical studies. It will yield visual absolute magnitudes with an uncertainty of ± 0.3 magnitude.

Concern is often expressed over our lack of knowledge of the true bolometric corrections for O and B stars. This lack of knowledge is not so serious as might at first be assumed, so long as the effective temperature and the bolometric correction which are adopted for each spectral type have been predicted from the same theoretical work. Model atmospheres without line blanketing certainly predict too bright a spectrum in the far ultraviolet for O and B type stars. However, this bright spectrum is used to derive both the effective temperature and the bolometric correction. So long as values from the same set of models are used together, the effect on the estimated visual absolute magnitude will not be so great as one might fear. It certainly is not so great as to invalidate this method of estimating absolute magnitudes for the early type stars. At least as much uncertainty is attached to the selection of systems such that the estimates of radius have validity. This point is discussed further in Chapter IX. The real problem is to decide upon the equivalent spectral type of a model atmosphere.

2. The calibration of the Victoria system of luminosity determination

2.1 *The A and B stars*

In two important papers Petrie (PETRIE and MAUNSELL, 1949, PETRIE, 1952) selected standard A and B type stars, determined their distances and plotted the resulting absolute magnitudes against the measured equivalent width of Hγ to derive an empirical calibration of the equivalent width of Hγ as a criterion of luminosity. These calibrations are valid from type A3 to B8, and from type B7 to type O9. One mean curve is given for each spectral range and correction terms are listed to allow for small changes due to spectral type.

Petrie considered that the distances of the selected standard stars could be determined in the following ways:

1. From trigonometric parallaxes. This method is useful only for A stars. The parallaxes of the nearest B type stars accessible from the northern hemisphere are too small to be individually of any weight.

2. From dynamical parallaxes, making use of binaries with known orbits.

3. From the spectroscopic parallaxes of G to M type companions of A and B stars. Petrie tests that each binary is a physical pair by examining the radial velocity of both

TABLE 2

Cluster distance moduli adopted by R. M. Petrie

Cluster or Group	True Modulus	Source for Modulus
Ursa Major nucleus	1.99	cluster motion
Hyades	3.29	cluster motion
Coma Berenices	4.27	trigonometric parallaxes
Pleiades	5.00	trigonometric parallaxes, dynamic parallaxes and spectroscopic parallaxes of stars of types later than A8, allowing for reddening
*Perseus moving cluster	5.23	cluster motion
Praesepe	5.62	spectroscopic parallaxes of 11 stars of types later than F0
*Perseus stream (α Per)	5.8	trigonometric and spectroscopic parallaxes (five A type stars)
I.C. 4665	6.9	spectroscopic parallaxes of three A stars
Orion cluster	7.8	trigonometric parallaxes and spectroscopic parallax of one B9 star
Messier 35	8.9	spectroscopic parallaxes of two G stars
h and χ Persei	10.7	spectroscopic parallaxes of three M stars allowing for reddening

* These appear to be different groups; the first group contains 2 stars which Petrie later (1964) assigned to the α Persei cluster.

stars. He adopts the Mount Wilson system of spectroscopic parallaxes for the G to M stars. This system is based on trigonometric parallaxes.

4. From eclipsing binaries by adopting effective temperatures and bolometric corrections from the compilation by KUIPER (1938), and using equation (III-4).

5. From galactic clusters the distances of which are known from geometric considerations. The clusters used by Petrie are listed in Table 2. Somewhat less than one-half Petrie's stars are selected from the groups given in Table 2 and from the Scorpio-Centaurus stream; the remainder are members of visual or eclipsing binary systems. The cluster moduli adopted by Petrie are listed as well as an indication of the source for the adopted distance of the group. Altogether there are 152 standard stars for calibrating the equivalent width of Hγ for the A stars and 99 standard stars for calibrating the B type relation. Seven of these stars are in the Scorpio-Centaurus stream, and Petrie adopts the absolute magnitudes for these stars deduced by BLAAUW (1946) from the motions of the stars.

The following relations are found:

Type A3 to B8

$$M = 2.09 - 13.25 \times 10^{-0.093W}, \tag{III-5}$$

Type B7 to O9

$$M = - 10.96 \times 10^{-0.145W}, \tag{III-6}$$

A correction depending upon spectral type must be added to each nominal absolute magnitude M. These spectral type corrections are listed in Table 3. Petrie estimates

TABLE 3

Spectral type corrections to Petrie's relations for absolute magnitude

A type calibration				B type calibration			
Sp Type	M_V	Sp Type	M_V	Sp Type	M_V	Sp Type	M_V
A3	+ 0.2	B8	+ 0.2	B7	+ 0.0	B2	+ 0.2
A2	+ 0.0			B6	+ 0.0	B1	+ 0.3
A1	− 0.4			B5	+ 0.1	B0	+ 0.3
A0	− 0.4			B4	+ 0.1	O9	+ 0.3
B9	+ 0.0			B3	+ 0.2		

that the probable error of an absolute magnitude determined from equations (III-5) and (III-6) is about ± 0.3 magnitude. Relations (III-5) and (III-6) were derived for use with the m_V system of apparent magnitudes. If they are to be used with the V system of magnitudes in order to derive distances, the values of M must be made brighter by 0.1 mag., since on the average for B stars the V magnitude is 0.1 mag. brighter than the m_V magnitude.

It is very difficult to find B type supergiants which can be used as luminosity standards. The clusters h and χ Persei are very important in this respect since they furnish most of the very luminous late B and A type standard stars used by Petrie. Petrie's adopted parallax for these clusters depends upon the absolute magnitude,

-4.0, which he adopts for the three M type supergiants HD 13136, HD 14270, and HD 14330 which are members of this cluster. Petrie estimates the absolute magnitudes of these stars by the method of line ratios, comparing single-prism spectra of the M stars with similar spectra of μ Cephei, α Orionis, 5 Lacertae, β Pegasi, and β Andromedae. Corrections for interstellar absorption are applied by means of colour excesses. The Orion cluster furnishes the early B type supergiants which determine the luminous end of relation (III-6). No stars from h and χ Persei are used in obtaining relation (III-6). Recently FEAST (1963) has added further important observational material concerning the luminosities of the B type supergiants, see section 5 of this chapter.

Relations (III-5) and (III-6) demonstrate an empirical qualitative relationship between electron pressure in an early type atmosphere and visual absolute magnitude, for the equivalent width of Hγ depends strongly upon the value of the electron pressure and only very slightly upon temperature (spectral type). This relation could be put in explicit form by using the relationship derived by UNSÖLD (1942b) between W(Hγ) and log P_e, but such a relation would have little real meaning, for the generalisations that are made to obtain Unsöld's formula are too sweeping to be valid in all early type stellar atmospheres.

More recently PETRIE (1964) has examined anew the problem of calibrating the equivalent width of Hγ in terms of absolute magnitude in the case of stars of spectral types O6 to A3. He has assembled a series of homogeneous measurements of W(Hγ) in the spectra of 166 early type stars belonging to the Pleiades, the α Persei cluster, NGC 2244, and to the association I Orionis and that around h and χ Persei. He also uses the components of six visual binary systems.

Petrie is able to show that for members of one cluster or association, the curve relating W(Hγ) to V_0, the visual magnitude corrected for interstellar reddening, has a shape which is independent of suspected evolutionary changes. Thus he is able to derive the shape of the relation between W(Hγ) and V_0 by sliding the curves for individual clusters and associations together. In this way he establishes the relative moduli of the above groups to the Pleiades. He fixes the zero point by means of the trigonometric parallaxes of 6 nearby stars of types A1 to A3, by means of 6 visual binaries which have late-type components which can be fitted to the standard colour-magnitude sequence, and by means of 9 eclipsing binaries which are considered to have well-established orbits. In this way Petrie finds that the true distance moduli for the calibrating groups of stars are as follows:

Pleiades	5.4 ± 0.07 mag
α Per cluster	6.0 ± 0.08 mag
NGC 2244	9.7 ± 0.10 mag
I Orionis	8.1 ± 0.08 mag
h and χ Per	11.8 ± 0.11 mag

The new calibration is given in the following table. The listed spectral type corrections must be added to the tabular M values in order to find M_V. This new calibration gives visual absolute magnitudes for the B type stars which are about 0.5 mag. brighter on the average than those obtained from the original calibrations (Table 3 and relations (III-5) and (III-6)).

Petrie's calibration of Hγ equivalent widths (1964)

$W(H\gamma)$ Å	M mag	$W(H\gamma)$ Å	M mag	$W(H\gamma)$ Å	M mag	$W(H\gamma)$ Å	M mag
1.56	−6.78	4.59	−3.09	9.26	+0.01	18.80	+1.49
1.84	−6.06	5.52	−2.37	11.20	+0.74	20.98	+1.65
2.43	−5.21	6.67	−1.52	13.86	+1.21		
3.53	−4.16	7.46	−1.23	16.02	+1.42		

Spectral type corrections to 1964 calibration of $W(H\gamma)$

Spectrum	Classes Ia, Ib	Classes II–V	Spectrum	Classes Ia, Ib	Classes II–V	Spectrum	Classes Ia, Ib	Classes II–V
O6		+1.0	B2	+0.0	+0.1	B8	−1.2	−0.2
O7		+0.8	B3	−0.2	+0.1	B9	−1.5	−0.1
O8	+0.5	+0.6	B4	−0.4	+0.0	A0	−1.8	+0.0
O9	+0.4	+0.5	B5	−0.6	−0.1	A1	−2.1	+0.1
B0	+0.3	+0.3	B6	−0.8	−0.1	A2	−2.4	+0.3
B1	+0.1	+0.2	B7	−1.0	−0.2	A3	−2.8	+0.5

2.2 *The O type stars*

It is difficult to determine the luminosities of the O type stars of spectral types earlier than O9. One cannot continue to use the equivalent width of Hγ as a criterion of luminosity because the measured equivalent width is influenced by the blend with He II λ4338.67. The strength of the He II lines varies strongly with spectral type. The higher Balmer lines H10 (λ3797.900), H11 (λ3770.632), and H12 (λ3750.154) are not seriously blended with each other nor with He II, since the He II 4–*n* series breaks off near the line 4–19 which is at λ3813.50 Å. However, they do not show a clear luminosity effect when the measured equivalent widths are plotted against the assigned absolute magnitude (UNDERHILL, 1955). The points scatter more or less uniformly over the diagram. This result was obtained from measures of equivalent width of H10, H11, and H12 in the spectra of 21 stars which have spectral types between O8.5 and O5 and for which absolute magnitudes could be estimated from their membership in a cluster or binary system containing B type stars. It is not certain whether this result is obtained because the spread in absolute magnitude among the available O type stars is small, or because the equivalent widths of the higher Balmer lines cease to be pressure sensitive in early O type stars.

There are 16 stars with spectral types between O5 and O8.5 to which standard absolute magnitudes fainter than −5.1 can be assigned. The mean absolute magnitude of this group is −4.2 and the scatter about the mean corresponds to a probable error for a single value of ±0.35 magnitude, which is close to the probable error expected from the uncertainties in the standard absolute magnitudes. Five early O type stars among the group studied appear to have brighter absolute magnitudes. They are

IC 1805 No. 5	O8.5	−5.5	variable radial velocity, ? dbl lines
HD 15558	O6	−6.5	variable radial velocity
HD 15570	O5.5f	−6.9	variable radial velocity
HD 15629	O7	−5.6	variable radial velocity
HD 17505	O7	−6.0	

The first four stars are close together in the sky and may belong to a cluster which lies in the nebulosity IC 1805. All have variable radial velocity or show double lines, thus the adopted absolute magnitudes should be reduced to allow for the light of the companion. The adopted distance for IC 1805 is 1.98 kpc, but this value is not certainly on the Victoria distance scale. It has not been confirmed by measurements of the strength of Hγ in the spectra of proved B type members of the cluster. The star HD 17505 is believed to belong to the cluster IC 1848 for which the estimate of distance is also not certainly on the Victoria distance scale since no confirmation by means of B type stars has been obtained. It is essential that the distances of these two groups be firmly established, for they offer the chief evidence of the existence of early O type stars with visual absolute magnitudes greater than −5.1 magnitudes. The spectra of these stars do not look appreciably different on low dispersion from the spectra of other early O type stars to which less luminous absolute magnitudes have been assigned.

Four of the early O type stars to which standard absolute magnitudes on the Victoria system were assigned by UNDERHILL (1955) are Of stars. These stars do not appear to be significantly brighter than the O stars in which the f emission features are not strong. It appears that the absolute magnitudes of early O type stars lie between −4.0 and −4.5 on the average. A few brighter stars, M_V about −6.0 or so, certainly do exist at spectral type O9 and later. The apparent non-existence of supergiants at early O spectral types may be due to the necessity for a large value of gravity to maintain a mechanically stable atmosphere against the large radiation pressure gradient which is expected for the hottest stars. Supergiants, by definition, have a low value of g ($g \leqslant 1000$) in their atmosphere. They seem to have luminosities brighter than −6.0 at all spectral types. Margherita Hack has re-examined the question of the spectral classification and the absolute magnitudes of the O stars (BOTTO and HACK, 1962; HACK, 1963). She has followed tradition in selecting the relative strength of the He I lines to the He II lines as a criterion for spectral type and has considered the following four criteria to be luminosity indicators:

(1) the equivalent width of Hγ,

(2) the ratio $r = \dfrac{W(\text{C III } 4650 + \text{N III } 4097 + \text{Si IV } 4088 + \text{Si IV } 4116)}{W(\text{H}\gamma + \text{H}\delta)}$,

(3) the ratio $R_c / W(\text{H}\gamma)$ where R_c is the central absorption of Hγ,

(4) the quantum number n_m of the last visible hydrogen line.

These criteria may all be used to isolate objects with extended atmospheres, but they do not clearly differentiate between shell stars and supergiants. This point will be discussed further in Chapter XV. The result of this choice of luminosity criteria is that the Of stars are assigned to luminosity class I or III, that is they are called supergiants or giants. The work of Margherita HACK (1963) purports to establish absolute magnitudes for each luminosity class among the O stars, but the problem of separating shell stars from supergiants is not considered and no independent new information about the absolute magnitudes of the O stars is obtained.

KOPYLOV (1958b) has studied the O stars and has attempted to assign absolute magnitudes to the O stars from consideration of membership in clusters and associations. He has also considered spectroscopic criteria for determining absolute magnitudes. According to HACK (1963), Kopylov's absolute magnitudes for main-sequence stars are systematically brighter than the absolute magnitudes assigned by UNDERHILL (1955).

One of the most serious problems to be solved before reliable absolute magnitudes can be assigned to the O stars from spectroscopic criteria is to find criteria that are surely unique indicators of effective temperature and of luminosity and of nothing else. Nothing that has been suggested so far is reliable in this respect.

3. Tests of the Victoria system

PETRIE (1956) and PETRIE and MOYLS (1956) have tested the internal consistency and the external consistency of the Victoria system. The internal consistency was tested in three ways. They measured the equivalent width of Hγ in the spectra of 53 B type stars for which trigonometric parallaxes are available and found that

$$\pi_{\text{sp}} - \pi_{\text{trig}} = +\,0\overset{''}{.}002 \pm 0\overset{''}{.}004\,.$$

Thus the original Victoria system agrees with the system of trigonometric parallaxes. Secondly they studied 25 eclipsing binary systems for which they felt the light curves and radial-velocity orbits were adequate to permit the deduction of visual absolute magnitudes and found

$$M_{\text{sp}} - M_{\text{ecl}} = -\,0.2 \pm 0.1\,\text{mag}\,.$$

Thirdly they measured the equivalent width of Hγ in the spectra of 8 stars of the Scorpio-Centaurus stream for which absolute magnitudes are known from the study of stellar motions (BLAAUW, 1946) and deduced absolute magnitudes finding that

$$M_{\text{sp}} - M_{\text{cl}} = +\,0.02 \pm 0.07\,\text{mag}\,.$$

They tested the external consistency of the original Victoria system by comparing the mean spectroscopic parallaxes of groups of stars with the mean parallaxes that may be estimated for these groups from proper motions, and by seeing if the average distances for the groups of stars agree with the distances that may be deduced from the average radial velocities assuming galactic rotation. The first method is most successful with nearby stars; the second method is most satisfactory for distant stars.

Petrie and Moyls computed mean parallaxes from the upsilon components of proper motion of 343 stars and compared these with the mean parallaxes which are deduced from Hγ measures. They divided their stars into three groups: dwarfs, which are defined as stars with $M_V \geqslant -2.0$, intermediates which are stars with $-2.0 < M_V < -4.0$, and giants which are stars with $M_V \leqslant -4.0$. From 177 dwarf B stars they found

$$\bar{\pi}_{sp} = 0''.0053 \pm 0''.0001 \qquad \text{and} \qquad \bar{\pi}_{\upsilon} = 0''.0051 \pm 0''.0003;$$

from 108 B stars of intermediate luminosity they found

$$\bar{\pi}_{sp} = 0''.0029 \pm 0''.0001 \qquad \text{and} \qquad \bar{\pi}_{\upsilon} = 0''.0019 \pm 0''.0002;$$

and from 58 intrinsically bright B stars they found

$$\bar{\pi}_{sp} = 0''.0013 \pm 0''.0000 \qquad \text{and} \qquad \bar{\pi}_{\upsilon} = 0''.0004 \pm 0''.0001.$$

For the stars brighter than -2.0, the results suggest that the spectroscopic distances are too small. However, this conclusion is not clear because the method of mean parallaxes begins to fail for the intrinsically bright stars because the measured proper motions are too small. The test is not adequate in this case.

In order to test the Victoria system by means of galactic rotation, Petrie and Moyls studied the radial velocities of 369 stars. They divided the stars into three groups according to luminosity as above. A solar motion of 20 km/sec was adopted together with a K-term of $+1.0$ km/sec, and the value of the galactic rotation constant A was put equal to 18.8 km/sec. The mean distances resulting from the analysis are the following. From 197 dwarf B type stars

$$r_{sp} = 290 \pm 3\,\text{pc} \qquad \text{and} \qquad r_{gr} = 250 \pm 46\,\text{pc};$$

from 111 B stars of intermediate luminosity

$$r_{sp} = 620 \pm 9\,\text{pc} \qquad \text{and} \qquad r_{gr} = 560 \pm 61\,\text{pc};$$

and from 61 stars in the giant category

$$r_{sp} = 1275 \pm 25\,\text{pc} \qquad \text{and} \qquad r_{gr} = 1170 \pm 82\,\text{pc}.$$

Petrie and Moyls conclude that the stars of the dwarf group are too close for galactic rotation to be an effective test of the system of absolute magnitudes.

Petrie and Moyls have compared their system of absolute magnitudes with other older systems such as that of WILLIAMS (1936b) and with the MK system. The chief discrepancy is with the absolute magnitudes suggested by calibrations of the MK system for main-sequence stars of spectral type B2 and earlier, the original Victoria absolute magnitudes being less bright by about 0.4 magnitude on the average.

Absolute magnitudes on the Victoria system have been assigned for 384 stars. Spectral types on the Victoria system have also been determined for these stars. The resulting Hertzsprung–Russell diagram is shown in Figure 13. The stars plotted in this diagram are the B type stars accessible from Victoria and brighter than the 8th magnitude. It is quite clear that the luminosities of the early B type stars spread more

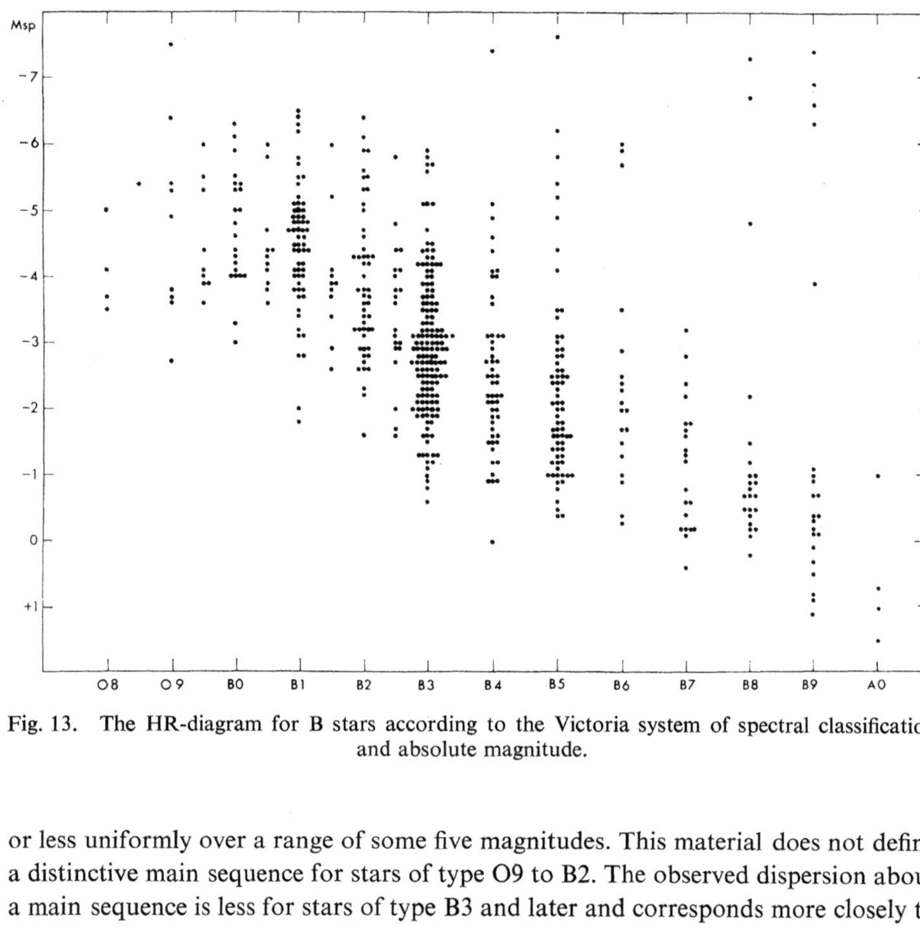

Fig. 13. The HR-diagram for B stars according to the Victoria system of spectral classification and absolute magnitude.

or less uniformly over a range of some five magnitudes. This material does not define a distinctive main sequence for stars of type O9 to B2. The observed dispersion about a main sequence is less for stars of type B3 and later and corresponds more closely to the dispersion expected according to the uncertainties of the method. Since the observing list is limited by apparent magnitude, the volume of space which has been searched is smaller for the late B type stars than it is for the early type stars. However, since the visual absolute magnitude of all B type supergiants appears to be about -6.5, such volume considerations do not apply to the supergiants. This material suggests that the space density of early B type supergiants is about two times greater than that of late B type supergiants. The spread in absolute magnitude that is observed among the early B type stars may be interpreted as a spread in evolutionary age (stage of development) for these stars. Apparently such a wide spread does not occur among the late B type stars. STRÖMGREN (1963) has shown that a similar spread does exist among the A and F type stars and has suggested that it is related to age.

4. The calibration of the MK luminosity classes for B type stars

A preliminary calibration of the MK luminosity classes was given by MORGAN and KEENAN (1951) where a tabulation is given of absolute magnitude *vs.* spectral class. They state that the values for spectral types A0 and later, luminosity classes II to V are derived from trigonometric parallaxes. For the stars of type B the listed absolute magnitudes are based partly upon the absolute magnitudes which were assigned by BLAAUW (1946) to some stars of the Scorpio-Centaurus stream from a study of space motions. Also mean parallaxes and absolute magnitudes were computed for stars of type B3 V and B5 V from their proper motions. No list is given of the stars so studied nor of the adopted proper motions. Finally it is stated that the luminosities for the supergiants depend upon members of the Orion cluster and of h and χ Persei, but it is not stated what distances were adopted for these clusters nor how these distances were determined.

JOHNSON and HILTNER (1956) gave a revised calibration of the MK system. Their idea was to fit the observed cluster sequences in a colour-luminosity diagram to a single zero-age main sequence. They emphasise the necessity of fitting the observed main sequences at a sufficiently low part that one can be sure that no evolutionary effects have moved the stars from the presumed well-defined zero-age sequence. This technique assumes that it is known that all the stars one observes are members of the cluster. The presence of a number of foreground or background stars could obscure the result. Although it is clearly essential that the colour and magnitude observations which are used refer only to bona-fide cluster members which are sufficiently close together in space that the spread Δm_v in apparent magnitude resulting from the true spread in distance $\Delta r/r$ is negligibly small, Johnson and Hiltner make no direct effort to verify this for their stars. Apparent propinquinity in the sky and a spectral type or colour close to that necessary to delineate a sequence in the colour-magnitude diagram is all that is required as proof of membership in the group.

Their method is the following. First they set up a standard main sequence valid from B8 to K7 by means of observations of nearby stars and clusters particularly the Hyades, Praesepe and the Pleiades. They use the theory of the evolution of stars to work back to zero age. They estimate that the probable error of the resulting zero-age main sequence is ± 0.2 magnitude. They then derive moduli for the Orion cluster, NGC 2362, and h and χ Persei by fitting the observations of stars believed to be members of these clusters to the adopted zero-age sequence in the range A0 to B8. Uniform reddening over NGC 2362 and over h and χ Persei is assumed. The interstellar absorption, A_V, is estimated by the relation

$$A_V = 3E(B\text{–}V), \tag{III-7}$$

where $E(B–V)$ is the observed colour excess in the $B–V$ colour. Each star in the Orion cluster is corrected separately using its measured colour excess.

In this way they find the following true moduli:

Orion cluster	8.0 mag
NGC 2362	10.9 mag
h and χ Persei	11.8 mag

Using these cluster distances and the spectral types of cluster members they calibrate the MK system in the range O5 to A5 and for M Iab stars. Their results for the O and B stars are given in Table 4. Nearly all of their early B type and O type stars are

TABLE 4
The Johnson and Hiltner calibration of the MK luminosity classes*

Type	V	IV	III	II	Ib	Ia
O5–O8	(− 5.1)					
O9	− 4.6	− 5.4	− 6.4			
O9.5	− 4.4	− 5.0	− 5.7	− 6.2	− 6.6	
B0	− 4.1	− 4.6	− 5.0	− 5.4	− 6.2	− 6.9
B0.5	− 3.6	− 4.2	− 4.7	− 5.2	− 5.8	− 6.9
B1	− 3.1	− 3.8	− 4.4	− 4.9	− 5.8	− 6.9
B1.5	− 2.6	− 3.4	− 4.0	− 4.8	− 5.8	− 6.9
B2	− 2.0	− 2.9	− 3.6	− 4.7	− 5.8	− 6.9
B3	− 1.5	− 2.5	− 3.3	− 4.6	− 5.8	− 6.9
B5	− 0.9					− 6.9
B6	− 0.5					− 6.9
B7	− 0.2					− 6.9
B8	+ 0.5					− 6.9
B9	+ 1.0					− 6.9
A0	+ 1.6					− 6.9

* The table gives M_V.

contained in the Orion association, which may well be much more extensive in depth than the Orion cluster, and in the so-called h and χ Persei association which includes, for instance, stars in the clusters IC 1805 and IC 1848.

Johnson and Hiltner note that their absolute magnitudes for the luminous stars in Orion and in h and χ Persei are about 0.7 magnitude brighter than those assigned by Petrie. This result is largely due to the different modulus adopted for h and χ Persei. For the main-sequence stars of type B3 and later there is essential agreement. It is chiefly the Orion stars and the stars of the so-called outer h and χ Persei association that define the luminosity calibration for the early B and O type stars. It is, therefore, essential to demonstrate that all these stars are at the distance of the B8 and A0 stars which are used to determine the distance to each association. If the resulting absolute magnitude calibration is to be accurate to within ± 0.2 magnitude the distances of the individual early type stars must be known to within 10 per cent. Since the Orion cluster is believed to lie at about 400 parsecs, all of the B8 to A0 *and* the early B type stars which are used should lie between 360 and 440 pc; all of the stars in the Perseus region should lie between 2000 and 2500 pc if the association has indeed a true modulus of 11.8 magnitudes.

The calibration of the MK system for the B stars is discussed further by JOHNSON and IRIARTE (1958) and by JOHNSON (1959). They note that the calibration given by Johnson and Hiltner is weak because of the poor junction between the stars observed in NGC 2362 and those in the Pleiades. They adopt MITCHELL's (1957) modulus of 6.1 for the α Persei group, and note that this permits them to join the colour-magnitude sequence of the α Persei cluster to that of the Pleiades in a satisfactory manner. They then fit all available UBV observations of galactic clusters containing B type stars to the α Persei cluster main sequence obtaining the distance moduli given in Table 5. Mitchell's result for the α Persei cluster depends upon photoelectric photometry of stars listed as members of this stream by HECKMANN, DIECKVOSS, and KOX (1956) from studies of proper motion. It is 0.3 magnitude greater than the value estimated by Petrie from trigonometric parallaxes and spectroscopic parallaxes of A type stars. Johnson and Iriarte give no discussion of membership in the clusters. They assume that all of the stars in the various photometric lists which they consult are sufficiently closely at the same distance that the resulting zero-age main sequence which they derive is accurate to within ± 0.1 magnitude. This point should be examined more carefully. Do the early type stars and the late B type or A type stars which are used to determine the modulus of the group really form a compact cluster?

Four of the clusters used by Petrie are also used by Johnson and Iriarte. In each case the modulus adopted by Johnson and Iriarte is larger than that adopted by Petrie. These differences in modulus are:

Pleiades	0.5 mag
α Persei	0.3 mag
I Orionis	0.3 mag
h and χ Persei	1.1 mag

The distances to these important groups of stars should not be a matter of opinion, but proved facts. Further information about these clusters is given in Chapter VI.

Johnson and Iriarte then use their distance moduli and all available measurements of the equivalent width of Hγ in spectra of stars they consider to belong to the clusters given in Table 5 to calibrate $W(\mathrm{H}\gamma)$ in terms of M_V. They derive a table of values of

TABLE 5

Cluster distance moduli adopted by Johnson and Iriarte

Cluster	True Modulus	Cluster	True Modulus
Pleiades	5.5	M 36	10.5
α Persei	6.1	NGC 2362	10.7
I Orionis	8.1	NGC 6530	10.7
M25	8.9	h and χ Persei	11.8
NGC 2264	9.4		

TABLE 6

The Johnson and Iriarte calibration of $W(H\gamma)$*

$W(H\gamma)$	0.0	0.1	0.2	0.3	0.4	0.5	0.6	0.7	0.8	0.9
0						− 8.5	− 8.4	− 8.2	− 8.0	− 7.8
1	− 7.6	− 7.5	− 7.3	− 7.2	− 7.0	− 6.8	− 6.6	− 6.4	− 6.3	− 6.1
2	− 6.0	− 5.9	− 5.7	− 5.6	− 5.4	− 5.3	− 5.2	− 5.1	− 5.0	− 4.9
3	− 4.8	− 4.7	− 4.6	− 4.5	− 4.5	− 4.4	− 4.3	− 4.2	− 4.1	− 4.1
4	− 4.0	− 3.9	− 3.7	− 3.6	− 3.5	− 3.3	− 3.2	− 3.0	− 2.9	− 2.8
5	− 2.7	− 2.6	− 2.4	− 2.3	− 2.2	− 2.1	− 2.0	− 1.9	− 1.8	− 1.7
6	− 1.6	− 1.5	− 1.4	− 1.3	− 1.3	− 1.2	− 1.1	− 1.0	− 0.9	− 0.9
7	− 0.8	− 0.7	− 0.7	− 0.6	− 0.5	− 0.5	− 0.4	− 0.4	− 0.3	− 0.2
8	− 0.2	− 0.1	+ 0.0	+ 0.0	+ 0.1	+ 0.1	+ 0.2	+ 0.2	+ 0.3	+ 0.3
9	+ 0.4	+ 0.4	+ 0.5	+ 0.5	+ 0.6	+ 0.6	+ 0.7	+ 0.7	+ 0.8	+ 0.8
10	+ 0.9	+ 0.9	+ 1.0	+ 1.0	+ 1.0	+ 1.1	+ 1.1	+ 1.2	+ 1.2	+ 1.2
11	+ 1.3	+ 1.3	+ 1.4	+ 1.4	+ 1.5	+ 1.5	+ 1.6	+ 1.6	+ 1.7	+ 1.7
12	+ 1.8	+ 1.8	+ 1.8	+ 1.9	+ 1.9	+ 2.0	+ 2.0	+ 2.0	+ 2.1	+ 2.1
13	+ 2.2	+ 2.2	+ 2.2	+ 2.3	+ 2.3	+ 2.4	+ 2.4	+ 2.5	+ 2.5	+ 2.5
14	+ 2.6	+ 2.6	+ 2.7	+ 2.7	+ 2.8	+ 2.8	+ 2.8	+ 2.9	+ 2.9	+ 3.0
15	+ 3.0	+ 3.0	+ 3.1	+ 3.1	+ 3.2	+ 3.2	+ 3.2	+ 3.3	+ 3.3	+ 3.4
16	+ 3.4	+ 3.4	+ 3.5	+ 3.5	+ 3.5					

* The table gives a nominal M_V; $W(H\gamma)$ is measured in angstroms.

TABLE 7

Spectral type corrections to Johnson and Iriarte's calibration of $W(H\gamma)$

Sp Type	Correction	Sp Type	Correction	Sp Type	Correction
O	0.0	B5	− 0.4	A0	− 1.5
B0	0.0	B6	− 0.5	A1	− 1.8
B1	0.0	B7	− 0.6	A2	− 2.0
B2	− 0.1	B8	− 0.9	A3	− 2.2
B3	− 0.2	B9	− 1.2	A5	− 2.6

TABLE 8

The Johnson and Iriarte calibration of the MK luminosity classes*

Type	V	IV	III	II	Ib	Iab	Ia
O5–O7	− 5.5						
O8	− 5.2						
O9	− 4.8		− 6.1	− 6.1	− 6.2	− 6.2	− 6.2
B0	− 4.4	− 5.0	− 5.4	− 5.7	− 6.2	− 6.3	− 6.6
B1	− 3.6	− 4.1	− 4.4	− 5.3	− 6.1	− 6.3	− 6.9
B2	− 2.5	− 3.3	− 3.6	− 4.9	− 6.0	− 6.3	− 7.1
B3	− 1.7	− 2.5	− 2.9	− 4.4	− 5.9	− 6.3	− 7.1
B5	− 1.0	− 1.8	− 2.2	− 4.0	− 5.8	− 6.3	− 7.0
B7	− 0.4	− 1.2	− 1.6		− 5.7	− 6.3	− 6.8
B8	+ 0.1	− 0.7	− 1.0		− 5.6	− 6.3	− 6.7
B9	+ 0.6	− 0.2	− 0.4		(− 5.5)	(− 6.3)	− 6.8
A0	+ 1.0	+ 0.7	(+ 0.6)		(− 5.3)		− 7.1

* The table gives M_V.

M_V *vs.* $W(\mathrm{H}\gamma)$ and give a table of spectral type corrections. This material is presented in Tables 6 and 7. Using this calibration of $W(\mathrm{H}\gamma)$ as a measure of absolute magnitude, they derive the visual absolute magnitudes of all possible MK B type standard stars from published measures of $W(\mathrm{H}\gamma)$. From this material they establish the calibration of the MK spectral types which is given in Table 8. Whether one considers this determination of the luminosities of the B type stars superior to any other that has been suggested depends upon how satisfied one is that the distance moduli given in Table 5 represent sufficiently accurately the distances of the stars used to calibrate the relationship between $W(\mathrm{H}\gamma)$ and M_V. JOHNSON (1959) has used the individual distances deduced by BERTIAU (1958) from space motions of stars of the Scorpio-Centaurus stream to confirm that the absolute magnitudes for main-sequence stars of types B2, B3, and later suggested by the data of Table 8 are correct. JOHNSON and IRIARTE (1958) also give a table of values relating $W(\mathrm{H}\gamma)$ to the intrinsic colour $(B-V)_0$ of an early type star.

No checks have been published of the external consistency of the calibration of the MK luminosity classes such as those attempted by Petrie and Moyls for the original Victoria system. Considering the rather superficial justification that has been given for membership of many of the stars in the clusters which are used to define the zero-age main sequence, particularly the stars of spectral type earlier than B3, it would be desirable to have such checks.

5. The absolute magnitudes of the B type supergiants

It is difficult to determine the absolute magnitudes of the B type supergiants because B type supergiants are rare and very few are known to be member of clusters the distances of which can be determined from members of later spectral type. None of the clusters are close enough that geometric methods of determining distance are useful. FEAST (1963) has assembled a list of 23 B type stars which fall in one or other subdivision of luminosity class I and for which magnitudes and colours on the UBV system exist. These stars are listed in Table 9. Two corrections have been made to Feast's list. The first column gives the *Henry Draper* number or the number in the *Henry Draper Extention*, or failing that, the BD number; the second column gives the cluster designation. The stars in *h* and *χ* Persei are numbered according to the list by OOSTERHOFF (1937); the spectral types are by JOHNSON and MORGAN (1955). FEAST (1958) identifies the stars in NGC 3292 and gives MK spectral types. Similar information for the stars of NGC 4755 is listed by FEAST (1963). The spectral types for NGC 6913 are taken from MORGAN and HARRIS (1956); the numbers are those of ZUG (1933). The spectral types for the remaining stars are from the catalogue by HILTNER (1956). They were assigned mostly by W. W. Morgan. The cluster numbers are from the catalogue by HOAG, JOHNSON, IRIARTE, MITCHELL, HALLAM, and SHARPLESS (1961). The absolute magnitudes listed in column 7 are those derived by Feast from the cluster moduli and average reddening corrections that he adopts. The modulus for each cluster was found by fitting the observed colour magnitude diagrams

TABLE 9

B type supergiants with MK spectral types and lying in clusters

HD Number BD Number	Cluster name	Spectral type	V	$B–V$	Source for Photo-metry	Feast M_V	Petrie	
							Type	M_V
91969	3293–4	B0Ib	6.52	+ 0.01	3	− 6.7		
227634	6871–5	B0Ib	7.92	+ 0.25	5	− 4.7		
229221	6913–6	B0:I:pe	9.20	+ 0.91	4	− 4.2		
91943	3293–3	B0.5Ib	6.72	+ 0.06	3	− 6.5		
229238	6913–3	B0.5Ib	8.86	+ 0.90	4	− 4.5		
13854	*h* and χ–16	B1Iab	6.49	+ 0.28	2	− 7.0	B0.5	− 6.2
229239	6913–1	B1Iab	8.97	+ 0.86	4	− 4.4		
14052	*h* and χ–662	B1Ib	8.18	+ 0.30	2	− 5.3		
111934	4755–I6	B1.5Ib	7.00	+ 0.18	1	− 6.3		
15690	957–1	B1.5Ib	8.01	+ 0.67	5	− 6.1	B2	− 5.5
14143	*h* and χ–1162	B2Ia	6.66	+ 0.51	2	− 6.8	B1	− 7.1
14443	*h* and χ–2227	B2Ib	8.05	+ 0.34	2	− 5.4		
13841	*h* and χ–3	B2Ib	7.40	+ 0.24	2	− 6.1	B0.5	− 5.2
111900	4755–C	B2Ib	6.80	+ 0.24	1	− 6.5		
111973	4755–B	B3Ia	5.94	+ 0.22	1	− 7.4		
14134	*h* and χ–1057	B3Ia	6.55	+ 0.47	2	− 6.9	B3	− 6.5
250290	2129–1	B3Ib	7.36	+ 0.61	5	− 6.3		
9311	581–1	B5Ib	7.26	+ 0.29	5	− 6.0	B5	− 4.7
+ 60° 339	663–1	B6Iab	8.42	+ 0.67	5	− 5.9		
7902	457–2	B6Ib	6.97	+ 0.40	5	− 6.7	B4	− 6.4
14542	*h* and χ–2621	B8Ia	7.00	+ 0.62	2	− 6.5	B8	− 5.6
+ 60° 333	663–2	B8Iab	8.87	+ 0.88	5	− 5.5		
111904	4755–A	B9Iab	5.75	0.32	1	− 7.5		

1. ARP, H. C. and VAN SANT, C. T.: 1958, *Astrophys. J.* **63**, 341.
2. JOHNSON, H. L. and MORGAN, W. W.: 1955, *Astrophys. J.* **122**, 429.
3. HOUCK, T.: 1956, *Thesis*, University of Wisconson.
4. MORGAN, W. W. and HARRIS, D. L.: 1956, *Vistas in Astronomy* **2** (ed. by A. BEER), Pergamon Press, London, p. 1124.
5. HOAG, A. A., JOHNSON, H. L., IRIARTE, B., MITCHELL, R. I., HALLAM, K. L., and SHARPLESS, S: 1961, *Publ. U.S. Naval Obs.* **17**, 439.

to the zero-age sequence recommended by JOHNSON and IRIARTE (1958). Petrie (PETRIE and MAUNSELL, 1949, PETRIE, 1956, PETRIE and MOYLS, 1956) has measured the equivalent width of Hγ for some of the stars. The spectral types and absolute magnitudes assigned by him are listed in columns 8 and 9. These absolute magnitudes have been found from relations III-5 and III-6.

Feast points out that the absolute magnitudes of the four stars which lie in the Cygnus Rift, that is 6871-5, 6913-1, 6913-3, and 6913-6, are rather fainter than the values estimated for the other supergiants. He concludes that this result is almost certainly due to underestimated distances for NGC 6871 and NGC 6913 owing to an abnormal ratio of total to selective absorption in the Cygnus Rift. This explanation may be valid, but it is not certainly so. The spectral classification of 6871-5 has been determined by only one observer and it should be confirmed. If it were in error by one luminosity class, there would be little reason to think that the distance to NGC 6871 had been seriously underestimated. Hiltner lists the spectral types B0II, B0I, and Bpe, respectively for 6913-1, 6913-3, and 6913-6 rather than the types listed in Table 9. Since the range in absolute magnitude between main-sequence stars and supergiants is only about two magnitudes at type B0, it is clear that one must be rather certain about the spectral type and luminosity class before concluding that the estimate of distance is in error owing to a faulty correction for interstellar absorption. It is difficult to obtain good classification spectra for these four stars on account of the severe reddening which distorts the appearance of the spectrum.

When the four stars in Cygnus are excluded, Feast finds the following average intrinsic colours for the B type supergiants:

Range of types	Mean type	No. of stars	Average $(B-V)_0$
B0 -B1	B0.5	4	−0.26
B1.5–B3	B2	9	−0.19
B3 -B9	B7	6	−0.07

The absolute magnitude does not change significantly with spectral type. From six stars of luminosity class Ia Feast finds that M_V is -7.0; from thirteen stars of luminosity class Ib he finds that M_V is -6.0.

The absolute magnitudes assigned by Petrie to eight of the stars in Table 9 are about 0.5 magnitude fainter on the average than the absolute magnitudes assigned by Feast. The discrepancy for HD 9311 is serious. Since the equivalent width of Hγ corresponds to an absolute magnitude near -4.7, the star cannot be at the assigned distance of the cluster. The spectral type B5Ib is probably correct, for the spectrum closely resembles that of 67 Ophiuchi, a standard B5Ib star.

Petrie uses spectrograms of higher dispersion than those used to obtain the spectral types quoted by Feast, thus it is not surprising that he assigns slightly different types, for different details may be seen on his spectrograms. The apparent systematic differences in absolute magnitude are more disquieting. However, before concluding that there is a systematic error in Petrie's calibration of $W(\mathrm{H}\gamma)$ for the most luminous stars, it would be advisable to check by some independent method that the supergiants of each cluster truly are at the distance found by fitting the adopted fainter members of the clusters to the zero-age main sequence. An error of 0.5 magnitudes corresponds to an error of 25 per cent in the distance.

6. A comparison of Victoria and MK luminosities for B type stars

Spectral types and luminosities from Hγ measures have been assigned by Petrie to many of the brighter B type stars which are accessible from the northern hemisphere. Spectral types and luminosity classes on the MK system are also available for some of these stars. By adopting a calibration of the MK luminosity classes one may obtain visual absolute magnitudes. Data for 131 stars are given in Table 10. The *V* magnitude for each star is listed if it was available; otherwise the m_v magnitude is given (in italics). The *V* magnitudes are probably correct to within about ± 0.02 mag (m.e.). It is rather disconcerting to find that some of the brighter stars appear with slightly different *V* magnitudes in different photometric lists. Since it was impossible to discover which value of *V* was correct, the first value encountered was entered in Table 10. The photometric lists did not differ regarding the MK spectral type. The corresponding MK absolute magnitude is taken from the calibration by Johnson and Iriarte.

The spectral types assigned by Petrie have been derived from the study of spectrograms of higher dispersion than those used to obtain the MK types. Since more fine detail is visible on these spectrograms, it is not unexpected that differences in type of 0.5 or 1 unit occur. It is more surprising to find that in nineteen cases the difference in spectral type is greater than or equal to 1.5 units. Usually Petrie assigns spectral types later than the MK types. Only for HD 37055 and HD 109387 does he assign an earlier type. The difference in type accounts for many of the positive residuals in the ΔM_V column. It is probable that Petrie's types are more nearly correct than the MK types, for the fine spectral detail may be seen better on his spectrograms. The spectral type differences among the early B stars, in particular, depend much upon the visibility of weak lines. Since the spectral type correction to the absolute magnitude deduced from the equivalent width of Hγ does not change by a significant amount with spectral type among the early B stars, uncertainty in the spectral type produces little uncertainty in the absolute magnitude.

Remarks are given at the end of Table 10 for all cases in which the differences in spectral type or luminosity appear to be larger than what may be expected from the uncertainty in assigning types and absolute magnitudes. The differences in assigned type make it impossible to make a strict, statistically valid comparison between Petrie's absolute magnitudes for a given type and those suggested by Johnson and Iriarte.

Johnson and Iriarte indicate that stars of luminosity class IV are about 0.8 mag. brighter than those of luminosity class V. It seems very doubtful that a clear systematic distinction between class IV and class V has been made in assigning the MK types. Some of the large positive differences in absolute magnitude between Petrie and the Johnson and Iriarte calibration of the MK luminosity classes occur because the star has been assigned to class IV when the Hγ strength suggests class V.

Setting aside doubts about it being possible to separate luminosity classes IV and V accurately by visual inspection of low dispersion spectrograms, the Victoria luminosities and the MK luminosities for main-sequence B stars from type B9 to B2 are in essential agreement. Petrie (1958a) has also demonstrated that the two systems of absolute magnitude are in essential agreement. He does not, however, list the individual stars upon which his comparison is made. On the other hand, the luminosities assigned to types BIV, B0V, and to the O stars by Johnson and Iriarte appear to be too bright. A study by Underhill (1955) has confirmed that the O stars, on the average, have visual absolute magnitudes in the neighbourhood of -4.2. Johnson and

TABLE 10

A comparison of spectral types and absolute magnitudes for 131 B stars

HD Number	Name	V	MK Type	MK M_V	Petrie Type	Petrie M_V	Re-marks	Δ Type P–MK	Δ M_V P–MK
886	γ Peg	2.83	B2IV	− 3.3	B2	− 2.6		0	+ 0.7
2905	κ Cas	4.15	B1Ia	− 6.9	B0.5	− 7.0		− 0.5	− 0.1
3360	ζ Cas	*3.7*	B2V	− 2.5	B2	− 2.4		0	+ 0.1
3369	π And	*4.4*	B5V	− 1.0	B5	− 0.8		0	+ 0.2
4180	o Cas	4.60	B2V	− 2.5	B5ne	− 2.2	1	+ 3	+ 0.3
4727	ν And	4.52	B5V	− 1.0	B5	− 1.2	dbl	0	− 0.2
9311		7.26	B5Ib	− 6.0	B5s	− 4.7		0	+ 1.3
13402		8.06	B0.5I	− 6.3	B0	− 6.2		− 0.5	+ 0.1
13841		7.40	B2Ib	− 6.0	B0.5	− 5.2	2	− 1.5	+ 0.8
13854		6.49	B1Iab	− 6.3	B0.5	− 6.2		− 0.5	+ 0.1
13866		7.52	B2Ib	− 6.3	B1	− 5.6		− 1	+ 0.7
14134		6.55	B3Ia	− 7.1	B3	− 6.5		0	+ 0.6
14143		6.66	B2Ia	− 7.1	B1	− 7.1		− 1	0.0
14357		8.53	B2II	− 4.9	B1.5	− 3.8	3	− 0.5	+ 1.1
14818	10 Per	6.25	B2Ia	− 7.1	B2	− 6.3		0	+ 0.8
14956		7.22	B2Ia	− 7.1	B2	− 6.5		0	+ 0.6
15497		7.03	B6Ia	− 6.9	B5	− 6.2		− 1	+ 0.7
15690		8.00	B1.5Ib	− 6.3	B2	− 5.5		+ 0.5	+ 0.8
22928	δ Per	3.03	B5III	− 2.2	B6	− 1.7		+ 1	+ 0.5
23180	o Per	3.82	B1III	− 4.4	B2	− 3.0	dbl	+ 1	+ 0.6
23288	16 Tau	5.45	B7IV	− 1.2	B7	− 0.7		0	+ 0.5
23302	17 Tau	3.69	B6III	− 2.0	B6	− 1.2		0	+ 0.8
23342	18 Tau	5.64	B8V	+ 0.1	B7	− 0.5		− 1	− 0.6
23338	19 Tau	4.29	B6V	− 0.7	B7	− 1.0		+ 1	− 0.3
23408	20 Tau	3.86	B7III	− 1.6	B6.5	− 1.2		− 0.5	+ 0.4
23432	21 Tau	5.75	B8V	+ 0.1	B8	+ 1.2	4	0	+ 1.1
23441	22 Tau	6.41	B9V	+ 0.6	B9	+ 1.0		0	+ 0.4
23480	23 Tau	4.16	B6IV	− 1.5	B6	− 1.1		0	+ 0.4
23630	η Tau	2.86	B7III	− 1.6	B7.5	− 1.8		+ 0.5	− 0.2
23850	27 Tau	3.62	B8III	− 1.0	B7	− 1.2		− 1	− 0.2
23923		6.16	B9V	+ 0.6	B8	+ 1.0		− 1	+ 0.4
24398	ζ Per	2.83	B1Ib	− 6.1	B0	− 4.6		− 1	+ 1.5
29763	τ Tau	*4.3*	B3V	− 1.7	B5	− 0.6	5	+ 2	+ 1.1
30614	α Cam	4.38	O9.5Ia	− 6.2	O9	− 6.9		− 0.5	− 0.7
30836	π^4 Ori	3.69	B2III	− 3.6	B2	− 2.8		0	+ 0.8
31237	π^5 Ori	3.70	B2IV	− 3.3	B2	− 2.8		0	+ 0.5
32630	η Aur	3.16	B3V	− 1.7	B5	− 0.8	5	+ 2	+ 0.9
33328	λ Eri	4.25	B2IV	− 3.3	B3	− 2.4	6	+ 1	+ 0.9
34085	β Ori	0.08	B8Ia	− 6.7	B8	− 6.1		0	+ 0.6
34503	τ Ori	3.58	B5III	− 2.2	B7	− 0.7	5	+ 2	+ 1.5
34748		6.30	B1.5V	− 3.0	B3	− 2.3	sb,5	+ 1.5	+ 0.7
34816	λ Lep	4.28	B0III	− 5.4	B1	− 2.1	3	+ 1	+ 3.3
35039	o Ori	4.72	B2IV	− 3.3	B2	− 2.0	6	0	+ 1.3
35149	23 Ori	4.96	B1V	− 3.6	B3	− 1.8	5	+ 2	+ 1.8
35299		5.71	B2V	− 2.5	B3	− 2.3		+ 1	+ 0.2
35411	η Ori A	3.33	B1IV	− 4.1	B2	− 2.3	dbl,7	+ 1	+ 1.8
35468	γ Ori	1.70	B2III	− 3.6	B2	− 2.0	3	0	+ 1.6
35497	β Tau	1.64	B7III	− 1.6	B8	+ 0.4	3	+ 1	+ 2.0
35762		6.74	B2V	− 2.5	B3	− 1.0	8	+ 1	+ 1.5
35777		6.60	B2V	− 2.5	B5	− 1.0	5	+ 3	+ 1.5

TABLE 10 (*continued*)

HD Number	Name	V	MK Type	MK M_V	Petrie Type	Petrie M_V	Re-marks	Δ Type P–MK	ΔM_V P–MK
35899		7.52	B5V	− 1.0	B4	− 0.7	sb	− 1	+ 0.3
36133		6.94	B2V	− 2.5	B5	− 0.8	5	+ 3	+ 1.7
36267	32 Ori	4.20	B5IV	− 1.8	B4	− 0.8	6	− 1	+ 1.0
36285		6.33	B1.5V	− 3.0	B3	− 1.6	5	+ 1.5	+ 1.4
36512	υ Ori	4.59	B0V	− 4.4	B3	− 2.4	5	+ 3	+ 2.0
36695	VV Ori	5.31	B1V	− 3.6	B2	− 2.1	5	+ 1	+ 1.5
36741		6.58	B2V	− 2.5	B5	− 1.0	5	+ 3	+ 1.5
36822	ϕ^1 Ori	4.50	B0IV	− 5.0	B0	− 2.9	8	0	+ 2.1
36824		6.69	B3V	− 1.7	B4	− 1.2		+ 1	+ 0.5
36959		5.67	B1V	− 3.6	B1	− 1.6	9	0	+ 2.0
36960		4.77	B0V	− 4.4	B0	− 2.5	10	0	+ 1.9
37016		6.22	B3V	− 1.7	B2	− 0.8	8	− 1	+ 0.9
37017		6.54	B1.5V	− 3.0	B2	− 1.0	8	+ 0.5	+ 2.0
37020	θ^1 Ori A	6.72	B0.5Vp	− 4.0	B2	− 2.5	5	+ 1.5	+ 1.5
37023	θ^1 Ori D	6.70	B0.5Vp	− 4.0	B0	− 1.8	11	− 0.5	+ 2.2
37041	θ^2 Ori A	5.07	O9V	− 4.8	B0	− 3.0	12	+ 1	+ 1.8
37042	θ^2 Ori B	6.38	B1V	− 3.6	B0	− 1.5	13	− 1	+ 2.1
37055		6.40	B3V	− 1.7	B1	− 0.9	14	− 2	+ 0.8
37128	ε Ori	1.71	B0Ia	− 6.6	B0	− 6.6		0	0.0
37303		6.02	B1V	− 3.6	B2	− 1.6	15	+ 1	+ 2.0
37397		6.83	B3V	− 1.7	B2.5	− 1.1		− 0.5	+ 0.6
37468	σ Ori	3.73	O9.5V	− 4.6	B0	− 3.8		+ 0.5	+ 0.8
37481		5.96	B1V	− 3.6	B3	− 1.7	sb,5	+ 2	+ 1.9
37742	ζ Ori A	1.72	O9.5Ib	− 6.2	O9.5	− 5.9		0	+ 0.3
37744		6.20	B1V	− 3.6	B3	− 1.5	5	2	+ 2.1
38771	κ Ori	2.04	B0.5Ia	− 6.7	B0	− 4.3	16	− 0.5	+ 2.4
38899	134 Tau	4.91	B9IV	− 0.2	B9.5	+ 1.3	6	+ 0.5	+ 1.5
39777		6.54	B2V	− 2.5	B3	− 1.1	5	+ 1	+ 1.4
40111	139 Tau	4.81	B1Ib	− 6.3	B0	− 4.4	16	− 1	+ 1.9
41117	χ^2 Ori	4.63	B2Ia	− 7.1	B2	− 6.3		0	+ 0.8
41753	ν Ori	*4.4*	B3V	− 1.7	B2	− 0.9		− 1	+ 0.8
42560	ξ Ori	*4.4*	B3V	− 1.7	B4	− 1.1		+ 1	+ 0.6
44743	β CMa	*2.0*	B1II	− 5.3	B1	− 4.0	3	0	+ 1.3
45542	ν Gem	4.15	B7IV	− 1.2	B6	− 1.1		− 1	+ 0.1
45546	10 Mon	4.98	B2V	− 2.5	B3	− 1.2	5	+ 1	+ 1.3
47105	γ Gem	1.95	A0IV	+ 0.7	A1	+ 0.9		+ 1	+ 0.2
48434		5.91	B0III	− 5.4	B0	− 5.4		0	0.0
48915	α CMa	− 1.47	A1V	+ 1.5	A1	+ 1.1		0	− 0.4
52918	19 Mon	*4.9*	B1V	− 3.6	B3	− 4.2	1	+ 2	− 0.6
74280	η Hya	4.29	B3V	− 1.7	B4	− 1.4		+ 1	+ 0.3
91316	ρ Leo	3.85	B1Ib	− 6.1	B0	− 5.6		− 1	+ 0.5
103287	γ UMa	2.43	A0V	+ 1.0	A1	+ 1.2		+ 1	+ 0.2
109387	κ Dra	3.84	B7V	− 0.4	B5	− 1.4	17	− 2	− 1.0
116658	α Vir	0.98	B1V	− 3.6	B2	− 2.7	dbl	+ 1	+ 0.9
120315	η UMa	1.86	B3V	− 1.7	B3n	− 1.1		0	+ 0.6
123299	α Dra	3.64	A0III	+ 0.1	A0	+ 0.8		0	+ 0.7
130109	109 Vir	3.75	A0V	+ 1.0	A0	+ 0.9		0	− 0.1
139006	α CBr	2.23	A0V	+ 1.0	A0n	+ 0.2		0	− 0.8
142096	λ Lib	5.02	B3V	− 1.7	B4	− 0.7	5	+ 1	+ 1.0
144217	β ScoAB	2.61	B0.5V	− 4.0	B0	− 3.5	dbl	− 0.5	+ 0.5
144218	β ScoC	4.92	B2V	− 2.5	B3	− 1.5	5	+ 1	+ 1.0

TABLE 10 (*continued*)

HD Number	Name	V	MK Type	MK M_V	Petrie Type	Petrie M_V	Re-marks	Δ Type P–MK	Δ M_V P–MK
144470	ω^1 Sco	3.97	B1V	− 3.6	B2	− 2.4	5	+ 1	+ 1.2
145502	ν^1 Sco	3.18	B2V	− 2.5	B3	− 1.5	5	+ 1	+ 1.0
147394	τ Her	3.89	B5IV	− 1.8	B6	− 0.6	6	+ 1	+ 1.2
148184	χ Oph	4.34	B1Vpe	− 3.6	B3	− 4.4	1	+ 2	− 0.8
149438	τ Sco	2.82	B0V	− 4.4	B0	− 3.4	8	0	+ 1.0
149757	ζ Oph	2.60	O9.5V	− 4.6	O9	− 3.9		− 0.5	+ 0.7
155763	ζ Dra	3.20	B6III	− 1.9	B7	− 1.4		+ 1	+ 0.5
160762	ι Her	3.80	B3V	− 1.7	B3	− 1.4		0	+ 0.3
161868	γ Oph	3.72	A0V	+ 1.0	A0	+ 1.5		0	+ 0.5
164284	66 Oph	4.84	B2V	− 2.5	B3n	− 1.6	5	+ 1	+ 0.9
164353	67 Oph	4.00	B5Ib	− 5.8	B4	− 5.1		− 1	+ 0.7
172167	α Lyr	0.03	A0V	+ 1.0	A1s	+ 1.2		+ 1	+ 0.2
176437	γ Lyr	3.23	B9III	− 0.4	B9s	− 0.7		0	− 0.3
180554	1 Vul	*4.6*	B3IV	− 2.5	B5n	− 1.6	1, 6	+ 2	+ 0.9
183913	β Cyg B	5.11	B8V	+ 0.1	B9	+ 0.6		+ 1	+ 0.5
187811	12 Vul	4.94	B3V	− 1.7	B4	− 1.3		+ 1	+ 0.4
196867	α Del	3.77	B9V	+ 0.6	B8.5	+ 0.8		− 0.5	+ 0.2
198478	55 Cyg	4.82	B3Ia	− 7.1	B2	− 7.2		− 1	− 0.1
199081	57 Cyg	4.64	B5V	− 1.0	B5	− 0.8	dbl	0	+ 0.2
202904	υ Cyg	4.45	B2V	− 2.5	B3ne	− 1.9		+ 1	+ 0.6
204172	69 Cyg	*5.8*	B0Ib	− 6.2	B0	− 5.6		0	+ 0.6
206165	9 Cep	4.72	B2Ib	− 6.0	B2	− 6.0		0	0.0
208947		*6.3*	B2V	− 2.5	B3	− 1.2	5	+ 1	+ 1.3
212120	2 Lac	*4.7*	B6IV	− 1.5	B5	− 0.9		− 1	+ 0.6
213420	6 Lac	4.50	B2IV	− 3.3	B3	− 3.0		+ 1	+ 0.3
214680	10 Lac	4.87	O9V	− 4.8	O9	− 3.0	8	0	+ 1.8
214993	12 Lac	5.25	B2III	− 3.6	B1	− 2.9		− 1	+ 0.7
216916	16 Lac	5.58	B2IV	− 3.3	B2	− 2.7	sb	0	+ 0.6
218440		*6.3*	B2V	− 2.5	B3	− 0.7	5	+ 1	+ 1.8
218674		6.73	B3IV	− 2.5	B3.5	− 1.6	sb	+ 0.5	+ 0.9

Sources for MK types and V magnitudes:

EGGEN, O. J.: 1961, *Roy. Obs. Bull.* No **41.**
JOHNSON, H. L. and BORGMAN, J.: 1963, *Bull. Astron. Inst. Neth.* **17,** 115.
JOHNSON, H. L. and HILTNER, W. A.: 1955, *Astrophys. J.* **122,** 429; 1956, *Astrophys. J.* **123,** 267.
JOHNSON, H. L. and MORGAN, W. W.: 1953, *Astrophys. J.* **117,** 313.
HARDIE, R. H. and CRAWFORD, D. L.: 1961, *Astrophys. J.* **133,** 843.

Sources for Petrie types and magnitudes:

PETRIE, R. M. and MAUNSELL, C. D.: 1949, *Publ. Dominion Astrophys. Obs.* **8,** 253.
PETRIE, R. M.: 1952, *Publ. Dominion Astrophys. Obs.* **9,** 251.
PETRIE, R. M. and MOYLS, B. N.: 1956, *Publ. Dominion Astrophys. Obs.* **10,** 287.
PETRIE, R. M.: 1956, *Vistas in Astronomy* **2** (ed. by A. BEER), Pergamon Press, London, p. 1346.

Remarks:

1. The Petrie type may be too late.
2. This is clearly a luminous star; the spectral type should be checked.
3. The luminosity class should be checked; the Hγ measures suggest a lower luminosity.
4. The luminosity assigned by Petrie from his adopted modulus for the Pleiades is + 0.6.
5. The strength of Hγ suggests that the later type is to be preferred.

TABLE 10 (*concluded*)

6. If the luminosity class were V rather than IV, there would be no discrepancy with the absolute magnitude assigned by Petrie.
7. The assigned MK spectral type of this double-lined star may be too early owing to the effects of a blended spectrum.
8. The Hγ absolute magnitude seems too faint for the type given by Petrie.
9. The luminosity assigned by Petrie from his adopted modulus for the Orion cluster is − 2.5.
10. The luminosity assigned by Petrie from his adopted modulus for the Orion cluster is − 3.4.
11. The luminosity assigned by Petrie from his adopted molduus for the Orion cluster is − 2.6.
12. The luminosity assigned by Petrie from his adopted modulus for the Orion cluster is − 3.1.
13. The luminosity assigned by Petrie from his adopted modulus for the Orion cluster is − 1.9.
14. The luminosity assigned by Petrie from his adopted modulus for the Orion cluster is − 1.7; his spectral type may be too early.
15. The luminosity assigned by Petrie from his adopted modulus for the Orion cluster is − 2.3.
16. The Hγ luminosity is in direct conflict with the assigned luminosity class, but it agrees well with the appearance of the spectrum at high dispersion.
17. This is a rapidly rotating Be star.

Iriarte's suggestion that the B1 and B0 main-sequence stars are brighter than this does not fit well with what may be deduced from the available information. At present there are too few B1, B0 and O stars known surely to be in clusters with securely established distances for the method of Johnson and Iriarte to be reliable for these stars. A study of Table 10 shows that a good many of the stars which are given types B1 or B0 on the MK system are given later types by Petrie. The Hγ absolute magnitudes are consistent with the later types.

The question of calibrating the luminosity classes of the MK types B1 and earlier requires further study. In particular, careful attention should be paid to proving that the early type stars under study are at the distance of the group to which they are supposed to belong and that the spread in depth of the group does not exceed fifteen or twenty per cent of the distance of the group. Unfortunately most of the stars of earliest type which are accessible from the northern hemisphere lie either in the extended Orion complex or in the so-called outer h and χ Persei association. Neither extended group is at a well-defined distance.

BLAAUW (1963) has discussed the question of calibrating luminosity criteria, and has demonstrated how the uncertainty in the modulus of a distant group may become quite large. He concludes that the modulus for the Orion cluster is 7.9 ± 0.2 mag, and that the modulus for h and χ Persei is 11.8 ± 0.3 mag. The quoted errors are probable errors. Petrie's estimate for Orion is in agreement with Blaauw's result; his estimate for h and χ Persei is significantly lower. The latter discrepancy, however, only affects slightly his estimates of the absolute magnitudes of a few A type supergiants. It is not the cause of the discrepancy between the absolute magnitudes suggested by Johnson and Iriarte for B1V and B0V stars and those suggested by Petrie. The Johnson and Hiltner absolute magnitudes for B1 and B0 stars are somewhat fainter than the Johnson and Iriarte values, and are in better accord with the results from Petrie's calibration of the equivalent width of Hγ.

WEAVER and EBERT (1964) have examined once more the question of the calibration of the MK luminosity classes for early type stars in terms of absolute magnitude. By adopting the distances established from UBV photometry and fitting to the zero-age main sequence of 35 clusters they derive the following absolute magnitudes for luminosity classes V to III. For luminosity classes II and I Weaver and Ebert conclude that the data are too few in number to provide a new calibration.

MK Type	V M_V	IV M_V	III M_V	MK Type	V M_V	IV M_V	III M_V
O5	−5.5			B1	−2.7	−3.8	−5.2
O6	−5.1			B2	−1.9	−3.1	−4.6
O7	−4.7			B3	−1.5	−2.5	−4.0
O7.5	−4.5			B5	−1.2	−1.9	−3.4
O8	−4.4			B6	−0.8	−1.4	−2.7
O9	−4.0	−5.0	−6.4	B7	−0.5	−1.0	−2.1
O9.5	−3.8	−4.7	−6.1	B8	−0.1	−0.5	−1.5
B0	−3.5	−4.4	−5.8	B9	+0.2	−0.1	−0.9
B0.5	−3.2	−4.1	−5.5	A0	+0.5	+0.4	−0.3

PETRIE's new (1964) calibration of $W(\mathrm{H}\gamma)$ in terms of M_V agrees quite well with that of Johnson and Iriarte, as may be expected since the new cluster distances adopted by Petrie are essentially the same as those adopted by Johnson and Iriarte. The absolute magnitudes for class V stars suggested by BLAAUW (1963) are a few tenths of a magnitude brighter than those suggested by the Hγ calibrations. Those of WEAVER and EBERT (1964) also differ by a few tenths of a magnitude.

An interesting question to ask is: why does the method used by Petrie in 1964 to calibrate $W(\mathrm{H}\gamma)$ in terms of M_V yield different results from that used by him in 1949 and in 1952, particularly for the main-sequence stars of types B0 to B3? The answer is found in the choice of standard stars which are used to define the shape of the $W(\mathrm{H}\gamma)$–M_V relation. In the original calibration the shape of the relation *and* the zero-point of the absolute magnitude scale were derived together by assigning absolute magnitudes to individual stars; in the second calibration the *shape* of the relation was found from cluster sequences and then the *zero-point* of the absolute magnitude scale was established by making a few stars with known absolute magnitudes fit the relation. In the first calibration quite a few eclipsing binaries and visual binaries were used as standard objects in addition to members of clusters and associations. In the second case the members of the clusters and associations were used only to establish the shape of the $W(\mathrm{H}\gamma)$–M_V relationship and the zero-point was established from a few carefully selected visual and eclipsing binaries and from a few stars with large trigonometric parallaxes. By this change of technique it was possible to determine the shape of the $W(\mathrm{H}\gamma)$–M_V relationship more accurately and the diffuseness due to the inherent difficulty of determining the distances of the B type stars was confined to the zero-point determination.

In the original calibration the position of the $W(\mathrm{H}\gamma)$–M_V curve was not well determined for the B0–B3 stars; the points from the Orion association tend to lie about 0.7 mag. higher than the points from the eclipsing and visual binaries for the same equivalent width of Hγ. The adopted curve is a compromise. If the curve had been drawn through the points from the Orion association only and little account taken of the points from the binaries, the revised results would have been obtained. Essentially the same absolute magnitudes are assigned to the Orion stars in both calibrations. In the new calibration, the part of the curve valid for main-sequence B0–B3 stars rests chiefly on the results from the Orion association, exactly as it does in the calibration of Johnson and Iriarte.

It is interesting to note that the six eclipsing binaries (σ Aql, TT Aur, AH Cep, Y Cyg, U Oph, AG Per) having spectral types B5 and earlier used by Petrie to set the zero-point of the new relationship, give a modulus 5.0 mag for the Pleiades exactly as adopted in Petrie's earlier work. When the three eclipsing pairs (AR Aur, β Aur, RX Her) of later type are used, the modulus of the Pleiades is 6.1 mag. Petrie's adopted value from all 9 pairs is 5.33 ± 0.15 mag. Only one visual binary of the six used by Petrie to estimate the zero-point is later than B5. It gives a modulus of 5.6 mag for the Pleiades; the other five stars yield a modulus of 5.16 mag. The seven stars with accurately known parallaxes are all A type stars. They yield a modulus for the Pleiades of 5.42 mag. Together the 11 stars of type B7 and later used by Petrie to set the zero-point of his new calibration of $W(\mathrm{H}\gamma)$ result in a modulus of 5.6 mag for the Pleiades while the 10 stars of types earlier than B7 yield a modulus of 5.1 mag. Clearly until this difference in zero-point depending upon spectral type is cleared up, no choice of the best calibration for the absolute magnitudes of the B stars can be made. None of the proposed calibrations differ systematically by more than 0.5 mag.

CHAPTER IV

SPECTRAL STUDIES BY MEANS OF PHOTOMETRY

1. Spectral classification and the UBV system of photometry

1.1 *The Q method of spectral classification*

The UBV broad-band system of colours and magnitudes was introduced by MORGAN and JOHNSON (1953). The properties of this photometric system and its relationships to earlier systems have been thoroughly studied, see for instance MORGAN and JOHNSON (1953) and CODE (1960). For B type stars, the V magnitude, effective wave length 5460 Å, is practically identical with the photo-visual magnitude system I_{pv}. We have

$$V = I_{pv} + (0.000 \pm 0.006) + (0.002 \pm 0.005)(B\text{–}V). \qquad \text{(IV-1)}$$

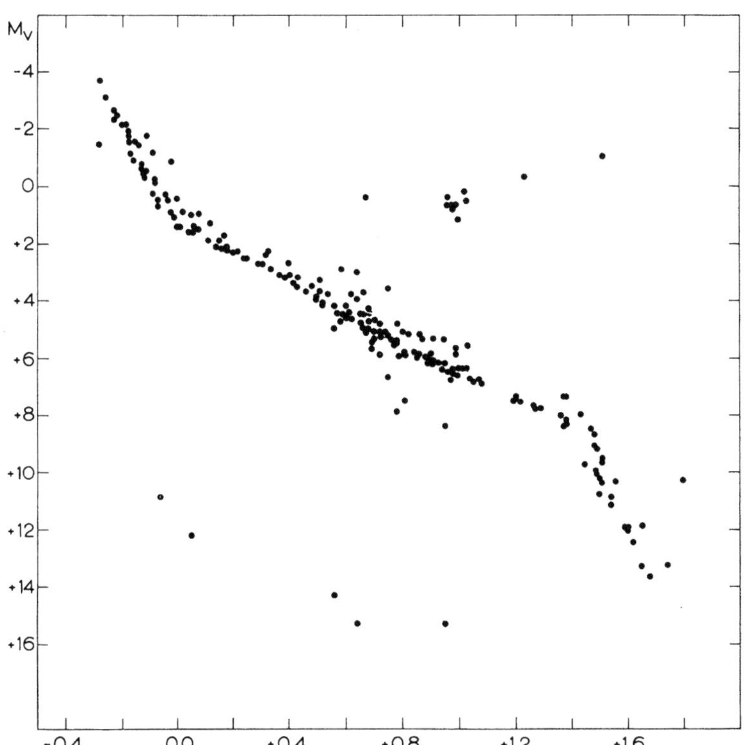

Fig. 14. The relationship between M_V and B–V for main-sequence stars according to MORGAN and JOHNSON (1953).

The B magnitude has an effective wave length of 4370 Å and the U magnitude has an effective wave length of 3460 Å. The width of each pass band is of the order of 1000 Å. The zero point for colour was established by setting

$$B–V = U–B = 0 \tag{IV-2}$$

for six A0V stars, namely α Lyrae, γ Ursae Majoris, 109 Virginis, α Coronae Borealis, γ Ophiuchi, and HR 3314. On this point, the UBV system is similar to the International system, which was based on photographic photometry, for the arbitrary ratio between the magnitudes measured in any two wave-length bands is fixed by stating that main-sequence A0 stars are equally bright in all colours. A list of photometric standard stars on the UBV system is given by JOHNSON (1955).

Using main-sequence stars with parallaxes greater than 0.″100, some stars in the Pleiades, in Praesepe, and in NGC 2362, Morgan and Johnson have plotted diagrams to illustrate the relationship between M_V and $B–V$ and between M_V and $U–B$. Their diagrams are reproduced in Figures 14 and 15. Their method is to establish the main

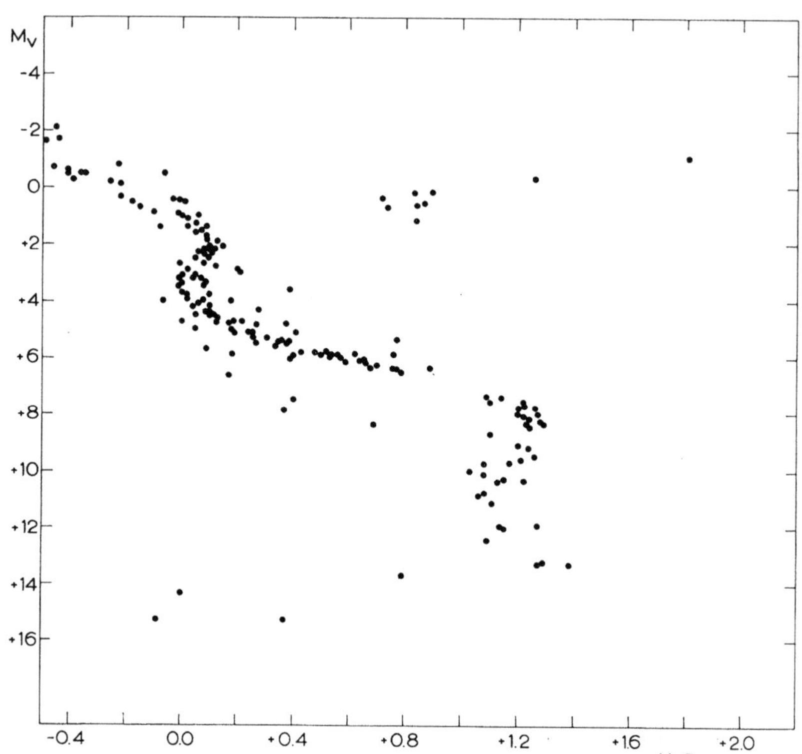

Fig. 15. The relationship between M_V and $U–B$ for main-sequence stars according to MORGAN and JOHNSON (1953). The abscissa is $U–B$.

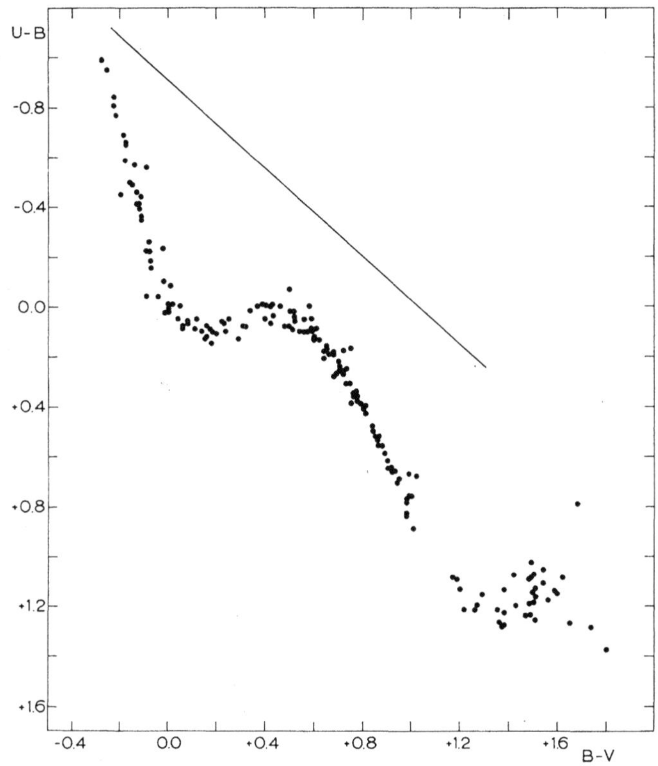

Fig. 16. The relationship between U–B and B–V for unreddened stars according to MORGAN and JOHNSON (1953).

sequence for stars of spectral type later than A0 using the stars within 10 parsecs, and then to fit the cluster colour magnitude diagrams to this sequence. The horizontal shift determines the reddening and the vertical shift gives the apparent distance modulus of the cluster.

Morgan and Johnson also plotted U–B against B–V and demonstrated the shape of the two-colour diagram for unreddened main-sequence stars. This relation is shown in Figure 16. The general trend upwards from right to left is due to the increasing temperature of the stars with smaller or negative B–V. The dip near U–B=0 and B–V=0 is caused by the blanketing effect of the strong Balmer absorption lines. Morgan and Johnson also indicate the position of the reddening path. The reddening path is the line on which B stars reddened by the selective interstellar absorption fall.

Colour excess is defined as the observed colour minus the intrinsic colour of any star, the intrinsic colour for each spectral type being estimated from the measured colours of nearby, and thus presumably unreddened, stars. Let the colour excess in the U–B range be designated by $E(U–B)$, that in the B–V range by $E(B–V)$. Morgan and Johnson found that the ratio $E(U–B)/E(B–V)$ was constant for the stars they studied and that it had the value 0.71 ± 0.03. This ratio is a measure of the wavelength dependence of interstellar reddening. They introduced the quantity

$$Q = (U-B) - \left[\frac{E(U-B)}{E(B-V)}\right](B-V), \tag{IV-3}$$

which is independent of reddening so long as the actual ratio of the colour excesses of the observed star is equal to the adopted average value. Morgan and Johnson set up an empirical relationship between Q and spectral type which is valid from the O types to type B9. The intrinsic colours for the B stars derived by MORGAN and JOHNSON (1953) and the Q-spectral type relation were later revised slightly by JOHNSON (1958). The revised values for main-sequence stars are given in Table 11.

TABLE 11
The intrinsic colours and Q values for main-sequence stars

MK Spectral type	$(B-V)_0$	$(U-B)_0$	Q	MK Spectral type	$(B-V)_0$	$(U-B)_0$	Q
O5	− 0.32	− 1.15	− 0.92	B2	− 0.24	− 0.86	− 0.69
O6	− 0.32	− 1.14	− 0.91	B3	− 0.20	− 0.71	− 0.57
O7	− 0.32	− 1.14	− 0.91	B5	− 0.16	− 0.56	− 0.44
O8	− 0.31	− 1.13	− 0.91	B6	− 0.14	− 0.49	− 0.39
O9	− 0.31	− 1.12	− 0.90	B7	− 0.12	− 0.42	− 0.33
O9.5	− 0.30	− 1.10	− 0.88	B8	− 0.09	− 0.30	− 0.24
B0	− 0.30	− 1.08	− 0.86	B9	− 0.06	− 0.19	− 0.15
B0.5	− 0.28	− 1.01	− 0.81	B9.5	− 0.03	− 0.10	− 0.08
B1	− 0.26	− 0.93	− 0.74	A0	+ 0.00	+ 0.00	+ 0.00

COUSINS, EGGEN, and STOY (1961) have measured colours and magnitudes for many southern B type stars on a photometric system similar to the original UBV system and they have established a relationship between Q_C, the value of Q formed from colours measured at the Cape Observatory and spectral type. They show that this relation yields spectral types on the MK system.

The use of the photometric quantity Q permits one to estimate spectral types for many early type stars which are not readily accessible for spectrographic study because of their faintness. However, these estimates cannot entirely supplant spectral studies, for it is implicity assumed that the differences in colour which are observed from star to star are representative of changes in the strengths of the features which are used for spectral classification and of the slope of the continuous spectrum. Shell stars, for instance or Be stars could give quite spurious results. Attempts have been made to use narrow-band photometry in order to overcome this inherent difficulty of spectral classification by means of photometry, but the problem of detecting peculiar stars remains. Is there enough redundancy in the photometric data alone to permit one to recognize peculiar stars? It is doubtful if the recognition of peculiar stars is possible when only UBV colours are available.

1.2 *The work of Wesselink on stars in the Magellanic Clouds*

The use of the wide-band photometric magnitudes U, B, and V to derive distances to clusters, associations and galaxies and to investigate the properties of interstellar material rests upon the assumption that the stars can be sorted uniquely into boxes defined by spectral type and luminosity class and that to each box there may be assigned a distinct set of values for the intrinsic colours $(B-V)_0$ and $(U-B)_0$. Experience has shown that this appears to be true for the early type stars within 2000 pc or so of the sun. WESSELINK (1962) observed stars in the direction of the Magellanic Clouds on the UBV system to test if the OB stars in the Magellanic Clouds gave results in accord with those found for stars in our galaxy.

Wesselink observed 48 foreground stars by which to tie his observations to the UBV system, 8 supergiant members of the Small Magellanic Cloud having MK spectral types between B0I and A0Ia and V magnitudes lying in the range $10.19 \leqslant V \leqslant 12.18$, and 12 members of the Large Magellanic Cloud having MK spectral types between O5f and A0Ia and V magnitudes lying in the range $10.53 \leqslant V \leqslant 12.60$. Wesselink derived the quantity

$$Q = U-B - 0.72(B-V) \qquad \text{(IV-4)}$$

and plotted it against the assigned spectral types (from spectra). The results are shown in Figure 17. The stars from the Large Magellanic Cloud are shown as filled circles,

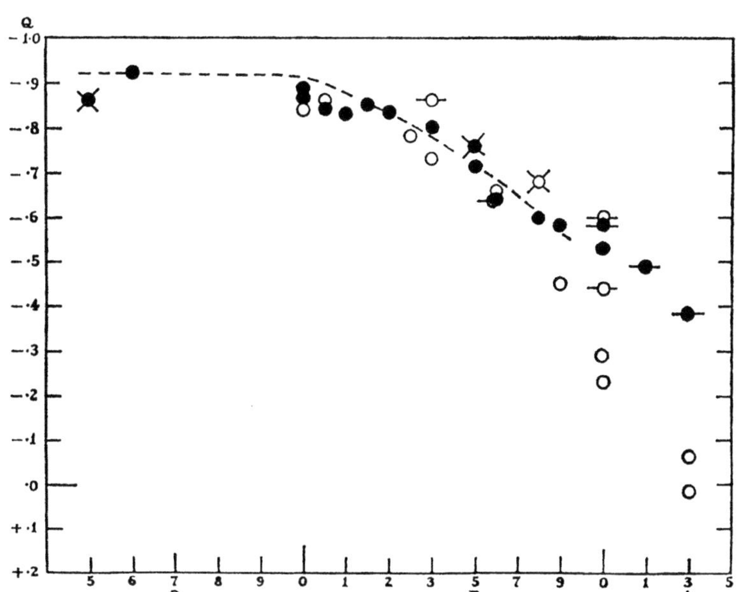

Fig. 17. The relation between Q and spectral type for stars in the Magellanic Clouds compared with that for stars in the neighbourhood of the sun. The latter relation is given by the dashed line. From WESSELINK (1962).

those from the Small Magellanic Cloud as open circles. Stars having, or suspected of having, emission lines are shown by other symbols. It is clear that the observed points fall along the line, given by JOHNSON (1958), relating Q to spectral type for supergiants. The observed supergiants in the Magellanic Clouds appear to be similar, photometrically, to the early-type supergiants in our galaxy.

Since Wesselink knows the spectral types of his stars, he is able to correct the observed colours of all the stars to those for type B0I. He then plots $(U-B)$ against

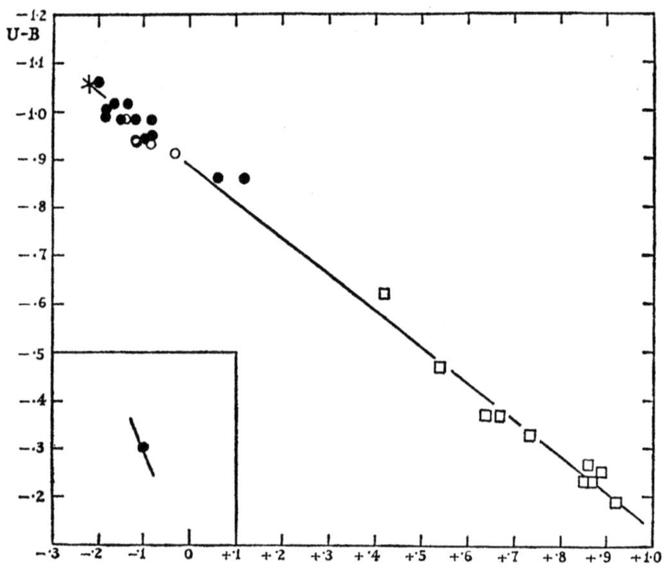

Fig. 18. The reddening path for B0I stars. Filled circles, Large Magellanic Cloud; open circles, Small Magellanic Cloud; squares, galactic supergiants; cross, intrinsic point for B0I stars according to Johnson. Inset: error due to an error of $\pm$ 0.1 spectral class. From WESSELINK (1962).

$(B-V)$ and shows that the points fall along the reddening line given by the galactic B0I stars. Wesselink's results are shown in Figure 18. Two conclusions may be drawn: (1) the law of reddening is the same in the Magellanic Clouds and in our galaxy, and (2) some reddening originates in the Magellanic Clouds, for the points for the stars of the Magellanic Clouds are distributed along the line.

2. Classification by means of narrow-band filters

Since spectral types are assigned by considering the multitude of detail in a stellar spectrum, particular attention being paid to the relative strengths of selected lines, photometry through two or three broad pass bands is not a very sensitive means of determining spectral types. Photometric measurements made through filters which transmit radiation in narrow regions stragetically placed can be related to spectral types and luminosity class in a more sensitive manner. B. Strömgren with his co-workers K. Gyldenkerne and D. L. Crawford has explored this method of obtaining quantitative measurements of the distribution with wave length of the intensity of radiation in stellar spectra. STRÖMGREN (1958, 1963) has published summaries of some phases of his work. He uses the six filters listed in Table 12. Filters *a*, *c*, *d*, and *e*

TABLE 12
The narrow-band filters used by Strömgren

Filter	Type	λ_{max}	Half-width	Remarks
a	Interference	5000 Å	90 Å	no H lines
b	Interference	4861 Å	35 Å	later replaced by a filter, 15 Å half-width
c	Interference	4700 Å	100 Å	no H lines
d	Interference	4500 Å	80 Å	no H lines
e	Interference	4030 Å	90 Å	no H lines
f	Glass	3600 Å	350 Å	

measure the brightness in regions between the hydrogen lines. In main-sequence A type stars and in late B type stars where the hydrogen-absorption lines may be very strong and wide, the pass band of filter *e* may include parts of the extended wings of Hδ and Hε.

Strömgren uses two indexes, the *c*-index which may be related to spectral type, and the *l*-index which is a measure of luminosity in the B and early A type stars. The *c*-index is defined as follows:

$$c = \text{const} + 2.5(2\log I_e - \log I_d - \log I_f). \qquad \text{(IV-5)}$$

It is thus a measure, on a magnitude scale, of the product of the relative brightnesses (I_{4030}/I_{UV}) and (I_{4030}/I_{4500}). The *l*-index is a measure, on a magnitude scale, of the amount of light removed by absorption in the Hβ line.

It is defined as follows:

$$l = \text{const} + 2.5(\tfrac{1}{2}\log I_a + \tfrac{1}{2}\log I_c - \log I_b). \qquad \text{(IV-6)}$$

CRAWFORD (1958, 1963a, 1963b) has published Hβ photometry and UBV photometry for many B type stars. Since the *l*-index is not sufficiently sensitive to the varying strength of the absorption line at Hβ, Crawford has introduced a new index, the β-index. He uses a beam-splitter to send one part of the light through a filter centered at Hβ which passes a beam of half-width 15 Å and the other part through a filter centered at Hβ and having a half-width of 150 Å.

He uses the resulting signals to form the index

$$\beta = 2.5\left[\log I(150\,\text{Å filter}) - \log I(15\,\text{Å filter})\right] + \text{const}. \qquad \text{(IV-7)}$$

The internal mean error in β from measures made on different nights is ± 0.006 mag. This photometer is very efficient and it can be used through clouds and on "non-photometric" nights.

Crawford selects the intrinsic colour $(U–B)_0$ as a measure of spectral type rather than $(B–V)_0$ because $(U–B)_0$ has the wider range for B type stars. However, the same relationship is not valid between $(U–B)_0$ and spectral type for stars of all luminosity classes. The $(U–B)_0$ scale is more compressed for supergiants. JOHNSON and IRIARTE (1958) give a table of the values valid for each luminosity class. Their correlation between $(U–B)_0$ and spectral type for class V stars differs only slightly from the relationship derived by Crawford. CRAWFORD (1958) has demonstrated that there is a linear relationship between $(U–B)_0$ and D, the measure of the intensity of the Balmer jump introduced by Barbier and Chalonge, and also between $(U–B)_0$ and Strömgren's c-index.

One may find $(U–B)_0$ from the observed UBV colours by the following iterative process. The method is based upon the observation (MORGAN and HARRIS, 1956) that for B type stars the intrinsic colours are in a constant ratio, thus

$$(B–V)_0 = 0.27\,(U–B)_0 \qquad \text{(IV-8)}$$

and the assumption that the interstellar reddening may be represented by

$$E(U–B)/E(B–V) = 0.70 + 0.06\,E(B–V)\,. \qquad \text{(IV-9)}$$

To begin with one assumes that the $E(U–B)/E(B–V)$ is equal to 0.70. In this case it follows that

$$(B–V)_0 = 0.332\,(U–B) - 0.232\,(B–V)\,. \qquad \text{(IV-10)}$$

With this estimate of $(B–V)_0$ one can use the observed colours and equation (IV-9) to calculate $E(U–B)$. With this new estimate of the $(U–B)$ colour excess and the observed colours one can then find $(U–B)_0$. This intrinsic colour can be inserted into equation (IV-8) to yield a new estimate of $(B–V)_0$. The process may be repeated until convergence is obtained.

In this way Crawford makes use of the published UBV colours for many O and B stars and of his β-index measurements to plot the stars in a β-$(U–B)_0$ diagram. Such a diagram is equivalent to an HR diagram. Crawford's results are shown in Figure 19. It is clear that there is a separation of the points according to luminosity class. Thus the β-index is truly a measure of luminosity. Crawford states that a spectral type may be predicted with a mean error of ± 0.01 spectral class when UBV colours and his β-index are used. This statement has little real meaning because spectral classes are not defined to such accuracy. The data of Table 10 illustrate the point that rather different spectral types may be assigned to the same star depending upon the dispersion of the spectrograms used.

The data shown in Figure 19 demonstrate a graphical relationship between the β-index and MK luminosity class. Crawford has made no attempt to relate the value of his β-index directly to visual absolute magnitude in the way Petrie related $W(H\gamma)$ to M_V.

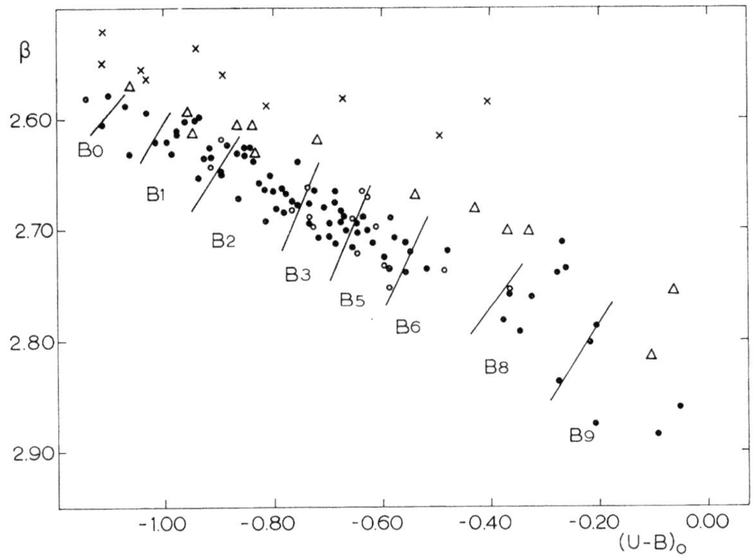

Fig. 19. The β–$(U–B)_0$ diagram for B stars with MK types. Dots indicate main-sequence stars; circles indicate stars whose $(U–B)_0$ disagree with the MK types; triangles are Class III stars; crosses are supergiants. From CRAWFORD (1958).

When Crawford plotted $(U–B)_0$ against β for all the main-sequence B type stars that were available to him, he found that for a given $(U–B)_0$, that is spectral type, there is a considerable scatter in β, see Figure 19. This scatter is precisely the same phenomenon that was demonstrated by PETRIE and MOYLS (1956). The B type stars in the neighbourhood of the sun do not fall on a narrow main sequence and on sequences for giant and supergiant stars. Crawford, following Strömgren, has interpreted the observed spread in absolute magnitude for a given value of $(U–B)_0$ to be an age effect. He implies that the more luminous stars at a given spectral type are older than the less luminous stars of the same type. The theoretical justification for such a conclusion comes from the theory of the evolution of stars at constant mass. The brighter stars, for a given spectral type, are presumed to be more massive than the fainter stars and to have left the main sequence somewhere to the left of their present position. They are thus at a later stage of development than the stars which are at present on the main sequence. To conclude that they really are significantly older than the stars of the same spectral type which are still on the main sequence requires that one have accurate estimates of the duration of each stage of development in a star's history. The theoretical indications are that the more massive stars evolve more quickly than the less massive stars. It seems somewhat hazardous, however, to conclude from the available data that the brighter stars for a given spectral type are indeed older, in the sense of having existed for a longer time, than the fainter stars. They may have merely passed through more stages of development in the same period of time. The fact that a considerable spread of absolute magnitude is observed for the early B type stars of a given spectral type suggests that the time to pass through the various stages of development from normal hydrogen burning on the main sequence to the giant and supergiant stages is of the same order of magnitude as the time since the stars were formed, and that possibly the formation of O and B stars within 2000 pc or so of the sun has been going on for a period of the same length, perhaps 10^7 to 10^8 years.

Bappu, Chandra, Sanwal, and Sinvhal (1961) have observed with narrow-band interference filters having a width at half-intensity of 45 Å to isolate regions centered at 4280 Å, Hγ, and 4410 Å, and have established a Γ-index which measures the strength of the Hγ absorption feature.

The Γ-index is defined as

$$\Gamma = -2500\left[\tfrac{1}{2}\log L + \tfrac{1}{2}\log S - \log\gamma\right], \qquad \text{(IV-11)}$$

where L is the deflection through the filter centered at 4410 Å, S is the deflection through the filter centered at 4280 Å, and γ is the deflection through the filter centered at Hγ. Bappu *et al.* observed 15 standard stars on many nights to establish a reliable scale and zero point for the Γ-index. They give results for about 50 bright field B type stars and for bright B type stars in the Orion association, the ζ Persei association, the northern part of the Scorpio-Centaurus association, the Ursae Major stream, the α Persei cluster, the Coma cluster, the Pleiades, and in Praesepe, h and χ Persei, NGC 2362, and NGC 2264. They list the Γ-index the spectral type, V_0 and $(U–B)_0$, the information other than the Γ-index being assembled from other sources.

Bappu *et al.* demonstrate that their Γ-index is indeed a measure of the strength of the Hγ absorption line by showing that a linear relationship exists between the equivalent widths of Hγ published by Petrie or Williams, for instance, and Γ. They plot Γ-$(U–B)_0$ and Γ-$(B–V)_0$ diagrams like the β-$(U–B)_0$ diagrams of Crawford. They show that stars of luminosity classes Ia, Ib, III, and V fall in typical parts of each diagram and they note that the MK spectral class is not a sharp means of selecting stars of a given absolute magnitude. Following Crawford they combine the Γ-$(U–B)_0$ diagrams for the several clusters and associations which they observe, and they find as Crawford did, that the stars in each group appear to fall along a different line in the Γ-$(U–B)_0$ diagram, the so-called older groups lying above the younger groups, particularly for values of $(U–B)_0$ which correspond to early B type stars.

There are two points upon which one would desire clarification before concluding that the positions of the apparent sequences of the individual groups, which are drawn by Bappu *et al.* and by Crawford, give true information about the ages of the stars. The first is to demonstrate clearly that the sequences are distinct and not due to the statistical distribution of the relatively few points which are obtained for each cluster or group; the second is to demonstrate that the observed range in the Γ-index, or β-index, for a given value of $(U–B)_0$ is due to changes in the stellar atmosphere which are caused by the age of the star and not by other means. It is well known that emission occurs in the Balmer series in the spectra of many B type stars. The presence of such emission would affect the Γ- and β-indexes and the $(U\text{–}B)_0$ colours. Furthermore, it appears that turbulence occurs in the atmospheres of early type stars (DE GROOT and UNDERHILL, 1963). This also can change the strength of some absorption lines in a spectrum relative to other lines thus making the relationship between colour index and spectral type somewhat diffuse. The presence of microturbulence implies shear motions in the atmosphere. These motions may possibly be due to differential rotation in the case of the early type stars. There seems to be little doubt that a flux of mechanical energy exists in the atmospheres of early type stars. It has not been demonstrated whether the presence of this mechanical energy is an age effect, or whether the state of motion in a stellar atmosphere is a remnant of the kinetic state of the gas from which the stars were formed. If the latter were the case, the observed distribution of stars in the Γ-$(U\text{–}B)_0$ or β-$(U\text{–}B)_0$ diagrams might be an indication of place of origin in the galaxy, rather than an age effect.

BORGMAN (1960) has given 7-colour photoelectric photometry of over 300 O and B stars using the narrow-band interference filters listed in the left side of Table 13. This work has been aimed chiefly at investigating the wave-length dependence of interstellar extinction and it will be discussed further in Chapter V. The prominent classification features for the OB stars which fall in the pass bands of the filters are noted in Table 13 as well as the wave length for maximum transmission and the

TABLE 13

The filters used by Borgman and by Johnson and Borgman

Filter	λ_{max}	Half-width	Remarks	Filter	λ_{max}	Half-width	Remarks
	Å	Å			Å	Å	
K	5880	215	Of emission	*L*	33700	10500	
L	5240	220		*K*	21400	6100	
M	4550	200	Of emission	*J*	11600	3800	He I 10830 Å
N	4055	200	wings Hδ, Hε	*I*	8400	2200	Paschen limit
P	3750	110	UV Balmer lines	*R*	6420	2100	Hα emission
Q	3560	90		*V*	5470	870	
R	3295	80		*B*	4340	980	Of, Hβ, Hγ emission
				U	3450	690	includes Balmer limit

half-width of the pass band. When narrow-band filters are used, variations in the spectra of the stars, such as the appearance of the Of emission lines, can contribute a few percent to the intrinsic scatter in the observed colours of stars of about the same spectral type. JOHNSON and BORGMAN (1963) have published more observations of O and B type stars with the set of filters listed in the right side of Table 13. Their purpose is to investigate interstellar reddening rather than to find photometric methods for classifying O and B stars.

3. Narrow-band photometry by interference techniques

To observe each colour separately for each star, as is done by the methods outlined in the previous sections is a rather inefficient use of observing time, and much observing time will be required for a program where many colours are observed. Furthermore there is the chance that transparency changes may occur during the observing time, or that the light of the star might change by a significant amount, if the star is a light variable. T. WALRAVEN and J. H. WALRAVEN (1960) have designed an instrument which uses interference techniques to divide the star light into several moderately narrow bands separated by dark spaces. Then each band of light is observed simultaneously by a bank of photometers.

After passing a wide entrance diaphragm the star light passes through an Iceland spar beam splitter which divides the light into beams in which the planes of polarisation make an angle $+45°$ or $-45°$ with the plane of symmetry of the spectrum analyser. The light then passes through a roof prism which restores parallelism to the two beams and through two equally thick slabs of quartz each of which has its axis in the direction of the optical axis. Consequently one beam is rotated clockwise and the other counter-clockwise, the amount of the rotation depending upon the thickness of the quartz and on the wave length. For certain wave lengths, depending upon the

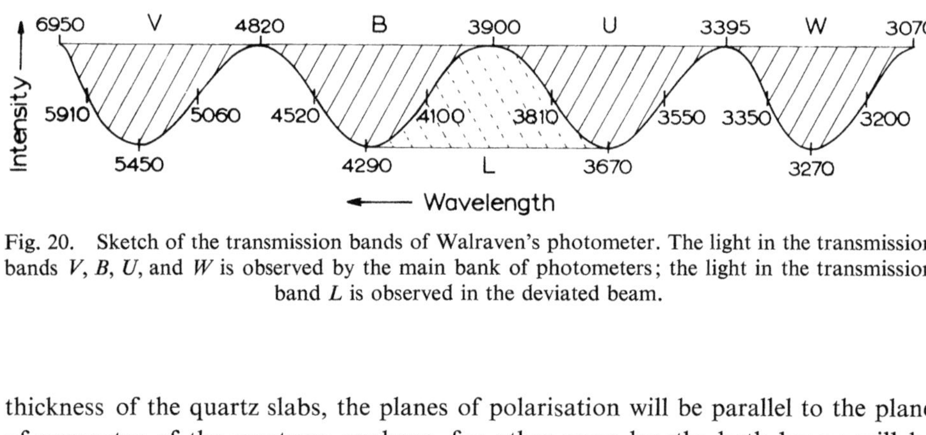

Fig. 20. Sketch of the transmission bands of Walraven's photometer. The light in the transmission bands *V*, *B*, *U*, and *W* is observed by the main bank of photometers; the light in the transmission band *L* is observed in the deviated beam.

thickness of the quartz slabs, the planes of polarisation will be parallel to the plane of symmetry of the spectrum analyser, for other wave lengths both beams will be polarised perpendicular to the plane of symmetry. The beams are passed through a second Iceland spar beam splitter which passes light polarised parallel to the plane of symmetry and deviates the light polarised perpendicular to the plane of symmetry. The transmission bands in the main beam of the instrument are shown schematically in Figure 20 by shaded areas. They are called *V*, *B*, *U*, and *W*. The pass band in the deviated beam which transmits light in the neighbourhood of the converging Balmer series is observed also. It is called *L*. After passing through a collimator the main beam containing the transmission bands *V*, *B*, *U*, and *W* goes through two quartz prisms to separate the individual pass bands which are then focussed on a bank of four photometers. In order to compensate for seeing, field lenses having their edges in the dark regions are placed before the photomultipliers. Then the shift of the spectrum due to image shift does not cause loss of light. The transmission coefficient of the entire instrument is estimated to be 70 per cent.

By varying the thickness of the quartz plates which do the rotation one can shift the transmission bands in the main beam to other parts of the spectrum. However, this is not easy to do. The system of colours must be selected in advance, and it cannot be changed so readily as with a conventional photometer where the pass band is determined chiefly by a filter which may be inserted into the single light beam of the instrument. The effective wave length of each pass band depends upon the spectral type of the star and upon the wave-length sensitivity of the photomultiplier. The resulting bands are intermediate in width between the UBV bands and the pass bands of the narrow-band systems described in the previous section.

T. Walraven and J. H. Walraven (1960) have observed more than 1000 O and B stars. They are able to show that for B type stars, their V and B magnitudes agree with the V and B magnitudes of the UBV system. However, their U magnitudes are different. They define the following indices

$$[B{-}U] = (B{-}U) - 0.66(V{-}B) \qquad \text{(IV-12)}$$

$$[U{-}W] = (U{-}W) - 0.55(V{-}B) \qquad \text{(IV-13)}$$

$$[B{-}L] = (B{-}L) - 0.40(V{-}B) \qquad \text{(IV-14)}$$

which are similar to the quantity Q of the UBV system of photometry and are thus essentially free from the effects of interstellar reddening. The numerical coefficients have been chosen to represent the average wave-length dependence of selective extinction as given by Divan (1954). In these equations the letters V, B, U, W, and L represent the transmission in the various pass bands expressed on a logarithmic scale. The zero point of the system was established by comparing the readings for standard stars with the given standard UBV magnitudes. A comparison with stars in the Scorpio-Centaurus cluster indicated that the zero point of the Walraven system was well chosen.

The Walravens demonstrate that for stars of types between B1 and A0 the $[B{-}L]$ and $[B{-}U]$ indices are equivalent to the β and $(B{-}U)_0$ indices of Crawford for separating supergiant stars from main-sequence stars and for spectral classification. By adopting the absolute magnitudes derived from Hγ by Johnson (1959) they are able to calibrate the index $[B{-}L]$ roughly in terms of absolute magnitude. For stars of type B0 and earlier there is no separation of stars in the $[B{-}L]-[B{-}U]$ diagram.

The $[U{-}W]$ index may also be used to separate supergiant B stars from main sequence stars. This index essentially measures the relative brightness of the region near 3670 Å and that near 3200 Å. This index is particularly useful for a luminosity separation of late B type stars.

The preliminary data published by the Walravens again demonstrate the fact that the early B type stars are not situated on a narrow main sequence and on narrow giant sequences, but that there is a wide spread of luminosity at any spectral type.

4. Observations of OB stars from space vehicles

Absorption by ozone in the earth's atmosphere completely cuts off stellar spectra at about 3050 Å, thus stellar spectra can be investigated at shorter wave lengths only by flying instruments in rockets or satellites above the ozone layer. The first successful detection of stars by photometers flown in rockets was done in 1955 and reported by BYRAM, CHUBB, FRIEDMAN, and KUPPERIAN (1957). Since then other successful flights have been obtained and results have been published by ALEXANDER, BOWEN, and HEDDLE (1963) and by CHUBB and BYRAM (1963). The apparently bright O and B stars, as expected, are the chief objects which are recorded.

The stellar photometry experiment reported by Chubb and Byram consisted of recording the responses from narrow-band photometers during the flight of Aerobee rockets. Each photometer consisted of a 4-inch or 6-inch paraboloidal mirror with an ultraviolet-sensitive gas-gain ionization chamber mounted at its prime focus. The photometers were mounted in the rocket so that they looked out perpendicular to the rocket spin axis and they scanned the sky during the roll and precession of the unguided rocket. The stars were identified from a study of the rocket trajectory and from knowledge of the precession and roll of the rocket.

Chubb and Byram give results obtained with a photometer having an effective wave length of 1427 Å which measured radiation between 1350 Å and 1550 Å and results obtained with a photometer which passed radiation between 1290 Å and 1350 Å and which had an effective wave length of 1314 Å. The average field of view was about 2°.5. They give absolute energy measurements based on laboratory calibrations of the ion chambers. Although they consider the absolute photometry of their flights to be of mediocre quality, they feel that it is unlikely that errors as great as a factor 2 have occurred. They list results with the 1427 Å photometer for 49 stars. The results with the 1314 Å photometer are divided into two classes: those in which the identification is good, 29 stars, and those for which there is some difficulty about the proposed identification, 75 stars.

In the wave-length bands 1290–1350 Å and 1350–1550 Å, the early type stars are the only stars seen and these stars radiate as point sources with an intensity between three and five times less bright than that predicted by model stellar atmospheres which take no account of line blanketing. This discrepancy is not unexpected when one recalls that the resonance lines of several abundant ionic species occur in the selected spectral regions. Owing to the small continuous opacity and to the sensitivity in the observed wave-length ranges of the Planck function to the changes in temperature which may be expected between the regions where the cores of the lines are formed and those where the extreme wings and the continuum are formed, these resonance lines may be expected to be wide and deep with central intensities of the order of 10 to 15 per cent. This is in contrast to the situation for O and B type stars in the usually observed spectral region. Here one is in the long wave-length tail of the Planck function and the emissivity differs only by a factor of the order of two between the line centre and the continuous spectrum. Consequently the strongest lines have central intensities near 0.5 and the lines are not particularly wide.

STECHER and MILLIGAN (1962) have published stellar ultraviolet photometry for seven stars obtained with an objective grating spectrograph mounted in a rocket. They were able to observe with a spectral purity of 50 Å over the range 1700 Å to 4000 Å. They list absolute intensities every 100 Å between 1600 Å and 4100 Å in the spectra of ε Canis Majoris, β Canis Majoris, γ Velorum, α Leonis, α Carinae, γ Orionis, and ι Orionis. From λ2600 to longer wave lengths the shape of the observed spectra is closely the same as that predicted from model stellar atmospheres. At wave lengths shorter than 2600 Å the observed fluxes are less than the predicted fluxes, the discrepancy being about a factor 3 at λ2000 for all the early type stars observed by Stecher and Milligan except for ε Canis Majoris. The observed shape of the spectrum of ε Canis Majoris is not the same as that for the other stars and it seems possible that there may be difficulties with the observations here. Possibly the spectra of other stars are blended with that for ε Canis Majoris, for the field of view of the photometer is one-half degree. It seems probable that the observed discrepancies between theory and observation are due to line blanketing. Further observations at higher resolution are needed to solve this problem.

GULLEDGE and PACKER (1963) have reported further photometric measurements from rockets of B stars in the ultraviolet. Measurements were made from an Aerobee rocket flown on 20 May 1963 using a photometer which consisted of an Ascop photomultiplier with solar blind response, an interference filter, and a 2.25-inch simple quartz objective. The solar blind multiplier was used to cut out a leak in the filter near 4000 Å. The response curve of this multiplier and filter peaked at 2120 Å and was 188 Å wide at the half-power points.

With this equipment results were obtained for nine stars having spectral types between B3 and O9.5 and having unambiguous identifications. The relative fluxes at λ2120 with respect to the fluxes at λ5560 (these fluxes are estimated from the assigned visual magnitudes and CODE's (1960) calibration of $m_V = 0$ in terms of absolute energy) are approximately 0.75 as large as those predicted by model stellar atmospheres without line blanketing. The ultraviolet fluxes measured by Gulledge and Packer are larger than those measured by Stecher and Milligan by about a factor two. Some measurements of the same stars with a photometer having an effective wave length at 2980 Å were obtained during a flight on 29 April 1963, but there is some doubt about the intensity calibration of this experiment. The indications are that B stars are not fainter near λ2000 than models corrected for line blanketing would predict.

The observations of ALEXANDER, BOWEN and HEDDLE (1963) were made with photometers having an effective wave length near 1900 Å and a half width of about 400 Å. The photometers were flown in 1961 on a Skylark rocket launched from the Woomera rocket range in Australia. The results confirm that B stars at λ1900 are only about one third as bright as model atmospheres without line blanketing predict. The measurements for individual stars have statistical value only owing to difficulties with the absolute energy calibration of the response of the photometers.

5. The classification problem

With a two dimensional classification system such as the MK system, the Victoria system of type and M_V assigned from the strength of the Hγ line, or the D–λ_1 system of Barbier and Chalonge, it is possible to classify almost all early-type stars and to arrange them in an HR diagram. Although the classification criteria were selected originally on an empirical basis, it has turned out that on the whole the stars are arranged according to the temperatures and pressures in their atmospheres. When one begins to inspect stellar spectra in detail, however, and to make quantitative measurements in the spectrum, differences are found between the spectra of stars which appear to be alike in gross traits. Factors in addition to temperature and pressure are important for determining the appearance of the spectrum.

In order to place apparently faint stars in the HR diagram which has been established from the study of spectra it is necessary to observe by photoelectric photometry. The photographic techniques of ordinary stellar spectroscopy are not sufficiently efficient for one to be able to record useable spectra of very faint stars. Thus a major problem of spectral classification is to find photometric criteria that permit one to arrange the stars in a two-dimensional array which is equivalent to the normal HR diagram. The selected classification colour indices must be equivalent to the criteria which are used to determine spectral type and luminosity. A spectral type and a luminosity class are assigned after considering a good many pieces of information, the absence of spectral features of a specified type being of equal significance to the presence of other features. With only a few narrow-band-pass filters, it is not easy, and may not be possible, to reconstruct the redundant information that is given by a stellar spectrum. Hence it is not surprising that a unique one-to-one correlation cannot usually be established between an HR diagram for a group of stars and a two-index diagram.

When one wishes to go further and to interpret the observed HR diagram, or its equivalent, in terms of a theoretical diagram giving the positions of model stars according to the coordinates log T_{eff} and luminosity, one must look carefully at the relationship between the empirically selected observational parameters and the theoretical parameters. It is necessary to check that this relationship remains valid and single-valued throughout the range of spectral type and luminosity that is being investigated. The empirically chosen criteria for determining spectral type do to a large extent vary in a systematic way with effective temperature, that is with the energy flux emitted per unit surface area of the star; also the empirically selected criteria for luminosity class on the whole vary systematically with the radius of the star. Nevertheless there are some indications that these general trends break down among the earliest spectral types.

For instance, in the O type stars the apparent relative strength of He II λ4541 with respect to He I λ4471 is an important criterion for determining the spectral type. The He II line comes from a level with a high excitation potential and its strength is strongly dependent on the temperature in the atmosphere. The He I line, on the other hand, arises from a level which is populated chiefly as a result of non-equilibrium effects (dilution effects). The ratio of the strengths of these He II and He I lines therefore is a measure of the effective temperature of the star *and* of the extent of atmosphere. The chief spectral type criterion at type O is not simply a measure of the effective temperature of the star, for a third parameter, the extent of the atmosphere, affects strongly the observed line ratio. This example illustrates the fact that if a simple physical meaning is to be attached to the value of each practical classification parameter, spectral type and luminosity, considerable redundancy is required in the classification criteria in order to be sure that some other factor such as chemical composition, amount of mechanical energy in the atmosphere, or lack of thermodynamic equilibrium does not condition strongly the adopted values of spectral type or luminosity. Since there are few strong spectral features that may be used for classification purposes with low dispersion spectra of early type stars, the danger of real misdirection in assigning types is not negligible. Attempts to use photometric measurements to replace spectral classification, particularly for the early type stars, are particularly sensitive to such misdirection, for in general few criteria are used and rather large superficial differences can and do occur in the spectra of early type stars that when looked at grossly have similar spectra. These differences may be symptoms of much deeper physical differences between the stars. The classification problem is to find useful criteria which are not sensitive to such misdirection.

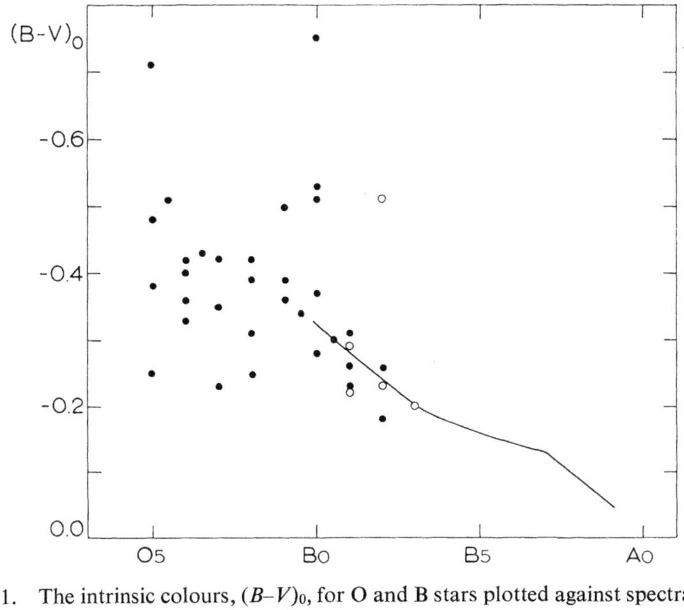

Fig. 21. The intrinsic colours, $(B–V)_0$, for O and B stars plotted against spectral type.

Wildey (1964) has made a detailed photometric study of the stars of the *h* and χ Persei association. In this region of the sky the reddening is very patchy, but Wildey has developed a technique which should enable one to make quite accurate corrections for the dimming of the stars by interstellar selective absorption. One of his achievements is to derive intrinsic colours, $(B-V)_0$, for the individual O stars of the association. His plot against spectral type of these intrinsic colours and of the intrinsic colours for the giant and subgiant B type stars in the neighbourhood is shown in Figure 21. It is quite clear that the intrinsic colours of the O stars spread through a range of at least 0.3 mag, as may be anticipated from spectral studies. This fact makes the O stars poor objects with which to investigate the law of interstellar reddening as described in Chapter V. The spread in intrinsic colour is much less for the B stars and confirms the generally adopted intrinsic colours for the B stars.

Greaves (1948), from the study of relative gradients in the continuous spectra of O stars, drew attention to what is, in effect, the same thing as the wide dispersion in intrinsic colour for the O stars which is demonstrated by Wildey's work. Greaves noticed that for some rather bright O stars the relative gradients at different wave lengths were much redder than for other O stars. He was of the opinion that it was more probable that these differences in colour were due to intrinsic reddening effects in the stars themselves than to extraordinary interstellar reddening. Greaves' point of view is justified by the measurements made by Wildey.

CHAPTER V

INTERSTELLAR MATERIAL AND THE OB STARS

1. The observation of selective extinction

It was early noticed that although most of the apparently bright O and B stars looked very blue to the naked eye, many of the fainter O and B stars looked quite yellowish. According to the ideas then in vogue, the most probable explanation of this observation was that the latter group of stars was significantly cooler than the first group. However, all these stars had spectra suggesting temperatures in the neighbourhood of 20000°. The solution to this riddle was not found until TRUMPLER (1930) from his study of open clusters realised that interstellar space was not transparent in all directions and that particularly in the plane of the Milky Way the line of sight often passed through layers of material which attenuated the light from the stars lying behind it. It was soon established that the attenuation was due to scattering by particles of interstellar dust and that the scattering was not equally efficient at all wave lengths, but varied about as λ^{-1}.

The variation with wave length of the interstellar extinction has been studied by photographic spectrophotometry, by spectrophotometry using a spectrum scanner, and by photometric work with wide-band and with narrow-band filters. The relative energy distributions in stars of the same spectral type but of differing amounts of reddening are compared and the variation of the interstellar absorption with wave length is derived. WHITFORD (1958) and SHARPLESS (1963) have summarized the first observational results. The results from the three methods of observation agree well. A sketch of the variation of interstellar absorption with λ^{-1} is shown in Figure 22.

The curve appears to be composed of three parts:

1. a linear portion in the range $3.0\mu^{-1} < \lambda^{-1} < 2.2\mu^{-1}$,
2. a second linear portion with a slightly greater slope in the range $2.2\mu^{-1} < \lambda^{-1} < 0.8\mu^{-1}$,
3. a curved portion in the infrared, $0.8\mu^{-1} < \lambda^{-1} < 0.0\mu^{-1}$.

This latter part has been extrapolated beyond $\lambda = 2\mu$ keeping in mind the shapes of theoretical curves for the scattering by small dielectric particles. There appears to be little doubt that the slope of the curve of interstellar extinction changes near $\lambda^{-1} = 2.2\mu^{-1}$ or $\lambda = 4500$ Å, and that the curve does not vary precisely as λ^{-1}. Although the zero-point of the interstellar extinction is a little uncertain owing to the extrapolation that is necessary to reach $\lambda = 0$, this uncertainty is only of the order of 0.1 or 0.2 mag. It is found that the interstellar absorption in the wave-length band of the V magnitude

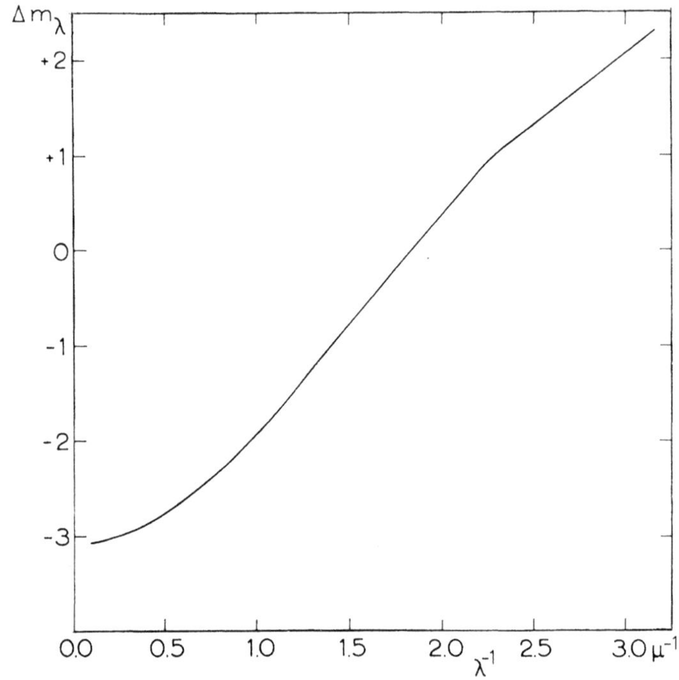

Fig. 22. The variation of interstellar absorption with λ^{-1}.

is related to the $(B–V)$ colour excess as follows:

$$A_V = 3.0\,E(B–V) \pm 0.2\,\text{mag}\,(p.\,e.). \tag{V-1}$$

Much interest attaches to the question of whether the reddening law is the same in all parts of the galaxy. Miss DIVAN (1954, 1956) from a careful spectrophotometric comparison of reddened and unreddened early type stars in all galactic longitudes concluded that there was no evidence for a real change of slope of the interstellar absorption curve with galactic longitude. WHITFORD (1958) also has concluded that this is so. However, the results of broad and narrow band photometry have been interpreted to indicate that the reddening law may differ in different directions in the galaxy. Some of these studies will be reviewed below.

There seems to be little doubt that the photometric colours of a few very early O type stars, such as Θ^1 Orionis C, which are imbedded in a dense emission nebula differ from those of normal, unreddened early O type stars and from the colours of normally reddened O stars. Some colour anomalies do exist. It is perhaps a matter of definition whether one considers these peculiar cases to represent abnormal interstellar reddening or abnormal stellar colours. It could well be in such cases that the star is surrounded by a circumstellar halo from which strong lines and continuous radiation are emitted as a result of the disturbance in the nebula caused by the stellar radiation. The resulting emission spectrum might distort the observed colours quite seriously. Clearly it would be very interesting to obtain high resolution spectra of the immediate neighbourhood of such stars in order to detect a possible circumstellar halo and to estimate how far such a particular halo might extend from the star.

When the colours of reddened and unreddened stars are used to derive information about the properties of interstellar space, it is assumed that the two stars which are being compared would have had identical colours if one had not been behind some interstellar dust. Because with low dispersion and low purity early B and O type spectra show little detail, stars with O and early B spectral types have been considered ideal objects with which to investigate the extinction properties of interstellar material. The photometric observers have compared the colour excesses of groups of stars of about the same spectral type at different galactic longitudes and have tried to discover if the law of interstellar reddening varied with galactic longitude. Some observers claim that the law of reddening is different in the Great Rift in Cygnus than elsewhere, but the results are not entirely convincing, partly because little attention has been paid to the problem of ascertaining whether it is reasonable to expect that the stars which are being compared form a homogeneous group with identical intrinsic colours, and partly because usually not enough stars are observed in any one area of the sky to determine the desired information with the necessary statistical accuracy.

Miss DIVAN (1954), by photographic spectrophotometry, compared the energy distributions between λ3130 and λ6100 of reddened and unreddened stars. She emphasised that it is necessary to compare stars of closely the same spectral type, particularly if it is wished to derive accurately the law of interstellar reddening in the ultraviolet. She concluded that the law of interstellar reddening was the same in Cygnus, the region where most of her stars occur, as elsewhere. In a later paper (DIVAN, 1956) she again drew attention to the fact that intrinsic differences in the energy distribution of O stars of neighbouring types do exist in the ultraviolet and that they should be taken into account in deriving the wave-length dependence of interstellar absorption.

RODGERS (1961) has observed 21 O stars chiefly in Cygnus and Cepheus with a Cassegrain photoelectric scanner attached to the 60-inch and 100-inch telescopes on Mount Wilson. The spectral types range from O5 to O9.5 and some of the stars are Of stars. He observed from 3400 Å to 6000 Å with a resolution of 20 Å. The observed fluxes were integrated over bands 100 Å wide in order to reduce the uncertainty introduced by the photon statistical noise. The average residual in the wave-length regions 3400–3700 Å and 5400–6000 Å was ± 0.03 mag. In the range 3700–5400 Å greater accuracy was possible, the average residual being ± 0.012 mag. The reddening law was found by subtracting the continuum of a moderately reddened star from that of a highly reddened star of the same spectral type. The interstellar absorption curve as a function of wave length was then normalised to an optical depth of 0.50 magnitudes between λ4250 and λ4550.

Separate mean interstellar absorption curves were found for the Cygnus region and for the other stars, most of which lie in Cepheus, and the results were compared with the interstellar absorption curve deduced by STEBBINS and WHITFORD (1945) from six-colour broad-band photometry. Rodgers found that the interstellar absorption curve from his stars in Cygnus was the same as that from the stars in Cepheus, and that both results agreed well with the interstellar absorption curve suggested by Stebbins and Whitford. A slight difference exists between the curves in the ultraviolet, but this difference is within the uncertainty introduced by observational errors. Rodgers' result indicates that if there is a real difference in the law of interstellar absorption in the Cygnus Rift (which appears to be unlikely), it is in the direction of producing less absorption in the ultraviolet in Cygnus than elsewhere.

Finally, using his mean interstellar absorption curve, Rodgers reduced all his observed spectral energy distributions to a constant amount of interstellar reddening, and he compared the energy distributions of the stars in Cygnus with those in Cepheus. He found that the Cygnus O stars probably have a stronger absorption in the Balmer continuum than the Cepheus stars. Since O stars appear to be surrounded by extended atmospheres or shells (see Chapter XI), it is quite probable that intrinsic differences in energy distribution of the sort noted by Miss DIVAN (1956) exist among the O stars. The fact that similar spectral types have been assigned from the study of low dispersion spectra is no guarantee that the stars will have precisely the same energy distribution over a wide range of wave length. Rodgers' contention that the very slight differences in colour that exist between his two groups of stars may be explained in terms of intrinsic differences between the stars is very reasonable. This work does not support the idea that the law of interstellar reddening is different in the Cygnus Rift.

The most detailed study of the shape of the interstellar reddening law in the Cygnus region is due to NANDY (1964). By spectrophotometric means he compared the apparent intensity distributions in pairs of reddened and unreddened stars of similar spectral type using objective prism spectra with a dispersion of about 1000 Å/mm at Hγ. His observations extend from 3480 Å to 8000 Å. All of his results and those of Miss Divan and of Rodgers can be brought to a single normalised reddening curve about which the scatter is due only to observational uncertainty.

Nandy has normalised his reddening curve so that the absorption is equal to unity when λ^{-1} is 2.22 μ^{-1} and the slope of the reddening curve is unity between λ^{-1} equal to 1.70 μ^{-1} and λ^{-1} equal to 2.28 μ^{-1}. Thus in this spectral region

$$\Delta m_\lambda = \lambda^{-1} - 1.22\,.$$

The section of the normalised reddening curve in the region $\lambda^{-1} > 2.30$ is given by

$$\Delta m_\lambda = 0.64\,\lambda^{-1} - 0.39\,.$$

UNDERHILL and WALKER (1966) have compared the relative intensity distributions found from spectrum-scanner observations by OKE (1964) and by WILLSTROP (1965) of O9 and B0 stars with the relative intensity distributions given by model atmospheres representing main sequence stars of types near B0 and O9. Oke's observations of ζ Ophiuchi O9V, and of ε Orionis, B0Ia, extend from 3390 Å to 10800 Å. These stars are lightly reddened, $E(B-V)$ being 0.36 and 0.18 respectively. Willstrop's observations of HD 154368, O9.5 Ia, and of HD 167971, B0V, extend from 4000 Å to 6500 Å. These stars are heavily reddened, $E(B-V)$ being 0.87 and 1.11 respectively.

Both sets of observations fall on a single normalized reddening curve. If the reddening curve is normalised so that the extinction occurring between the V point ($\lambda^{-1} = 1.83$) and the B point ($\lambda^{-1} = 2.29$) is one magnitude, the reddening curve is

$$\Delta m_\lambda = 2.23\,\lambda^{-1} - 4.46\,, \qquad \text{for} \qquad 1.45 < \lambda^{-1} < 2.25$$

and

$$\Delta m_\lambda = 1.42\,\lambda^{-1} - 2.65\,, \qquad \text{for} \qquad 2.25 < \lambda^{-1} < 2.50\,.$$

These lines are identical with those of NANDY (1964) and they confirm the existence of the knee in the reddening curve at about $\lambda^{-1} = 2.25$ (λ 4450). This knee cannot be due to stellar features, for it would then disappear in the normalising process that has been used.

Although the stars ζ Ophiuchi, ε Orionis, HD 154368, and HD 167971 are in parts of the sky quite separated from the Cygnus area, they give the same reddening law that Nandy found for the Cygnus area. It appears that the extinction caused by interstellar grains has the same wave-length dependence in large parts of the galaxy, if not all over.

MORGAN and JOHNSON (1955) using a colour-difference method similar to that introduced by BECKER (1938, 1948) investigated the interstellar absorption by means of UBV observations of 32 O stars. When they plotted the stars in a two-colour diagram, $U-B$ *vs.* $B-V$, they found that seven stars lying in the Great Rift in Cygnus did not fall on the same reddening path as the other stars. The ratio $E(U-B)/E(B-V)$ was 0.72 for most of the stars, but it was about 0.81 for the stars in Cygnus. They suggested that the difference was due to the Cygnus O stars not being so closely associated with the dust clouds as in the case of the other stars, thus that the interstellar material had a different character in Cygnus than elsewhere.

HILTNER and JOHNSON (1956) assembled UBV photometric measurements for 262 O stars and they examined the distribution of points in the $U-B$ *vs.* $B-V$ diagram. They found that most of the points lay along a curved line in accord with BLANCO's (1956) remark that a large amount of interstellar reddening will cause the effective wave lengths of the U, B, and V filters to change, and thus that one may expect the colour excesses to be related by an equation of the form

$$E(U-B)/E(B-V) = a + bE(B-V)\,. \tag{V-2}$$

Hiltner and Johnson suggested that $a = +0.72$ and $b = +0.05$. However, they concluded that the 50 stars in Cygnus, that is the stars lying between $l^{\mathrm{I}} = 40°$ and $l^{\mathrm{I}} = 50°$ and between galactic latitude $b^{\mathrm{I}} = -3°$ to $+5°$, fell along a different line which was not curved. They felt that the difference was most probably due to a difference in the interstellar absorption law and not to intrinsic differences among the stars, but they did not give the equation of this different reddening path. In addition they noted that the scatter in the two-colour diagram tended to increase with increased reddening. This result they interpreted as evidence for variations in the reddening law. It could be due to increased uncertainties in the measurements of the faint stars, for the heavily reddened stars are also the faintest stars which are observed.

The two-colour diagram established by Hiltner and Johnson is shown in Figure 23. The stars in Cygnus are indicated by open circles. The line drawn corresponds to

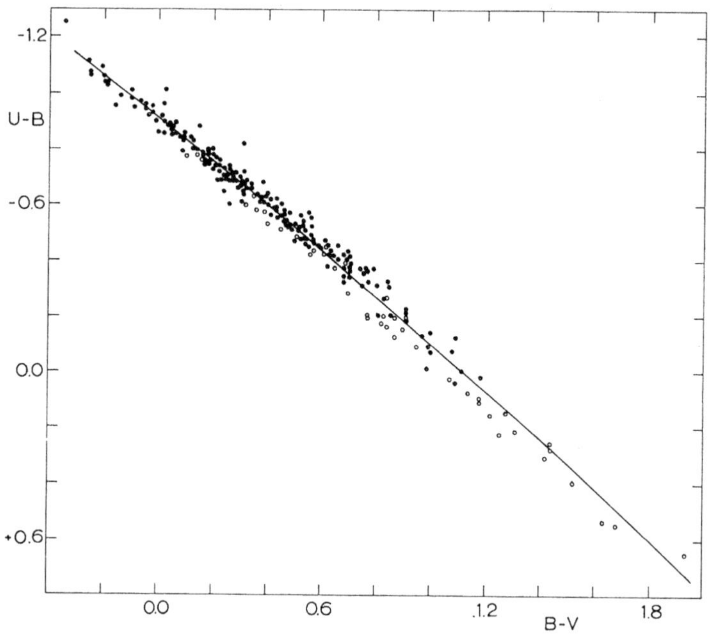

Fig. 23. The two-colour diagram for O stars. The abscissa is B–V. From HILTNER and JOHNSON (1956).

$a = +0.72$, $b = +0.05$. It is clear that the points from the Cygnus stars tend to fall below this line, and that in the neighbourhood of $B-V = +1.0$ the vertical scatter among the points is somewhat greater than that near B–$V = 0.0$.

Hiltner and Johnson do not say how they found the constants a and b which they adopt. Clearly an impersonal method is required. Therefore the data from Table 1 of Hiltner and Johnson have been divided into two groups, *Rift stars* which are stars 176 to 226, and *field stars* which are the remaining stars in the table, and a least squares solution has been done to determine for each group the constants a and b of equation (V-2). Finally a solution was done for all the stars together. Each star is given unit weight. The intrinsic colours of JOHNSON and IRIARTE (1958) for O stars were adopted. Two stars were rejected because their spectra are abnormal, namely No. 51 (X Per) and No. 233 (BD+28° 4211). The O7 intrinsic colours were adopted for No. 217 which is given no subtype, only the letter O. The results are as follows:

Group	No. Stars	*a*	*b*
Field stars	209	$+0.719 \pm 0.016$	$+0.018 \pm 0.016$
Rift stars	51	$+0.750 \pm 0.063$	$+0.097 \pm 0.035$
All stars	260	$+0.740 \pm 0.018$	$+0.034 \pm 0.013$

Statistically these solutions cannot be distinguished the one from the other. One cannot say that the reddening path has a different shape in Cygnus than elsewhere. In passing, it may be noted that the solution for the Cygnus stars gives the largest curvature term, *b*. This result is in direct contradiction to the statement of HILTNER and JOHNSON (1956) who, presumably from visual inspection of the data, concluded that there was no curvature of the reddening path in Cygnus. It seems best to adopt the results from the solution which uses all the stars. Thus the reddening path determined from O stars is

$$E(U\text{–}B)/E(B\text{–}V) = +\,0.740 \pm 0.018 + (0.034 \pm 0.013)\,E(B\text{–}V)\,. \qquad \text{(V-3)}$$

BLANCO (1956) adopted the interstellar absorption curve proposed by WHITFORD (1948), which is similar to that shown in Figure 22, and found that the UBV colour excesses should be related as follows:

$$E(U\text{–}B)/E(B\text{–}V) = +\,0.76 + 0.09\,E(B\text{–}V) + \cdots \qquad \text{(V-4)}$$

This expression is derived by taking account of the change in effective wave length of the UBV pass bands caused by interstellar reddening. The close similarity between equations (V-3) and (V-4) indicates that the UBV photometry of the O stars gives no sound reason to suspect that the properties of interstellar dust in all parts of the Milky Way are not well represented by WHITFORD's interstellar absorption curve. Refinement of this conclusion can only be attempted when it is possible to make a more detailed spectral classification of each star and to take into account the small differences in intrinsic colour, particularly in the ultraviolet, that result from the intrinsic differences among the O stars.

SERKOWSKI (1963) has considered statistically all available UBV observations of O and B stars in order to investigate if there is any dependence of the interstellar reddening law on the polarisation caused by the interstellar dust particles. He is able to show that the slope of the reddening path is independent of the ratio of polarisation to interstellar reddening and therefore that the alignment of the interstellar dust grains cannot be responsible for the regional variations in the reddening law that have been suggested.

Serkowski demonstrates that the curvature constant, *b*, cannot be well determined from the available observations and he suggests that the value +0.06 be adopted. This conclusion is not contrary to the results found here for O stars. In addition Serkowski demonstrates, as first shown by LINDHOLM (1957), that the constant *a* (*X* in his notation) is dependent on the spectral type thus on $(B\text{–}V)_0$. Clearly the colour excesses are measuring more than the interstellar reddening law. Lindholm has suggested that the variation in a results from the change in effective wave length as the equivalent widths of the Balmer lines changes with spectral type. The question once more arises: Are the adopted spectral types and their related intrinsic colours $(B-V)_0$ and $(U-B)_0$ sufficiently accurate for one to be able to isolate small differences in the interstellar absorption law? The answer to this question would appear to be No. When early B type and O type spectra are compared with the spectra computed from a homogeneous set of models, it is found that many of the spectral classification criteria represent the outer halo or shell surrounding the star rather than the star itself, and that the stars grouped in any one spectral class are an inhomogeneous lot (UNDERHILL, 1965). This subject will be discussed further in Chapter XI. At present it seems best to be cautious in attributing slight variations in the broad-band colours of early type stars to variations in the law of interstellar reddening rather than to intrinsic differences among the stars of each spectral type. Serowski's claim that the deviations in $E(U-B)$, which remain after using improved intrinsic colours and accounting for the apparent variation of a with spectral type and for the curvature of the reddening path, are irrefutable evidence of the existence of variations in the interstellar reddening law is valid only in so far as one has confidence that the adopted intrinsic colours are valid for all stars of an assigned spectral type.

The curvature in the reddening path, which is described by the parameter b of equation (V-2), is due to the width of the pass bands in UBV photometry. By observing with narrow pass bands one should be able to eliminate this term and write

$$E(1)/E(2) = c \qquad \text{(V-5)}$$

where $E(1)$ is a colour excess corresponding to one pair of narrow pass bands and $E(2)$ is an excess corresponding to another pair of narrow pass bands. If both pairs of pass bands fall on one or other side of the knee in the interstellar absorption curve at about λ 4500, then the constant c should reflect only the slope of the appropriate linear part of the interstellar absorption curve. If one of the pairs straddles λ 4500 the value of c will be a function of the slope of each linear section. However, if the interstellar reddening law is constant in space, c should be constant and independent of spectral type provided that the filter pass bands have been so chosen that they do not fall on any strong stellar spectral features that are sensitive to spectral type and provided that the intrinsic colours of the stars are well known.

In order to investigate the question of interstellar reddening WAMPLER (1962) observed O stars, and B0 and B1.5 supergiants known to suffer differential amounts of reddening. He used the following interference filters:

λ_{max}	half-width	transmission at λ_{max}
3310 Å	75 Å	28 per cent
4500	150	64
5500	200	62

He considers two colours:

$$\mathfrak{U} = m_{3310} - m_{4500},$$

and

$$\mathfrak{G} = m_{4500} - m_{5500}.$$

The probable error of $\mathfrak{U}$ is ± 0.020 mag. while that of $\mathfrak{G}$ is ± 0.011 mag. From equation (V-5) it follows that a linear relationship should exist between $\mathfrak{U}$ and $\mathfrak{G}$, thus

$$\mathfrak{U} = c\mathfrak{G} + d, \tag{V-6}$$

where

$$d = \mathfrak{U}_0 - c\mathfrak{G}_0. \tag{V-7}$$

Here $\mathfrak{U}_0$ and $\mathfrak{G}_0$ are the intrinsic colours for each spectral type.

The results of fitting Wampler's observed data to equation (V-6) by the method of least squares are given in Table 14. The stars near $l^{II}=0°$ are included in the group $l^{II}=17°$; the star BD+61° 411 is omitted because it is anomalous. If there is a real difference in the reddening law which depends upon galactic longitude, then the quantity c should differ from group to group. The results are not statistically different the one from the other, particularly when it is realised that the quoted probable errors reflect only the uncertainty in fitting a linear relationship to the observed data and that they do not take into account the uncertainties in measuring $\mathfrak{U}$ and $\mathfrak{G}$. Wampler considers that the data in Table 14 indicate that gross variations exist in the law of interstellar reddening. However, a more just interpretation of the data would be to conclude that there is no conclusive evidence for variations in the reddening law. In assessing the data of Table 14 it should be borne in mind that the average declinations of the stars observed in the group at $l^{II}=17°$ is $-12°$ and that the observations were made from the McDonald Observatory. These stars therefore had zenith distances of the order of 35° towards the south at meridian passage. The other groups respectively had zenith distances of about 8° and 25° towards the north at meridian passage.

Borgman (1960, 1961) has used seven-colour photoelectric photometry of O and B stars through narrow pass band interference filters to investigate the law of interstellar reddening. The wave lengths of peak transmission and widths at half power of his filters are given in Table 13. The zero-point of his system of colours is defined by making all colour indices of 15 S Monocerotis equal to zero. Borgman investigates the

TABLE 14

Wampler's solutions for the reddening law with indices $\mathfrak{U}$ and $\mathfrak{G}$

Region l^{II}	O stars			B stars		
	No.	c	d	No.	c	d
17°	9	1.32 ± 0.02	2.29 ± 0.01	15	1.38 ± 0.02	2.42 ± 0.01
78°	16	1.48 ± 0.02	2.50 ± 0.01	9	1.44 ± 0.02	2.52 ± 0.01
133°	8	1.37 ± 0.03	2.42 ± 0.01	8	1.23 ± 0.08	2.36 ± 0.04

properties of interstellar matter by plotting his various measured colour indices against the index [N–M] which measures the slope of the spectrum between λ 4055 and λ 4550. His results for the index [M–L] are shown in Figure 24. It is clear that the points fall more or less along a straight line, just as one would expect if equation (V-5) were

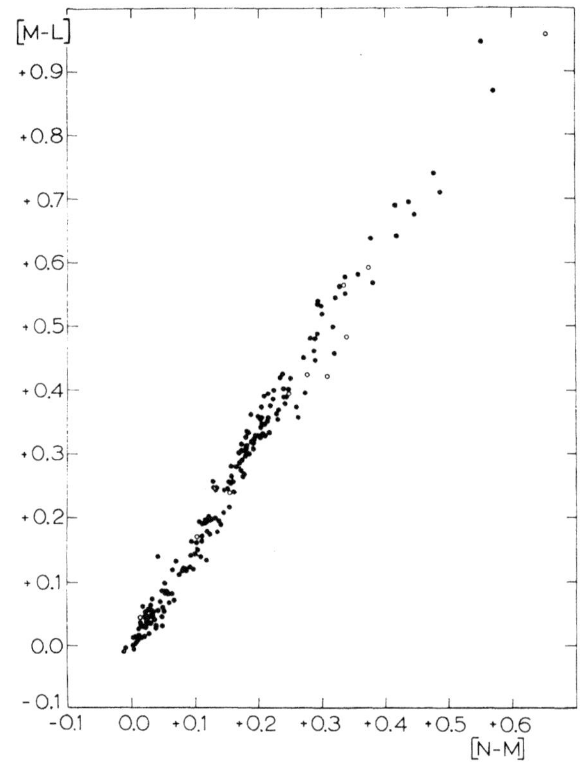

Fig. 24. Borgman's [M–L] *vs.* [N–M] diagram for early B and O type stars. The Of stars are indicated by circles.

valid. There is an increase in the scatter as the amount of reddening increases, that is towards more positive indices. Borgman considers this scatter to be more than the accuracy of the observations will explain, and he concludes that it is an indication of a variable law of interstellar reddening. He is able to show that although the strength of the broad interstellar feature λ 4430 is different in the various stars and thus will introduce some scatter in the [M–L] *vs.* [N–M] diagram, the scatter due to this cause is not enough to account for the observed distribution of points. Many of the stars observed by Borgman are O stars and some are well known Of stars. The pass band of filter M includes the well known Of emission lines. Variation in the intensity of the Of emission would displace the points in the [M–L] *vs.* [N–M] diagram to the lower right. It seems very probable that a major part of the scatter of points in Figure 24 is due to this cause. BORGMAN also examines the behaviour of the colour indices [N–P], [Q–N], [R–Q], and [I–K]. He concludes that his observations indicate that there is no unique law of interstellar reddening, but that variations occur which are connected with regions in the vicinity of very hot stars. This conclusion is based on the assumption that the spectral groups which he considers together have identical intrinsic colours. This point should be investigated further, particularly in connection with the pass bands which cover lines known to appear in emission in Of stars or in supergiants.

JOHNSON and BORGMAN (1963) using wide-band filters which cover the range from 3540 Å to 3.50 μ have continued this type of investigation. They plot two-colour diagrams for O stars, and from the scatter of points conclude that there is no unique reddening law. Their method of analysis is based on the assumption that the intrinsic colours of the O stars are pretty well alike and thus that any deviations from the adopted colours must be due to interstellar selective extinction. This work would be greatly strengthened if it were accompanied by a careful spectrophotometric investigation of the spectra of some of the most divergent stars. Most of the weight of the argument for variable interstellar selective extinction seems to be based on the divergence of the series of measurements for Θ^1 Ori C. Possibly BORGMAN's (1961) conclusion that most of the variation occurs in the vicinity of hot stars should be interpreted to mean that the measured colour variations originate in the atmospheres of these stars.

JOHNSON and BORGMAN (1963) have concluded that in certain dense emission nebulae the interstellar reddening law is such that the ratio

$$R = A_V / E(B-V)$$

is equal to 7.0 rather than about 3.0. They point out that if this is so the distance moduli for certain clusters should be revised. In particular they note that adopting R equal to 7 for the region of the Trapezium in Orion would imply that the distance modulus of the Trapezium stars is 7.0 mag. rather than about 8.0 as has been suggested by astrometric work and by spectroscopic studies of the stars in the Orion association.

The spectra of the stars of the Trapezium are typical of stars on the lower part of the main sequence (MORGAN and LODEN 1966). A modulus of 7.0 for this group is inconsistent with the measured visual magnitudes and the appearance of the spectra of the stars; it is also inconsistent with the astrometric measurements for this group of stars (see Chapter VI). Thus it is not possible to agree with JOHNSON and BORGMAN that $R = 7.0$ in the Orion Nebula. The usual value, $R = 3.0$, is consistent with the absolute magin tudes deduced from the appearance of the spectra and those deduced from an adopted modulus of 8.0 mag. and the measured V magnitudes and $E(B-V)$ values.

Johnson and Borgman suggest that R should be 7.0 because their measured J and K magnitudes for the Trapezium stars are too bright to be consistent with R equal to 3.0 and the generally accepted V magnitudes. It is very possible that their infrared observations are affected by strong nebular emission. Their pass bands are not particularly narrow (see Table 13). It is known that [S III] lines near 0.9 μ arise from the Orion Nebula and that the He I line at 1.083 μ is strong in emission (YESSIPOV,

1962). Other forbidden and permitted emission lines from atoms and ions may be present. Furthermore, the (2,0) and (1,0) bands of the quadrupole radiation from the H_2 molecule fall near 1.2 μ and 2.2 μ respectively. It is most necessary to observe the Orion Nebula in the infrared with sufficient resolution to determine if the J and K magnitudes of JOHNSON and BORGMAN for the Trapezium stars are vitiated by strong nebular emissions which have not been properly compensated by the technique used to subtract out the nebular and sky background.

JOHNSON (1965) has published wide-band photometry between 3500 Å and 10 μ for 104 stars including some early type stars. In this article he reviews the problem of determining the ratio R from wide-band photometric measurements. He demonstrates how R may be found for a group of stars which are all at the same distance if the MK spectral type is available for each star. It is hazardous to assume that all the stars in each group that Johnson discusses are closely at the same distance, and furthermore no account is taken of the fact that many of the stars will be members of spectroscopic binaries, nor of the possibility that in some cases the photometry may be disturbed by nebular emissions or by peculiarities of the stellar spectrum. These factors will cause a dispersion in the results. Johnson lists the following results:

Region	R	Region	R
h and χ Persei	3.0	Orion Belt	4.8
III Cepheus	5.4	Orion Sword	5.7
I Ara	6.6	NGC 2244	6.0

A critical examiniation of the data for each star is required in order to ascertain the true uncertainty in these estimates of R.

From his measurements Johnson is able to estimate new intrinsic colours, particularly in the infrared, for spectral types O5 to A0. He then uses his new observational data to derive extinction curves in various regions of the galaxy. All departures from the adopted intrinsic colours are attributed to selective extinction by the interstellar dust. The normalised reddening curves obtained in this way have a variety of shapes and they cannot generally be approximated by two straight lines as NANDY (1964) and UNDERHILL and WALKER (1966) found. Johnson draws the conclusion that the ratio R is variable around the galaxy and that the value 3.0 is a minimum value.

What Johnson does not consider is the possibility that the anomalous "extinction" curves which he finds could be the result of emission from gas in the neighbourhood of the stars. It seems much more probable that nebular emission, continuous and line emissions, or emission from an extended atmosphere affects the wide-band colours than that the sizes and composition of the interstellar grains are so diverse in the various parts of the galaxy as to produce greatly differing laws of interstellar extinction. It is particularly difficult to achieve different shapes for the reddening curve in the infrared because in the infrared (if our present ideas about interstellar dust have any validity at all) the particles are small with respect to the wave length of light and Rayleigh scattering should predominate.

2. The diffuse interstellar absorption features

Interstellar material leaves its mark on B type spectra by producing absorption lines as well as by causing selective extinction. The very sharp lines produced by transitions in atoms, ions, and molecules are discussed in Section 3. In addition a few diffuse interstellar absorption lines are known. Their origin is unknown, though it is suspected that they result from optical interaction between star light and interstellar dust grains.

The first diffuse interstellar line to be recognised and the strongest is the broad band centered at 4430 Å. This feature, which looks like a thumb mark on low dispersion spectra of highly reddened B type stars, may have a central depth up to about 15 per cent absorption, and it has very extensive wings which can be traced to $\pm$ 75 Å. Typical observed profiles in highly reddened stars are shown in Figure 25. These were obtained by R. WILSON (1958) from spectrograms obtained at the Dominion Astrophysical Observatory. The profile of this diffuse absorption lines makes it very difficult to estimate accurately the level of the continuous spectrum between Hγ and λ 4500.

Other weaker diffuse interstellar absorption lines are known. Most of them were discovered by MERRILL (1936) and by MERRILL and O. C. WILSON (1938). Further broad shallow absorption features have been noted by R. WILSON (1958). These features may be classed as interstellar absorption bands or as diffuse interstellar

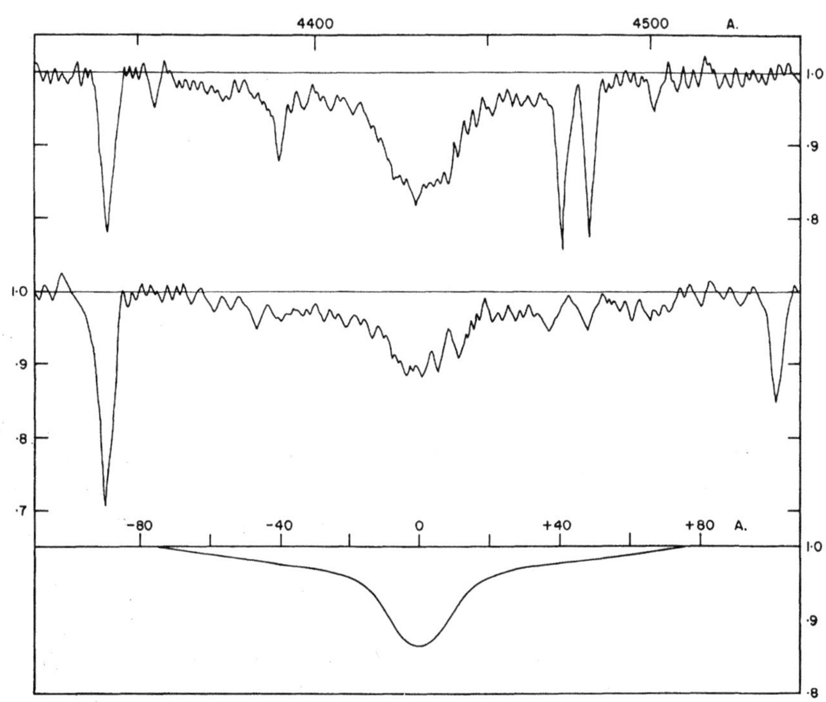

Fig. 25. Observed profiles for interstellar λ 4430. From WILSON (1958).

absorption lines. The latter group of features have a width at half intensity of a few angstroms. The features recognised at present occur at the following wave lengths:

Diffuse absorption lines	Diffuse absorption bands
5780 Å	4430 Å
5797	4760
6203	4890
6270	6180
6284	
6614	

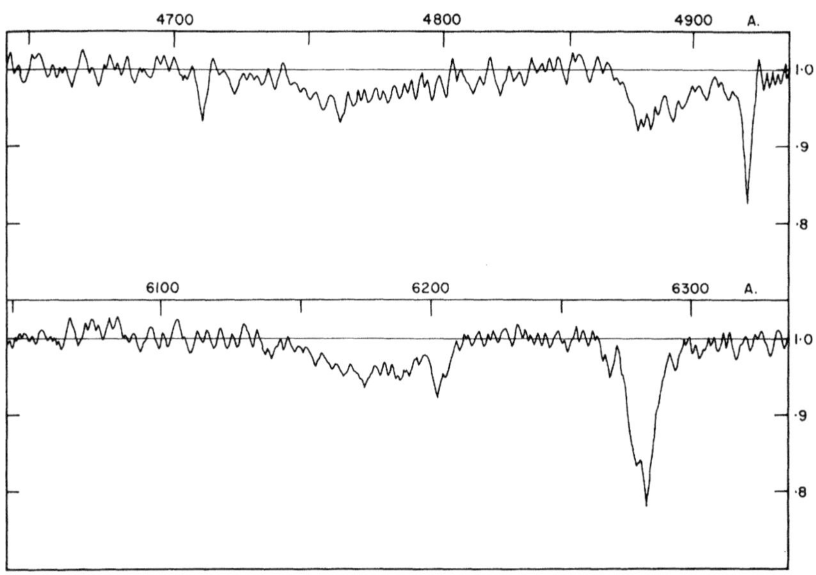

Fig. 26. Observed profiles for diffuse interstellar lines and bands. From WILSON (1958).

Some of these lines are shown in Figure 26, which is due to R. Wilson. The star HD 183143 is a heavily reddened supergiant with spectral type B7Ia, $E(B-V) = +1.36$.

The strength of λ 4430 correlates fairly well with the amount of interstellar extinction, see, for example, GREENSTEIN and ALLER (1950), DUKE (1951), UNDERHILL (1956a), WALKER (1963), but the relationship does not seem to be unique. The same strength of λ 4430 can occur with quite widely different amounts of interstellar extinction. It is thought that interstellar polarisation is also caused by the grains of interstellar dust, but the available measurements of the strengths of the diffuse interstellar absorption lines and bands do not correlate well with measured amounts of interstellar polarisation. R. WILSON (1960) has studied the correlation between the strengths of the two strongest diffuse interstellar lines λ 6284 and λ 5780 and the ratio $p/E(B-V)$, where p is the amount of polarisation caused by the interstellar dust, and finds support for the DAVIS-GREENSTEIN (1951) theory for the alignment of interstellar grains in a magnetic field. Wilson makes the remark that it is better to use the strengths of the diffuse interstellar *lines* as criteria for the mass of grains in the line of sight than the strength of λ 4430 because the equivalent widths of these lines can be measured more accurately since these lines have sharper profiles than λ 4430 has. Unfortunately very few quantitative measurements have been made of the strengths of diffuse interstellar absorption lines and bands other than λ 4430. WALKER (1963) has not been able to confirm Wilson's conclusions regarding polarisation.

It is not known what is the origin of the diffuse interstellar bands and lines. HERZBERG (1954) has given reasons for believing that if these lines are due to molecules, the molecule must be of a type HXY where X and Y are heavy elements. Diatomic molecules would not produce features of the observed diffuse wide character. HERBIG (1963) has suggested that the λ 4430 band might arise from absorption due to metastable H_2 molecules trapped on interstellar grains. However, MALVILLE (1964) has shown that it is very unlikely that an appropriate density of H_2 molecules in the metastable $c^3\Pi_u$ state can be maintained in interstellar space. UNSÖLD (1963) has noted that although the general properties of interstellar extinction and reddening can be explained by dielectric particles of a suitable size, colloidal particles of metals may well be the origin of the λ 4430 absorption feature. However, there is insufficient information available at present to indicate which metal is most important. STOECKLY and DRESSLER (1964) have shown that λ 4430 is weaker for a given colour excess in the spectra of those stars where the radial velocity shown by the interstellar Ca lines is more than 10 km/sec more negative than the stellar velocity. They interpret this to mean that these stars are on the average of earlier spectral type and that the negative velocity is evidence of a stellar radiation field blowing the interstellar dust and gas away from the stars and causing the evaporation of small dust particles. It is implied that λ 4430 is absorbed preferentially by small dust particles. They throw out the suggestion that the line may be due to Ca, but give no supporting physical arguments for this idea. WHITNEY (1964) has given arguments to suggest that λ 4430 is due to metallic Na particles covered by an ice.

WALKER (1963) using spectral scans and narrow-band filters, has made a very extensive series of measurements of the strength of λ 4430 in the spectra of O and B stars in the southern hemisphere as well as in the northern hemisphere. He confirms the conclusion of GREENSTEIN and ALLER (1950) that the strength of λ 4430 is very closely correlated with the strength of the diffuse interstellar lines λ 6284 and λ 5780. Furthermore he shows that the slope of the line correlating the strength of λ 4430 with the selective extinction, measured for instance by the colour index $E(B-V)$, appears to depend upon galactic longitude. The data on this point are not absolutely convincing, but they are too suggestive to be entirely ignored. One difficulty is in being sure that the effects of the underlying stellar spectrum in each case are sufficiently accurately removed by the adopted intrinsic colours and by the measurements made in the comparison bands for λ 4430. The distribution of spectral types is not the same in each group of stars that is investigated, thus one cannot count upon such effects cancelling out on a statistical basis.

WAMPLER (1963) has also measured the strength of λ 4430 by narrow-band filters in a few stars, but he gives no catalogue of results. He compares the strength of λ 4430 with $E(B-V)$ and with the excess in his 𝔘𝔊 colour system, and also suspects that the relationship between the strength of λ 4430 and selective extinction varies with galactic longitude. His results are not so detailed as those of Walker, but they are generally the same. Walker concludes that λ 4430 is weak relative to reddening in the southern Milky Way, and towards Cygnus. He finds that it is particularly weak towards the galactic centre but strong towards Monoceros. HUTCHINGS (1964) has reported that the λ 4430 absorption is found in the spectra of reddened stars of the Magellanic Clouds.

3. The sharp interstellar lines

For many years the H and K lines of calcium discovered by HARTMANN (1904) and the sodium D lines discovered by Miss HEGER (1919) were the only lines in stellar spectra known to be of interstellar origin. These absorption lines are very sharp and their displacements give the radial velocity of the clouds in interstellar space which lie between the star and the earth. When the coudé spectrograph of the Mount Wilson 100-inch telescope went into service, more sharp interstellar lines were discovered. BEALS (1942) has summarised the information which rapidly became available. Lines from atomic spectra and from molecular spectra are found, but in every case only those lines are found which arise from the ground level of the term. In the case of molecules only lines from a few of the lowest rotational levels of the ground state of the molecule are observed. HERZBERG (1954) has summarised the information about

TABLE 15
The sharp interstellar lines of atomic origin

Spectrum	λ (Å)	Transition	J – J′
Na I	3302.34	$3^2S - 4^2P^0$	$\frac{1}{2} - 1\frac{1}{2}$
Na I	3302.94	$3^2S - 4^2P^0$	$\frac{1}{2} - \frac{1}{2}$
Na I	5889.95	$3^2S - 3^2P^0$	$\frac{1}{2} - 1\frac{1}{2}$
Na I	5895.92	$3^2S - 3^2P^0$	$\frac{1}{2} - \frac{1}{2}$
K I	7664.91	$4^2S - 4^2P^0$	$\frac{1}{2} - 1\frac{1}{2}$
K I	7698.98	$4^2S - 4^2P^0$	$\frac{1}{2} - \frac{1}{2}$
Ca I	4226.73	$4^1S - 4^1P^0$	$0 - 1$
Ca II	3933.66	$4^2S - 4^2P^0$	$\frac{1}{2} - 1\frac{1}{2}$
Ca II	3968.47	$4^2S - 4^2P^0$	$\frac{1}{2} - \frac{1}{2}$
Ti II	3072.97	$a^4F - z^4D^0$	$1\frac{1}{2} - \frac{1}{2}$
Ti II	3229.19	$a^4F - z^4F^0$	$1\frac{1}{2} - 2\frac{1}{2}$
Ti II	3241.98	$a^4F - z^4F^0$	$1\frac{1}{2} - 1\frac{1}{2}$
Ti II	3383.76	$a^4F - z^4G^0$	$1\frac{1}{2} - 2\frac{1}{2}$
Fe I	3719.94	$a^5D - z^5F^0$	$4 - 5$
Fe I	3859.91	$a^5D - z^5D^0$	$4 - 4$

TABLE 16
The sharp interstellar lines of molecular origin

Spectrum	λ (Å)	Transition and line		
CH	4300.32	$A^2\Delta - X^2\Pi$	(0,0)	$R_2(1)$
CH	3890.21	$B^2\Sigma^- - X^2\Pi$	(0,0)	$P_2(1)$
CH	3886.41		(0,0)	$Q_2(1)$
CH	3878.77		(0,0)	$R_2(1)$
CH	3146.00	$C^2\Sigma^+ - X^2\Pi$	(0,0)	$P_2(1)$
CH	3143.17		(0,0)	$Q_2(1)$
CH	3137.53		(0,0)	$R_2(1)$
CN	3875.77	$B^2\Sigma^+ - X^2\Sigma^+$	(0,0)	P(1)
CN	3874.61		(0,0)	R(0)
CN	3874.00		(0,0)	R(1)
CH^+	4232.57	$A^1\Pi - X^1\Sigma^+$	(0,0)	R(0)
CH^+	3957.71		(1,0)	R(0)
CH^+	3745.30		(2,0)	R(0)
CH^+	3579.04		(3,0)	R(0)
CH^+	3447.08		(4.0)	R(0)

the molecular lines. The sharp interstellar lines are listed in Table 15. Lines which are found only in diffuse nebulae are not included. The lines of molecular origin are listed in Table 16. DOUGLAS and MORTON (1960) identified the line at 3579 Å and HERBIG (1964) has identified further lines in the extreme part of the accessible ultraviolet.

The radial velocities of some of these lines, particularly of the Ca II lines, have been measured in order to study the motions of the interstellar gas, see for example ADAMS (1949), MÜNCH (1957), and PETRIE and PEARCE (1962), and measurements have been made of the equivalent widths of the stronger lines. Some of these studies are summarised by Beals. However, since we are not concerned here with the dynamics of the interstellar material, these results will not be reported.

In 1951 neutral hydrogen gas was detected in interstellar space by means of the radiation of 21 cm wave length (EWEN and PURCELL 1951, MÜLLER and OORT 1951, and PAWSEY, 1951). The latest component of the interstellar gas to be detected is OH (BARRETT, MEEKS, and WEINREB, 1964 and PENZIAS, 1964). Observations are made of radiations at 1667.357 Mc and at 1665.402 Mc. Some lines from the (0, 0) vibrational band of the $A^2\Sigma^+ - X^2\Pi$ transition of OH which have rotational quantum number 1 occur between 3062.52 Å and 3090.47 Å. High resolution stellar spectra might reveal these lines.

4. Interstellar polarisation

It was discovered in 1949 that the dust in interstellar space polarised the light from B stars. HALL and MIKESELL (1950) and HILTNER (1956) have published extensive catalogues giving the degree of polarisation as well as information regarding the position angle of the electric vector for many O and B type stars. Observations in various wave-length bands have indicated that the degree of polarisation depends upon the wave length. There is some evidence that the degree of polarisation reaches a flat maximum near 6500 Å and that it is less at shorter and at longer wave lengths. This subject has been studied from the observational side chiefly by GEHRELS (1960), by GEHRELS and TESKA (1963), and by Behr and his colleagues, see SCHMAHL (1963). The character of the interstellar particles and the manner in which they interact with light is still a problem in which much is unknown. In 1960 a conference was held to try to consolidate the available information. A report, with contributed papers by workers active in the field, has been prepared by HALL (1960). There is some possibility that some of the observed polarisation may originate in the immediate neighbourhood of the hotter stars, but no truly general conclusions can be drawn. The instrumental difficulties of making reliable polarisation measurements are great and they restrict severely the amount of available information about the variation of polarisation in different wave-length bands.

CHAPTER VI

THE DISTRIBUTION OF THE O AND B STARS

1. The general distribution of the early type stars

The O, B, and Wolf-Rayet stars are closely confined to the plane of the galaxy and many of them are immersed in dust and gas. The distribution of the O and B stars may be readily visualised from the *Atlas Eclipticalis* and the *Atlas Borealis* (BEČVÁŘ, 1958, 1962). The *Atlas Eclipticalis* covers the declination zones $-30°$ to $+30°$ and the *Atlas Borealis* covers the zone $+30°$ to the pole. The positions of the stars are plotted for the equinox 1950.0 and all stars to the 10th magnitude for which positions are listed in the *Yale Catalogues of Proper Motion* or in the *General Catalogue* are plotted, including variable stars. Different colours are used to indicate spectral types, the OB stars being violet, the A stars blue, the F stars green, the G stars yellow, the K stars orange, and the M, N, R, and S stars red. The magnitude is indicated by the size and shape of the symbol representing the star. The boundaries of the constellations are indicated and also whether the stars are known to be binaries or variable in light. In addition to being strongly concentrated to the Milky Way, the O and B stars tend to occur in open clusters and in associations. A few O and B stars appear to be "unattached".

1.1 *Open Clusters*

Open clusters are recognised by a local increase in the apparent density of stars over the general background. Nearby clusters may be distinguished easily. Distant clusters which are projected against the Milky Way are more difficult to see. Helen Sawyer HOGG (1959) has prepared a *Catalogue of Star Clusters* with an extensive bibliography. A comprehensive list of clusters and associations has also been prepared by ALTER, RUPRECHT, and VANÝSEK (1958) together with a bibliography. Supplements to this catalogue appear at regular intervals. The early type stars occur in galactic or open clusters and in associations but not in globular clusters.

The open clusters form a flattened disk-like system which is about 1000 pc thick and which has a diameter of about 10000 pc. The plane of symmetry of this system is inclined $2°.3$ to the galactic plane and has a pole at $12^h50^m.4$, $+27°.7$ (1950). The system is concentrated towards a point about 350 pc distant at galactic longitude 247°, near the rich open cluster NGC 3532. TRUMPLER (1930) showed that the space distribution of the open clusters outlines spiral arms.

1.2 *Associations*

The concept of associations as significant building blocks of the galaxy was introduced by AMBARTZUMIAN (1949). He defined *O associations* to be stellar systems where the partial density of O to B2 stars is larger than the average field density of these stars. The accepted O associations cannot be explained by chance fluctuations. It is necessary that O or B0 stars are present. (There are also T associations which contain T Tauri stars.) The radii of O associations range up to 200 pc, thus O associations are much larger than open or globular clusters. The star density is small in an O association. An O association may contain open clusters which contain O and B stars. ALTER, RUPRECHT, and VANÝSEK listed 46 associations in 1958. Twenty-seven associations were listed by MORGAN, WHITFORD, and CODE (1953) shortly after the introduction of the concept by Ambartzumian.

2. Comprehensive lists of O and B stars

Much work has been done on O and B type stars because of their importance as spiral-arm tracers. Consequently a number of publications exist which list the early type stars which have been selected for one or other type of study. A few of these sources are given here as a convenient starting point for finding OB stars. BOK (1964) has emphasised once again how important it is to have relatively simple and efficient methods for recognising OB stars among the multitude of faint stars which may be seen in any Milky Way field. He finds the use of *U* and *B* magnitudes obtained by a photographic process very convenient. From double exposure photographs it is easy to recognise those stars for which the *U* magnitude is brighter than the *V* magnitude. Almost all of these stars when examined in more detail turn out to be OB stars.

Comprehensive objective prism surveys designed to discover OB stars have been carried out at the Hamburg-Bergedorf Observatory and at the Tonanzintla Observatory, and the astronomers at these Observatories have prepared many lists of faint OB stars in certain areas of the sky.

MCCUSKEY and HOUK (1964) have studied the distribution of 159 OB stars in an area of 15.5 sq deg centered at $1^h\,40^m$, $+60°$ by photographic photometry. The spectral types of the stars are found from objective prism plates. McCuskey and Houk find that the OB stars are concentrated for the most part in a region between 1 and 3 kpc from the sun at about the same distance as the five galactic clusters NGC 581, NGC 654, NGC 659, NGC 663, and Tr 1. The peak space density of OB stars appear to be between 4 and 6 per 10^6 cubic parsecs at about 1800 pc from the sun. The OB stars delineate the position of the Perseus spiral arm in this area.

The classic paper on the colours of the OB stars is that by STEBBINS, WHITFORD, and HUFFER (1940). This paper gives the colours of 1332 B stars on the c_1 photoelectric system. Later long lists of colours, magnitudes, and spectral types for OB stars were given by MORGAN, CODE, and WHITFORD (1955) and by HILTNER (1956). Hiltner's colours are on the UBV system and his spectral types are on the MK system. Magnitudes, colours, and spectral types of nearly 5000 southern stars including many B stars have been published by COUSINS and STOY (1963). M. L. WOODS (1955) and A. DE VAUCOULEURS (1957) have published lists of southern B type stars with spectral classifications.

Radial velocities of O and B stars north of $-11°$ and brighter than 7.5 mag have been published by PLASKETT and PEARCE (1930), while PETRIE and PEARCE (1962) have published radial velocities for northern B stars between magnitudes 7.5 and 8.5. Thackeray and colleagues at the Radcliffe Observatory have published radial velocities for many southern B type stars. Much of the material about the space velocities of the OB stars within 3 kpc of the sun has been collected and discussed by RUBIN, BURLEY *et al.* (1962) who give a catalogue of 1440 OB stars. BEER (1961, 1964) has also assembled a comprehensive list of B stars in order to investigate the distribution of B type stars that may be deduced from absolute magnitudes derived from Hγ measures. ROBERTS (1962) has compiled a list of 123 Wolf-Rayet stars and has discussed their distribution.

3. Some important groups of O and B stars

In this section information will be given about a few groups of early type stars which play an important role in our studies of O and B stars. For more complete information the reader is referred to the catalogues mentioned in Section 1 and to the reports of Commission 37 in the *Trans. I.A.U.*

3.1 *The Pleiades*

The earliest spectral type is B6 and there are many late B and early A type stars in the cluster. This cluster is very important for determining the absolute magnitudes of the late B type stars. The distance to this cluster, which is imbedded in gas and dust and in which reflection nebulae are visible near the brighter stars, is not known precisely even yet. There is about 0.04 mag reddening in the blue. BINNENDIJK (1946), from a comprehensive discussion of all available parallax measurements for the group, adopted a group parallax of $0''.010 \pm 0''.003$. PETRIE and MAUNSELL (1949) and PETRIE (1952) discussed the available information and adopted the same value, pointing out that one must allow for the reddening when converting measured visual magnitudes to visual absolute magnitudes by means of the adopted modulus of 5.0 mag. GÜNTHER (1949, 1950), from photographic photometry, came to essentially the same conclusion regarding the distance of the cluster, and so did MICZAIKA (1950) who used observed magnitudes and an adopted relationship between M_V and the number of visible Balmer lines to find the distance to the Pleiades. MITCHELL and JOHNSON (1957) deduce a larger modulus, namely 5.5 mag. They fit their observed lower part of the main sequence in the colour-magnitude diagram to an adopted zero-age sequence. Their corrections for interstellar reddening are about one-half to one-third those adopted by the other investigators. PETRIE (1964) has rediscussed the absolute magnitudes of the B stars and finds a modulus of 5.4 ± 0.7 mag for the Pleiades. He makes no use of trigonometric parallaxes for stars of this group.

STROM (1964) has measured the spectral distribution of eight of the brighter stars in the Pleiades with a spectral scanner and he has used this material to relate these stars to a series of model atmospheres. He has thereby derived effective temperatures for stars of types B6.5 to B8.5.

HERBIG (1962) has pointed out that there are no T Tauri stars in the Pleiades and has suggested that this means that star formation in this cluster occurred over a long period of time.

3.2 *The Orion association*

There are at least three clusters in the Orion association, all apparently close together in the sky. They differ in declination more than in right ascension. Thus we have the λ Orionis cluster at a declination of about $+10°$, the Belt group at about $-1°$, and the Trapezium cluster at $-5°$. Since the bright stars are embedded in nebulosity, it is difficult to find faint members of the clusters and it is difficult to estimate the reddening accurately. The strong emission lines from the Orion nebula contaminate spectra of the Trapezium stars in particular.

STRAND (1958) has studied the proper motions of the stars near the centre of the Orion Nebula cluster (Trapezium cluster). He has derived an expansion age for the cluster of 3×10^5 years. From his new proper motions and from the radial velocities available to him Strand found a distance of 520 pc for the group, thus a distance modulus of 8.6 mag. The most recent study of the proper motions in the region of the Orion nebula is that of MEURERS and SANDMANN (1963). Their proper motions agree well with those of Strand, but they consider that there is no conclusive evidence for an expansion of the Trapezium or of the Orion group.

SHARPLESS (1952, 1954) has made a spectroscopic and photometric study of the Orion association; he lists V magnitudes, $B–V$ and $U–B$ colours, and spectral types. He suggests that the distance modulus of the aggregate is 8.5 mag. The distance estimates of Strand and of Sharpless are somewhat greater than the value 8.0 mag which is the mean of the values resulting from the attempts to use the Orion association as a standard group for determining the luminosities of the O and B stars.

MCNAMARA and LARSSON (1962) and ABT and HUNTER (1962) have measured the rotational velocities of 70 stars in the Orion association; the two sets of results are in good agreement. The spectral types covered are from O9.5 to B5, most of the stars being of types B0 to B3. The projected rotational velocities ($v \sin i$) of the Orion stars are on the average smaller than those of the general field stars of the same spectral type. If a random orientation of axes of rotation is assumed, then $\bar{v}=135$ km/sec for the B0–B3 stars in Orion, but 186 km/sec for similar stars in the general field. Abt and Hunter point out that there is a greater frequency of binaries in clusters where the average velocity of rotation is low than in clusters such as the Pleiades where the stars rotate more rapidly than the average. McNamara and Larsson point out that the He I lines appear to be unusually weak in some of the sharp-lined Orion stars in comparison to what would be expected for the measured $(U–B)_0$. Clearly the atmospheres of these stars are anomalous and deserve further study.

3.3 *The Scorpio-Centaurus association*

This group of relatively nearby B type stars (the distance is approximately 200 pc) is very important for furnishing standard stars whose luminosities are known because their distances have been found by geometric means. In an important paper BLAAUW (1946) used the proper motions and radial velocities available to him at that time to derive individual distances of the stars by considering the relation of the individual proper motions to the mean stream motion. Later BERTIAU (1958) made a definitive study using revised proper motions, improved radial velocities, and new photometry and he derived visual absolute magnitudes for the stars which he considered were members of the association. Bertiau concluded that the best way of handling the observations was to treat the stars as if they formed an expanding cluster. Finally Bertiau was able to calibrate the MK spectral classes from B0 to B3, luminosity classes IV to V, in terms of absolute magnitude.

PETRIE (1961) discussed revised radial velocities for stars of the Scorpio-Centaurus association and re-discussed the space motion of the group. Petrie makes the point that the radial velocities play a critical role in the assessment of the nature of the Scorpio-Centaurus group and that further observations of radial velocities are highly desirable for the stars of this group. He was able to confirm the well-known general recession of this group of stars from the sun, but he found that the moving cluster hypothesis is not tenable. This point is important when the motions of the stars are to be used to yield distances and individual absolute magnitudes, for the distances are found by assuming that one can compute the tangential component of motion of each star from its radial velocity. This procedure is valid only if the space motions of the stars are truly equal in amount and direction. This state of affairs is far from being established for the stars of the Scorpio-Centaurus group. One can compute a mean parallax for the group which gives a modulus of 5.8 mag. The result indicates that the absolute magnitudes suggested by Bertiau for spectral types B0 to B3 are too bright by about 0.4 mag. PETRIE (1964) has found a modulus of 6.1 ± 0.10 mag for the group from spectroscopic absolute magnitudes.

HARDIE and CRAWFORD (1961) have done UBV photometry for 80 stars and Hβ photometry for 50 stars in the northern part of the Scorpio-Centaurus association. They investigate stars to spectral type A0 and they investigate carefully the corrections for interstellar reddening in various parts of the association. By adopting Bertiau's distances for certain members of the group they are able to use their measured values of V_0 to derive more precise visual absolute magnitudes than were available with the previous photometry.

Spectral types and luminosity classes essentially on the MK system have been derived by Pamela MORRIS (1961) for 6 O stars, 201 B stars and 13 A stars in the region of the Scorpio-Centaurus association. She shows that many of the early type stars which she examines lie beyond the Scorpio-Centaurus association and that some of the late B type and A type stars are in the foreground. BUSCOMBE (1962) has published radial velocities for 70 stars in the area.

3.4 *The Cassiopeia-Taurus group*

There are a considerable number of apparently bright B type stars in the constellations Cassiopeia and Taurus, and BLAAUW (1956) suggested that they showed a community of motion. He considered the Cassiopeia-Taurus group to be identical with the extended Perseus stream which had been recognised by earlier investigators but which later was subjected to doubt. Blaauw analysed the motions of the group assuming an approximately linear expansion and found individual parallaxes and absolute magnitudes for the stars.

PETRIE (1958b) has determined new radial velocities for 40 of the 49 stars of the Cassiopeia-Taurus group and has rediscussed the space motion of the group. He finds that there is no evidence for community of motion and that the stars are a sample of the general population possessing normal solar motion plus the average random motion. Consequently individual parallaxes cannot be derived using moving cluster techniques. According to Petrie the stars do not show an expansion from a common origin. EGGEN (1961) has given reasons for considering the Cassiopeia-Taurus stars and the Scorpio-Centaurus stars to form a local association some 1000 pc in diameter, less than 100 pc thick and inclined at an angle of about 15° to the galactic plane.

The above discussions have been based on photometric data from a variety of sources. In order that the data for these well known stars may be homogeneous, CRAWFORD (1963) has provided UBV and Hβ photometry for these stars. Spectral types on the MK system have been given by Blaauw; types on the Victoria system and absolute magnitudes from Hγ measures are given by Petrie. From a consideration of the photometric data, Crawford supports Petrie's conclusion that the full group of Cassiopeia-Taurus stars does not form a cluster. However, Crawford does not exclude the possibility that some members may form a physically real, small moving cluster.

3.5 *The ζ Persei association*

From a study of the proper motions of 17 stars in an area covering 7° in declination and 40^m in right ascension around the B1Ib star ζ Persei, BLAAUW (1952) showed the group to be at a distance of about 300 pc, modulus 7.4 mag, and that the group was expanding. The expansion is near the limit of detectability. The systematic errors in the proper motions are crucial for assessing the validity of Blaauw's conclusion; accurate radial velocities are required to confirm or disprove the hypothesis of expansion.

This association is under observation for radial velocity at the Dominion Astrophysical Observatory and at the Royal Greenwich Observatory. JOHNSON (1957) has adopted a modulus of 7.7 mag for this group from UBV photometry, and PETRIE (1964) has derived a modulus of 7.7 ± 0.18 mag. BAPPU *et al.* (1961) have given Γ-index measurements for this group and have assembled the existing UBV photometry.

3.6 *The h and χ Persei association*

By the *h* and χ Persei association is meant the group of early type stars lying between about 1^h50^m and 2^h43^m in right ascension and between about +53° and +61° in declination. This large area of the sky contains a considerable number of O stars and supergiants as well as the clusters *h* and χ Persei.

One of the first spectroscopic studies of the double cluster in Perseus was made by BIDELMAN (1943, 1947) who gave spectral types for 153 stars in the vicinity of the double cluster and attempted to derive the distance of the clusters using spectroscopic parallaxes. Bidelman noted that the interstellar reddening in this area of the sky is very patchy and difficult to evaluate. However, he was able to show that the clusters are at a distance of about 2000 pc and that the radius of each nucleus is about 7 pc. There are thirteen M type supergiants in the neighbourhood of h and χ Persei and Bidelman showed that these stars are very probably closely associated with the early type stars forming the clusters. As described in Chapter IV, Petrie, Hiltner, and Johnson have used the stars of the double cluster as standards for deriving the luminosities of the B stars. The consensus of opinion is that the double cluster has a modulus of 11.8 mag.

A comprehensive photometric study of the stars in h and χ Persei and in the surrounding association has been made by WILDEY (1964). This complements the massive photographic study by OOSTERHOFF (1937). Wildey gives MK types and UBV photoelectric photometry for 21 O stars, 51 supergiants, and 17 early type giants and subgiants in the area as well as photographic magnitudes and colours for 298 stars in the nucleus of h Persei and for 296 stars in the nucleus of χ Persei. Finding charts are given for the stars. Wildey makes a particular effort to determine accurately the reddening in the neighbourhood of the O stars and of the B type supergiants by measuring the colours of faint early type companions to these stars. Since the intrinsic colours of early type main-sequence stars are fairly well known, Wildey is able to estimate rather closely the reddening near each luminous star. He draws attention to the great patchiness of the reddening in the Perseus association and he plots diagrams showing curves of equal reddening. From his measured colours and his estimates of the local reddening, Wildey is able to derive intrinsic colours for the O stars and for the supergiants.

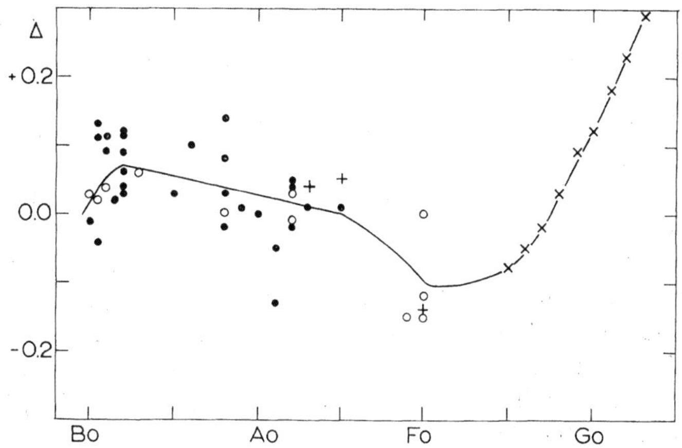

Fig. 27. Observed differences in $(B-V)_0$ between supergiants and main-sequence stars as a function of spectral type. From WILDEY (1964).

It is quite clear that the intrinsic colours of early type supergiants are not the same as those of early type main-sequence stars. The differences which are found are shown graphically in Figure 27. They probably arise because of the differences in line strength between supergiants and main-sequence stars. WILDEY's intrinsic colours for the O stars and for the B type stars of classes II, III, and IV in the h and χ Persei association are shown in Figure 21. The luminosity class II stars are plotted as open circles; the other stars are plotted as filled circles. The relation between $(B-V)_0$ and spectral type for main-sequence B stars which has been selected by ARP (1958) from a survey of all material available to him is indicated by a line. The striking thing in this diagram is that the range in colour for a given spectral type among the O stars is large, greater than 0.3 mag, whereas the range in colour among the B stars is quite small. Very little of the range in colour among the O stars can be due to incorrect estimates of interstellar reddening. This range is intrinsic to the stars and bears out the conclusion, reached in Chapter V, that the O stars are not a homogeneous group and that they are not suitable objects for investigating the interstellar reddening law.

Wildey finishes his paper with a discussion of the distribution of the stars of the h and χ Persei association in the colour-magnitude diagram and of the meaning of this distribution in terms of current theories of stellar evolution. The results appear to signify a non-unique age, three pronounced maxima of star formation having occurred. There is no significant correlation between spatial position and the suggested epoch of star formation.

3.7 *The I Lacertae association*

From a study of proper motions and of spectra BLAAUW and MORGAN (1953) derive a distance of 460 pc for the concentration of O–B5 stars in the constellation Lacertae. From the proper motions in declination they conclude that the association is expanding; the proper motions in right ascension cannot be used because of the systematic errors in the catalogues. WOOLLEY and EGGEN (1958) have concluded that there is no sound evidence for expansion. They show that the radial velocities give no support to the conclusion that expansion exists and that the existing small variations in the proper motion catalogues may well be errors of a cyclic nature. HARRIS (1955) has given UBV photometry for the early B stars of the Lacertae aggregate, while HARDIE and SEYFERT (1959) have given UBV photometry and spectral types for stars of types B5 and later. They estimate a modulus of 8.4 mag or a distance of 480 pc for the group.

CRAWFORD (1961) has published Hβ photometry for stars of the I Lacertae association to type A0 and he determines a modulus of 8.9 mag. (600 pc) by fitting to the V_0–$(U-B)_0$ curve found by Hardie and Crawford for the northern part of the Scorpio-Centaurus association. He corrects for a uniform absorption which is estimated from the reddening. From spectroscopic absolute magnitudes, PETRIE (1964) finds a modulus of 8.7 ± 0.13 mag.

ABT and HUNTER (1962) have derived rotational velocities for the brighter stars in the I Lacertae association. They note that the frequency of binaries is greater in groups of stars like I Lacertae, I Orionis, and the α Persei group which have lower than usual rotation speeds.

3.8 *NGC 2264*

This is a well-known group of bright B stars close to the O7 star 15 S Mon. The B stars show a community of motion which is not the same as that of the O star. The radial velocities of the B stars are near +20 km/sec, but the well determined radial velocity of S Mon is +33.2 km/sec. The observed radial velocities are chiefly due to reflex solar motion. At present there is no satisfactory explanation for the difference in radial velocity between S Mon and the B stars. The observed velocity shift does not appear to be due to incorrect wave lengths for the O type spectrum. The explanation that it is a gravitational red shift leads to an unlikely mass for S Mon.

M. WALKER (1956a) has done photometry of many stars in the neighbourhood of the B stars, and he has studied some spectra. He estimates that the cluster is at a distance of about 800 pc. HERBIG (1954) has shown that the stars are projected against a dark cloud. Since the area is full of nebulosity, it is very difficult to do precise photometry in this area. Many T Tauri stars are found in this neighbourhood and Walker has suggested that the colour-magnitude diagram of the group of stars studied by him is typical for a cluster so young that the less massive stars are still in the contracting phase. Much work must be done to eliminate more surely foreground stars of type F and later from the suspected cluster members. It is by no means certain that all the stars considered by Walker are in the close neighbourhood (within about 20 pc) of the group of B stars. Accurate radial velocities are needed to make the decision about membership in the cluster, but these are difficult to obtain.

Spectroscopic absolute magnitudes, whether found from Hγ measures or from other spectral characteristics, usually are uncertain by about ±0.3 mag. This results in an uncertainty in distance of the order of 15 per cent. Thus at a distance of 800 pc, one can only hope to place the stars within a band some 200 pc deep when using spectroscopic parallaxes. Such an uncertainty in depth might serve to define an association, but it hardly suffices for defining a cluster, for the diameter of a cluster is usually not greater than 20 pc. Possibly in NGC 2264 we are viewing an O association which is elongated in the line of sight, rather than a normal moving cluster. The density of the dark cloud limits the length of our line of sight to about 800 pc.

3.9 *NGC 3293 and NGC 4755*

These clusters are important because like *h* and χ Persei they contain B type supergiants and M supergiant stars. An A type supergiant is also possibly a member of NGC 4755. FEAST (1958, 1963) has studied these clusters and has compared them with *h* and χ Persei. He gives MK spectral types for 35 stars and radial velocities for 26 stars in NGC 3293 and spectral types for 35 stars, radial velocities for 31 stars in NGC 4755. NGC 4755 is also known as κ Crucis. A comprehensive photometric study of NGC 4755 has been done by ARP and VAN SANT (1958). Feast quotes unpublished photometric work by HOUCK for NGC 3293.

FEAST adopts a modulus of 12.3 mag for NGC 3293 and 11.9 mag for NGC 4755. He finds the distance to NGC 3293 by correcting the observed magnitudes for reddening using the intrinsic colours of the OB stars given by HILTNER (1956) and the calibration of the MK system by JOHNSON and HILTNER (1956). The distance to NGC 4755 is found from 19 early-type members of luminosity classes III to V using the photometry of Arp and Van Sant, adopting the intrinsic colours of Johnson and determining mean reddening corrections. Feast adopts the standard relationship

$$A_V = 3.0\,E(B\text{–}V)$$

in order to estimate the interstellar absorption.

Once the distances to NGC 3293, NGC 4755, and h and χ Persei are known it is possible to plot a combined colour-magnitude diagram for the three clusters. Feast notes that all three clusters have a vertical sequence at spectral type B1 and that there is a gap between the evolved main sequence which ends at about $M_V = -5.5$ and the supergiant branch which lies near $M_V = -6.0$. Feast suggests that this gap is caused by the stars entering a period of rapid gravitational contraction of the stellar core following hydrogen exhaustion. Rapid evolution from A type supergiants to M type supergiants is expected to occur during helium burning.

In addition Feast has looked at the relative radial velocities of stars of different luminosity classes in NGC 3293, NGC 4755, and in h and χ Persei. He concludes that the supergiants tend to have positive velocities with respect to the other stars in the cluster. More observations are needed to clarify this problem.

3.10 *Regions of origin of O and B stars*

Most O and B stars are found in clusters or in associations. In fact associations are defined by the presence of O stars. Essentially all known O stars are either in clusters or in associations. All known B stars are not in these groups. Can the observed number of B stars be interpreted as a result of the longer life of B stars than O stars? That is, is it reasonable to suggest that all B stars were formed in O clusters or in O associations and that the unattached B stars are from broken-up clusters, the O stars having evolved into something else?

ROBERTS (1957) has given an affirmative answer to this question. From the known number of O clusters and O associations, from the estimated ages of O and B stars, and from the average number of O and B stars per cluster, Roberts estimated that there should be between 1.2×10^6 and 3.0×10^6 B stars in the Galaxy. Since it is known from surveys that there are about 2×10^6 B stars in the Galaxy, Roberts concluded that all B stars were formed in O clusters or in O associations. In addition he interprets the fact that some B stars are not observed in clusters as strong evidence in favour of the concept of expanding clusters and of the destruction of clusters by the peculiar motions of the members.

BLAAUW (1961) has investigated "run-away" stars, that is O and B stars of large space motion for which the direction of the motion is sufficiently accurately known and for which the distance the star has travelled since it left the association is still small enough to allow the identification of its origin. Blaauw lists 19 run-away stars of spectral types O6 to B5 and for 13 of them he is able suggest an association from which they probably came. This list is by no means complete, for the necessary information for finding such stars is not available over the whole sky.

The group of run-away stars has two properties different from that of OB stars in general. The first is the absence of double or multiple stars which contrasts strongly with the high frequency of duplicity or multiplicity among the low-velocity objects. The second is the great preponderance of O stars among the group. The relative number of O stars to B stars in the group is ten times what it is in the general field. Blaauw proposes that the run-away stars were the secondary components of massive binary systems which have dissolved as a result of the primary star loosing much of its mass in a sudden event. He gives reasons for believing that the ejection of mass by the primary and the consequent freeing of the companion may be identified with supernova events of Type II.

Blaauw considers the following stars to be run-away stars: λ Cep, ζ Per, HD 157857, HD 152408, 68 Cyg, α Cam, AE Aur, ζ Oph, μ Col, HD 151397, HD 149363, 53 Ari, HD 97991, HD 197419, 72 Col, HD 214930, HD 216534, HD 4142, and HD 201910.

CHAPTER VII

THE HELIUM SPECTRUM

1. Spectroscopic notation and the energy level diagram of helium

The presence of absorption lines from the spectrum of neutral helium is a defining characteristic of spectral type B. A summary of the chief properties of the He I spectrum will be given here as an introduction to the astrophysical study of B type spectra and also for the purpose of reviewing spectroscopic notation. The lines from the spectrum of ionized helium, the He II spectrum, appear in O type stars in addition to the He I lines. The He II lines are hydrogen-like and present few special points of interest.

Since in neutral helium atoms there are two electrons outside the helium nucleus, the spectral terms of He I can be either singlets, in the case in which the individual spins, s, of the two electrons are opposed, or triplets when the two spins are aligned. The multiplicity is $2S+1$, where S is the vector sum of the spins. A spectral term, which represents a possible energy state of the atom is given the designations S, P, D, or F according to whether the vector sum, L, of the individual orbital angular momentum quantum numbers is 0, 1, 2, or 3. The total angular momentum of the state is designated by J which is the vector sum of L and S.

Permitted transitions in the case of electric dipole interactions are those in which $\Delta S=0$, $\Delta L=\pm 1$, and $\Delta J=0, \pm 1$. The transition $J=0$ to $J=0$ is forbidden. The change in the value of the chief quantum number, n, can have any value. These rules are known as Russell-Saunders or LS coupling and they govern the production of spectral lines from simple atoms. For a permitted transition the parity must change. The parity of a term is denoted by a superscript to the right of the symbol representing the term. The parity is odd (superscript small o) if the scalar sum of the orbital angular quantum numbers is odd, and even (no superscript) if the sum is even.

A level diagram of some of the lowest terms of He I is shown in Figure 28. The ground level of He I is 1^1S and it lies 19.81 volts below the first excited level, 2^3S, and 20.61 volts below the second excited level, 2^1S. No strong intercombination lines between the singlet and triplet terms are known in the case of He I. The 2^3S and the 2^1S states are metastable; they do not combine strongly with each other or with the ground state. The fact that these levels are metastable plays an important role in the interpretation of the strength of the He I lines in some O and B type stars. It is to be noted that the excited levels of He I lie much closer to the ionization continuum than to the ground state. A line representing the leading member of each series is drawn.

Long series of He I absorption lines may be observed in B type spectra. They arise

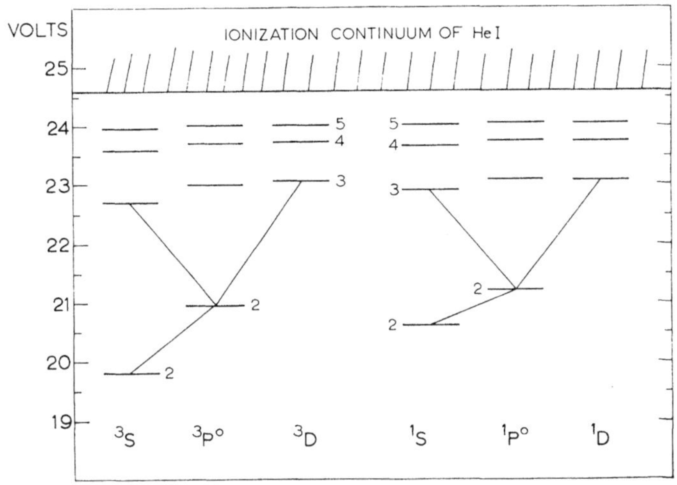

Fig. 28. An energy level diagram for He I.

as a result of transitions from the 2^1S, 2^1P°, 2^3S, and 2^3P° terms to upper terms. Some of the members of the most well known series are listed in Table 17 together with their *gf* values. The wave lengths have been taken from the paper by MARTIN (1960). When there is more than one component in the multiplet, the component of shortest wave length is usually the strongest line. The lines of the 1^1S–n^1P° series fall in the wave-length range between 584.334 Å and the Lyman limit of He I at 504.3 Å, or 24.58 volts.

2. Stark broadening and the forbidden lines

When helium atoms are in an electric field, the bound electrons are perturbed and the energy levels are shifted. Transitions occur which are forbidden according to the LS coupling rules. These forbidden lines are electric quadrupole radiation and they form series such as 2P–*nP* and 2P–*nF*, both for the singlet and for the triplet terms. The forbidden lines are well known in the spectra of main-sequence B stars. Some of the strongest forbidden lines of He I are listed in Table 18. The wave length is given only to the nearest tenth angstrom because the separation of the energy levels depends upon the field strength. Both the permitted and the forbidden lines are shifted in an electric field. One of the luminosity effects in B type stars is the appearance of Stark broadening wings on the *2P–nD* series lines and the presence of strong, somewhat diffuse lines of the *2P–nF* series. The forbidden lines at 4469 Å and at 4025 Å are particularly prominent in main-sequence stars; these lines are not visible in supergiant spectra.

GRIEM, BARRANGER, KOLB, and OERTEL (1962) have studied the Stark broadening of the helium lines in a plasma and have given formulas for the shape of lines which

TABLE 17

Wave lengths and f values for some lines of He I

n	Wave length (Å)	f	Source for f	n	Wave length (Å)	f	Source for f
	2^3P^0–n^3D Series				2^1P^0–n^1D Series		
3	5875.966, .621	0.6234	1	3	6678.151	0.7253	1
4	4471.682, .479	0.1229	1	4	4921.931	0.1205	1
5	4026.359, .191	0.04668	1	5	4387.929	0.04308	1
6	3819.758, .607	0.02472	2	6	4143.761	0.0225	2
7	3705.148, .005	0.01412	2	7	4009.268	0.0126	2
8	3634.369, .232	0.00849	1	8	3926.534	0.00740	1
9	3587.405, .270	0.00578	2	9	3871.791	0.00501	2
10	3554.547, .415	0.00408	1	10	3833.554	0.00351	1
11	3530.621, .491	0.0030	4	11	3805.740	0.0026	4
12	3512.640, .512	0.0023	4	12	3784.862	0.0019	4
13	3498.772, .645	0.0018	4	13	3786.784	0.0015	4
14	3487.850, .723	0.0014	4	14	3756.107	0.0012	4
	2^3P^0–n^3S Series				2^1P^0–n^1S Series		
3	7065.707, .190	0.07639	1	3	7281.349	0.04640	1
4	4713.376, .146	0.01159	1	4	5047.738	0.00824	1
5	4120.992, .815	0.00518	2	5	4437.551	0.00341	2
6	3867.630, .475	0.00256	2	6	4168.967	0.00151	1
7	3733.010,2.865	0.00134	2	7	4023.973	0.00087	1
8	3652.130,1.990	0.00071	1	8	3935.912	0.00055	1
9	3599.448, .314	0.00048	2	9	3878.181	0.00037	4
10	3562.979	0.00034	1	10	3838.100	0.00026	4
	2^3S–n^3P^0 Series				2^1S–n^1P^0 Series		
2	10830.337,.248,9.088	0.60021	1	2	20581.30	0.3578	1
3	3888.648	0.05705	1	3	5015.678	0.1646	1
4	3187.745	0.02083	1	4	3964.729	0.0508	1
5	2945.106	0.0114	3	5	3613.643	0.0239	2
6	2829.076	0.0061	3	6	3447.586	0.01257	1
7	2763.804	0.0038	3	7	3354.550	0.0066	3
8	2723.191	0.0026	3	8	3296.773	0.0044	3

1. TREFFTZ, E., SCHLÜTER, A., DETTMAR, K. H., and JÖRGENS, K.: 1957, *Z. Astrophys.* **44,** 1.
2. Interpolated linearly in $n^3 f_n$ using values of Trefftz *et al.*
3. DALGARNO, A. and KINGSTON, A. E.: 1958: *Proc. Phys. Soc. A.* **72,** 1053.
4. From a graphical extrapolation of $n^3 f_n$ using values of Trefftz *et al.*

TABLE 18

Forbidden lines of He I

n	2^3P^0–n^3F^0	2^3P^0–n^3P^0	2^1P^0–n^1F^0	2^1P^0–n^1P^0
	Å	Å	Å	Å
3	–	6067.0	–	6631.9
4	4469.9	4517.4	4920.4	4910.7
5	4025.5	4045.2	4387.6	4383.3
6	3819.2	3829.5	4143.5	4141.3

occur as a result of transitions between well separated levels. The Stark effect of the complex of lines from levels which lie close together, such as the $2P$–nD and the $2P$–nF lines in both singlets and triplets, cannot be handled in a satisfactory manner by the above theory in the case of a stellar atmosphere where the electric fields arise from the close passage of ions and electrons.

Foster and Douglas (1939) made the first systematic investigation of the Stark effect of He I in B stars and they attempted to interpret the blended profiles of lines from the $2P$–nD and the $2P$–nF series quantitatively using experimental measurements of the Stark effect of He I in constant electric fields and a simple intuitive theory of line formation in stellar atmospheres. They noted that on this basis it was not possible to explain the apparent relative strengths and shapes of line pairs such as λ4471–λ4469 and λ4922–λ4920. They suggested that cyclic transitions occur in the He I spectrum which tend to form extra absorption at the zero field position of the $2P$–nF lines and tend to weaken the leading member of the series by emission. Further observations which appeared to support this suggestion were made by Underhill and W. Petrie (1944).

All series of equivalent width measurements of the He I lines in main-sequence and supergiant B stars give similar relative intensities to those reported by Foster and Douglas and by Underhill and Petrie. However, it now seems that the interpretation made in 1939 of the observations is too simple. Theoretical studies of line broadening in plasmas have indicated that Stark broadening in constant electric fields is not the only important broadening agent for the selected lines but that collisional broadening due to electrons is also important. In addition it is now known that strong spectral features, such as the He I lines in early B type spectra are formed through a great depth of atmosphere. Simple theories of line formation are not adequate for interpreting the *whole* line profile. The forbidden lines and the Stark broadened and collision broadened wings of the $2P$–nD lines are formed in the much deeper layers of the atmosphere than the layers in which the cores of the $2P$–nD lines are formed. It is not to be expected that a one-layer theory and simple theories of line formation will be adequate to explain both the observed equivalent widths of the leading members of the diffuse series (for which Griem *et al.* have given a useful theory of line broadening) and the equivalent widths of the complex line groups formed by the higher members of the $2P$–nD series and the $2P$–nF series. The Foster-Douglas cycle probably does not occur. The observed line strengths in B type stars cannot be taken as evidence in favour of it. The argument used in 1944 was that since the leading members of the diffuse series, λ5876 and λ6678, were much weaker in main-sequence stars than in supergiants of the same spectral type, these lines must be weakened by re-emission in the main-sequence stars. This argument overlooks the fact, not realised then, that lines in supergiants are generally much stronger than the lines in main-sequence stars of the same spectral type as a result of turbulent broadening. One cannot expect supergiants and main-sequence stars of the same spectral type to have lines of the same absolute strength. The fact that the spectral type is the same only means that certain characteristic lines have the same relative strength in both objects.

3. Non-equilibrium effects due to the presence of metastable levels

In order to interpret theoretically the strengths of stellar absorption lines from excited levels, it is generally assumed as a first approximation that the levels are populated according to the Boltzmann law, that is as if the gas were in thermodynamic equilibrium at the temperature assigned to the layer of the atmosphere. However, in general this is not the case in the outer parts of a stellar atmosphere. In particular, striking departures from the equilibrium populations can occur if there are some metastable levels in the atom. In this section we shall look at this problem with special reference to the He I spectrum and to spectra that have a similar electronic structure.

3.1 *Effects due to the metastability of the* 2^1S *and the* 2^3S *levels of* He I

Struve and Wurm (1938) in an important paper discussed the problem posed by the strengthening of the He I triplets with respect to the singlets in stars of type B2 and earlier and by the remarkably great strength of the lines from only the 2^1S and 2^3S levels in certain shell stars where the underlying star is of spectral type B1 and earlier. The strengthening of the diffuse series triplets with respect to the singlets as the spectral type becomes earlier is shown in Figure 29. This diagram has been prepared from the equivalent widths measured by the Edinburgh observers. The equivalent width of λ4471, $W(4471)$, is taken to be representative of the strength of the triplet

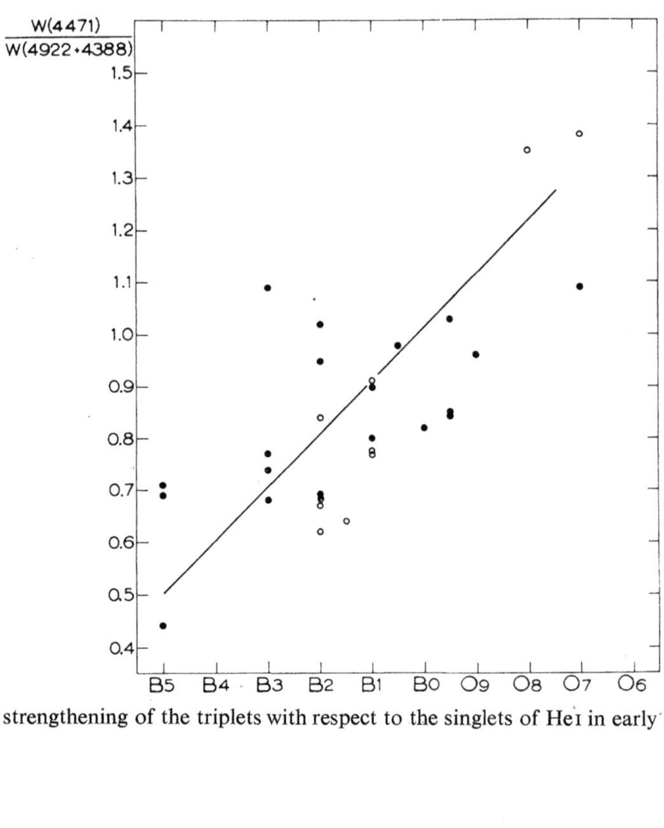

Fig. 29. The strengthening of the triplets with respect to the singlets of He I in early spectral types.

series; the sum of the equivalent widths of λ4922 and of λ4388, $W(4922+4388)$, represents the strength of the singlet series. Main-sequence stars are represented by filled circles and giant and supergiant stars are represented by open circles. It is clear that the ratio $W(4471)/W(4992+4388)$ increases steadily from about 0.5 at type B5 to approximately 1.3 at type O7. The points for two stars fall outside the diagram. Namely that for λ Cep, O6f, which falls at 1.83 and that for 68 Cyg, O8II, which falls at 1.61.

Struve and Wurm noted that when the absorbing gas is some distance from the surface of the star, as it is in a shell, the atoms will be subjected to a diluted radiation field, for radiation will not impinge on the atoms from all directions as in an optically thick layer, but only from the cone filled by the star's surface when viewed from the position of the atom. In this case the distribution of the atoms among the various states will not be that which is valid in thermodynamic equilibrium. One can state, however, that statistical equilibrium will be reached in the level populations. This means that on the average the number of atoms entering a level by all possible methods must equal the number leaving the level by all possible methods.

Let us consider level j of an atom. The levels below level j may be indicated by the letter l and the levels above level j by the letter u. The atom is situated in a radiation field of intensity I. The radiation intensity at a frequency corresponding to a transition to a lower level, j–l, is I_{jl} and that at a frequency corresponding to a transition to an upper level, j–u, is I_{ju}.

The number of atoms leaving the level j per unit time by spontaneous emission will be $\sum N_j A_{jl}$. Here N_j is the population of level j and the sum is to be taken over all possible lower levels. The quantity A_{jl} is the Einstein spontaneous transition coefficient. The number of atoms leaving the level by emission in the presence of the radiation field will be $\sum_l N_j I_{jl} B_{jl}$. Here B_{jl} is the Einstein impressed emission coefficient and the sum is carried out over all possible lower levels. The number of atoms leaving the level as a result of absorption of radiation will be $\sum_u N_j I_{ju} B_{ju}$, where B_{ju} is the Einstein absorption coefficient and the sum is carried out over all possible upper levels. In addition atoms may leave the level j as a result of collision with perturbers of type p. The collisions may lead to upper levels or to lower levels. The number of atoms leaving level j due to collision with all possible types of perturber p is thus $N_j \sum_l \sum_p C_{jl}(p) N_p + N_j \sum_u \sum_p C_{ju}(p) N_p$. Here the $C_{jl}(p)$ and $C_{ju}(p)$ are collision cross-sections.

Similarly atoms may enter level j from levels above j as a result of spontaneous emission, as a result of impressed emission and as a result of collisions. They may enter level j from levels lying below j as a result of absorption and as a result of collision. The principle of statistical equilibrium states that the total rate of atoms entering level j must equal the total rate of atoms leaving level j. Thus the following equation results:

$$N_j \Big[\sum_l A_{jl} + \sum_l B_{jl} I_{jl} + \sum_u B_{ju} I_{ju} + \sum_l \sum_p C_{jl}(p) N_p + \sum_u \sum_p C_{ju}(p) N_p\Big]$$
$$= \sum_u N_u (A_{uj} + B_{uj} I_{uj}) + \sum_l N_l B_{lj} I_{lj} + \sum_l \sum_p C_{lj}(p) N_p + \sum_u \sum_p C_{uj}(p) N_p . \qquad \text{(VII-I)}$$

There is a similar equation for each level of the atom. The whole set of equations must be solved for the level populations, N_j.

In order to make a start on this problem, for the case of He I, Struve and Wurm neglected the collision terms and they considered a six level atom, namely the ground state, 1^1S, the first four excited states 2^3S, 2^1S, 2^3P, and 2^1P, and the ionized state. Furthermore they represented the radiation field by $W.B_\nu(T)$ where W is the dilution factor (a free parameter in their work), and $B_\nu(T)$ is the Planck function for a temperature T. In order to calculate the degree of ionization they assumed that the electron temperature was equal to the radiation temperature and that Saha's law was valid. In this way they found the following relative populations for the first four excited states of He I in the case that the temperature is 10^4 degrees. The relative populations according to the Boltzmann law (thermodynamic equilibrium) are given for comparison with the case $W=1$.

Level	Boltzmann	$W=1$	$W=0.1$	$W=0.02$
2^1S	100	100	100	100
2^1P	150	152	11	5
2^3S	753	756	814	1140
2^3P	610	602	75	30

The 2^1P and 2^3P levels are depopulated as the radiation field becomes more dilute and the 2^3S level is overpopulated. The atoms tend to accumulate in the metastable levels. If one defines b_j as the ratio N_j/N_j(l.t.e.), then one can show that as W becomes small the factor $b_j \sim W^{-1}$ for 1^1S, 2^1S, and 2^3S. The atoms concentrate in these three levels. This is the dilution effect in the He I spectrum and it serves to explain in broad outline the observed strengthening of lines from 2^1S and 2^3S in objects with extended atmospheres.

Struve and Wurm also discussed qualitatively what collisions with electrons would do to the level populations. They concluded that a strong preference for the triplet levels will occur when collisions become comparable in effect to dilution. This tends to happen at high temperatures and moderate dilutions, that is in the O stars.

Wellmann (1951) rediscussed the problem including levels to $n=4$ explicity in his work and allowing for the electron temperature, which enters the calculation of the degree of ionization, to be different from the radiation temperature. His results in general parallel those of Struve and Wurm, but give more detail. Wellmann did not include collisions in the equations of statistical equilibrium. Ghobros (1962) has calculated more b_j values for He I following Wellmann's exposition of the problem. Mathis (1957) has discussed the He I spectrum in connection with departures from local thermodynamic equilibrium in a gaseous nebula. Here the dilution factor is very low. In this case two-photon transitions can be important for depopulating 2^3S.

3.2 *Effects in other spectra of the same structure*

In any atom or ion where two electrons lie outside a tightly bound nucleus one will obtain singlet and triplet terms. The probability of intercombinations between the two sets of terms will be small so long as the inner electrons and the nucleus form a tightly bound unit. Consequently the lowest triplet levels will be metastable and dilution effects exactly like those observed in the He I spectrum may be expected in the case of stars with extended atmospheres. Since the ions of the He I iso-electronic sequence produce lines in the far ultraviolet, they are not interesting so far as interpreting ordinary stellar spectra is concerned. However, C III, N IV, and O V have helium-like terms (singlets and triplets), the electron configuration of the ground state being $1s^2 2s^2$. Since the $1s^2$ electrons are tightly bound, they may be considered to form a core with the atomic nucleus. No strong intersystem-combination lines have been discovered for the C III, N IV, and O V spectra, thus the lowest triplet level in each case is metastable just as in the case of He I.

A similar situation exists for Si III. In this case the configuration of the ground state is $1s^2 2s^2 2p^6 3s^2$. The first ten electrons together with the atomic nucleus form a tightly bound unit. The spectroscopic terms are singlets and triplets. Again it is very doubtful if any strong intersystem combination exist, thus the lowest triplet term is metastable.

The theory that has been worked out for He I indicates that in extended stellar atmospheres one may expect that the absorption lines arising from the metastable triplet levels and from triplet levels which are connected by strong transitions to these levels will be strengthened by dilution effects. Relevant information concerning the triplet levels of He I, C III, N IV, O V, and Si III is given in Table 19. The lines of the He I

TABLE 19

Energy levels concerned in dilution effects

Spectrum	He I	C III	N IV	O V	Si III
Ground state	$1s^2\ ^1S^0$	$1s^2 2s^2\ ^1S^0$	$1s^2 2s^2\ ^1S^0$	$1s^2 2s^2\ ^1S^0$	$1s^2 2s^2 2p^6 3s^2\ ^1S^0$
Lowest triplet terms built on ^2S term of next ion	$1s2s\ ^3S$ $1s2p\ ^3P^0$ $1s3p\ ^3S$ $1s3p\ ^3P^0$	$2s2p\ ^3P^0$ $2s3s\ ^3S$ $2s3p\ ^3P^0$ $2s3d\ ^3D$	$2s2p\ ^3P^0$ $2s3s\ ^3S$ $2s3p\ ^3P^0$ $2s3d\ ^3D$	$2s2p\ ^3P^0$ $2s3s\ ^3S$ $2s3p\ ^3P^0$ $2s3d\ ^3D$	$3s3p\ ^3P^0$ $3s3d\ ^3D$ $3s4s\ ^3S$ $3s4p\ ^3P^0$
Multiplet showing dilution effects	2^3P^0–4^3D λ 4471	3^3S–3^3P^0 λ 4650	3^3S–3^3P^0 λ 3478	(3^3S–3^3P^0) λ 2790	4^3S–4^3P^0 λ 4552
Multiplet connecting lowest level	2^3S–2^3P^0 λ 10830	2^3P^0–3^3S λ 538	2^3P^0–3^3S λ 322	2^3P^0–3^3S λ 215	3^3P^0–4^3S λ 997

multiplets arising from the 2^3P° term are well-known to be strengthened in high-temperature extended atmospheres. Extra strong absorption in the C III λ 4650 multiplet and in the N IV λ 3478 multiplet is observed in Wolf-Rayet spectra. It is likely that similar extra strong absorption will be observed in the corresponding O V multiplet at 2790 Å when spectra of very hot stars have been obtained in this wave-length region. The abnormal strengthening of the Si III multiplet 4^3S–4^3P°, which has lines at 4552 Å, 4568 Å, and 4574 Å, in absorption in objects of about type B1 having extended atmospheres (for instance P Cygni and ε Persei, see UNDERHILL, 1963a) can be attributed to the same type of dilution effect. In fact, in any spectrum where metastable levels exist because the level systems of different multiplicity are not strongly connected by line transitions, similar dilution effects may be expected.

The cases mentioned here are the most striking found to date. According to the present analysis the multiplet which connects the lower level of the multiplet showing dilution effects to the metastable level can be expected to be very strong in absorption in those objects which are observed to show dilution effects. The connecting multiplet is available for observation from the surface of the earth only in the case of He I, and even this multiplet is not easy to observe owing to the lack of sensitivity of the presently available recording devices at 10830 Å.

CHAPTER VIII

WAVE-LENGTH STUDIES OF EARLY TYPE STARS

1. Measurement for wave length

Measuring wave lengths on stellar spectrograms is essentially a question of determining accurately the position on the spectrogram of the stellar lines with respect to the position of lines in the comparison spectrum and interpolating for wave length. The plate to be measured is set on the table of a measuring machine and either viewed with a microscope or viewed by projection. A magnification of 20 to 50 is usually required. Position along the plate may be measured by counting the rotations of a long screw, 1 mm or 0.5 mm pitch, and using a divided head so that one can read to 0.001 unit, or by measuring against a fixed scale, as in an Abbe comparator.

With prismatic spectrograms the Hartmann formula is used for interpolation. Thus

$$\lambda = \lambda_0 + \frac{C}{s_0 - s}. \qquad \text{(VIII-1)}$$

Here s is the setting for a line of wave length λ. The three constants s_0, C, and λ_0 are characteristic of the spectrograph and measuring machine, and they must be found for each spectrograph by measuring the settings s for three lines of known wave length and using equation (VIII-1) three times to obtain equations for the three unknowns. Equation (VIII-1) may be used for interpolation over a range of about 200 Å with any one set of constants.

With grating spectrograms the linear dispersion is nearly constant over a considerable wave-length range. Within this range linear interpolation may be used. Thus

$$\lambda = \lambda_0 + D(s - s_0). \qquad \text{(VIII-2)}$$

Here s is the setting for a line of wave length λ, and λ_0, D, and s_0 are constants characteristic for the spectrogram and the measuring machine. It follows that D is the linear dispersion expressed as angstroms per unit setting.

Neither equation (VIII-1) nor equation (VIII-2) fit a typical stellar spectrogram exactly over more than a few angstroms. However, these equations and a single set of constants may be used over something like 200 Å by making use of a correction curve. Thus from the settings for three lines of known wave length, usually selected to fall at either end of the region and near the middle of the region, a solution is made for the three constants λ_0, C, and s_0 or λ_0, D and s_0. Then the positions s_i are calculated for a number of standard wave lengths in the comparison spectrum. Let the measured positions of these lines be s_i'. The s_i' will not in general be exactly the same as the computed s_i. A correction curve is found by plotting $s_i - s_i'$ against s_i'. Usually this curve is nearly a straight line for prismatic spectrograms and a parabola for grating spectrograms. Once this curve has been found for a spectrogram, it is a simple matter to read off the correction $s_j - s_j'$ for every stellar line λ_j lying at setting s_j'. The corrected setting, s_j, of the stellar line is found by adding the correction to the measured setting s_j'. With s_j thus determined on the scale which was used to find the constants, either interpolation formula (VIII-1) or (VIII-2), as appropriate, may be used to find λ_j, the apparent wave length of the stellar line.

By using this technique, one set of constants may be used for reducing to wave length the settings made on all spectrograms taken with a given spectrograph. It is not necessary to solve for new constants each time. In each case the zero point of the setting scale must be adopted by making a selected setting s_i correspond to a selected wave length λ_i in the comparison spectrum. It is customary to select a well-defined comparison line in the middle of the spectrogram for this purpose.

2. Standard wave lengths

In order to perform the interpolation to obtain wave lengths on stellar spectrograms, one must have available a list of known, stable, accurate wave lengths in the comparison spectrum. Such standard wave lengths are called *Secondary Standards*. They are assigned to Class A or to Class B, depending upon the accuracy of the set of values.

Class A secondary standards are highly reproducible standards which have been directly compared with the primary standard in several laboratories with an accuracy comparable to that of the primary standard, that is, to better than 0.0001 Å. Class A secondary standards may serve as substitutes for the primary standard. The following lines have been selected. The unit is 10^{-10}m or 1 Å.

Kr^{86}	Hg^{198}	Cd^{114}
6458.0720	5792.2683	6440.2480
6422.8006	5771.1983	5087.2379
5651.1286	5462.2705	4801.2521
4503.6162	4359.5624	4679.4581

Class B secondary standards permit the routine measurement of spectral lines with an accuracy better than 0.01 Å. There must be many of these standards at more or less uniformly distributed wave lengths. These wave lengths are known with an accuracy of 0.001 Å or better, but they need not be of the high accuracy of Class A secondary standards.

The lines of the iron arc in air are commonly used as secondary standards of Class B. Commission 14 of the International Astronomical Union has prepared lists of standard wave lengths in several spectra. A list of the standard lines in the Fe I spectrum between 2084 Å and 11973 Å is given in *Trans. I.A.U.* **9** (1955) 220.

The lines of the open iron arc, however, are rather broad and they are liable to displacements by pole and pressure effects. The Fe lines emitted in hollow cathode discharges or in high-frequency discharges through iron halide vapours using Ne, Ar, or He as the carrier gas have been found to be greatly superior to the Fe lines emitted by open arcs and they are to be preferred when an adequate light source of this type is available. There are some appreciable differences in wave length between the two types of source. Commission 14 is looking into the question of providing a list of suitable standard wave lengths from hollow cathode discharges. A preliminary list of wave lengths is given in *Trans. I.A.U.* **11A** (1961) 100.

The Fe I spectrum does not have a satisfactory distribution of lines in the yellow-red spectral region for use as a comparison spectrum. Other sources of comparison spectra are required. One of the most convenient is Ne I. A list of lines in the region 5852 Å–7032 Å is given in *Trans. I.A.U.* **8** (1952) 190, and further lines are listed in *Trans. I.A.U.* **9** (1955) 205. Commission 14 is continuing to look into the problem of providing secondary standard wave lengths for the ultraviolet and for the infrared spectral regions.

The *Primary Standard* of wave length is a line from the first spectrum of Kr^{86}. The wave length in vacuum is 6057.80210_5 Å. This wave length defines the angstrom unit (Å) as 10^{-10}m exactly. The wave length in air is 6056.12525_3 Å. This value is obtained from λ_{vac} by adopting the formula for the refractivity of air proposed by EDLÉN (1953). The above wave length in vacuum is the legal international standard of length and it replaces the meter bar at Sèvres.

3. The refractivity of air

The wave length in vacuum is related to the energy of the light quantum. Thus the energy of a photon is

$$E = h\nu = hc/\lambda_{vac}\,. \tag{VIII-3}$$

All theoretical formulas which concern radiation are expressed in terms of the frequency ν, or of the wave length in vacuum, λ_{vac}. However, all experimentally determined wave lengths greater than 2000 Å are usually measured in air, and it is customary to list λ_{air} in wave-length tables. The two wave lengths are related as follows

$$\lambda_{vac} - \lambda_{air} = \lambda_{air}(n - 1)\,. \tag{VIII-4}$$

Here n is the index of refraction of air and it varies with the wave length and with the temperature and pressure of the air.

Commission 14 has discussed the question of the refractivity of air. By 1952 it was apparent that the then available information on the dispersion of air was inadequate and that the available formulas gave results that differed by amounts that could not be tolerated. The dispersion of air is found by converting observed wave lengths to wave numbers, $\sigma = 1/\lambda$, and comparing the results with the values found in complex spectra from the combination principle using the term values of well established energy levels. The subject was reviewed by EDLÉN (1953) and he derived a new formula for the refractivity of air. This formula has been adopted by Commission 14. It is

$$(n-1)\,10^7 = 643.28 + \frac{294981}{146-\sigma} + \frac{2554.0}{41-\sigma^2}. \qquad \text{(VIII-5)}$$

Here σ is the wave number in vacuum expressed in reciprocal μ.

Tables of the vacuum correction $(\lambda_{vac} - \lambda_{air})$ corresponding to this formula have been computed at the Physics Department, University of Lund. The correction from air to vacuum is not negligible, as the following abstract from the Lund Tables shows:

λ_{air} Å	$\lambda_{vac} - \lambda_{air}$ Å	λ_{air} Å	$\lambda_{vac} - \lambda_{air}$ Å
4000	1.131	5000	1.395
4200	1.183	5500	1.528
4400	1.236	6000	1.662
4600	1.289	7000	1.932

COLEMAN, BOZMAN, and MEGGERS (1960) have published a table of wave numbers 2000 Å (.01) 10 000 Å based on Edlén's formula for the refractivity of air. They also give vacuum corrections.

4. The correction of measured wave lengths to absolute values

The apparent wave lengths that are derived from a stellar spectrum by interpolation in the comparison spectrum are not the same as the true wave lengths in a star. The true wave length is displaced by the motion of the star in the line of sight, that is by the star's radial velocity, and by the component of motion of the earth in the line of sight. In the case of prismatic spectrograms there is a further displacement due to the curvature of the lines.

Spectral lines obtained with a prismatic spectrograph are curved because the light from each end of a long slit goes through a longer path in the prism than does the light from the centre of the slit. Thus these off-axis beams are deviated more than the on-axis beams. The result is that each image of the slit has a parabolic shape, the parabola opening towards the violet. This subject has been discussed, for instance by LORD (1897) and by ADAMS (1900). Since stellar spectrographs are so arranged that the stellar spectrum uses the central, on-axis part of the slit and the comparison arc goes through the outer parts of the slit, the centre of gravity in the direction of the dispersion of the images of the comparison lines is displaced to the violet from the centre of the stellar line.

The curvature correction is best found empirically by measuring spectra taken with a long slit at different places along the slit to find the shape of the image of the slit. It is found in some cases that the curvature correction in Å is nearly proportional to wave length. This means that in velocity units the correction is nearly independent of wave length, for

$$v = \Delta\lambda \cdot \frac{c}{\lambda}. \qquad \text{(VIII-6)}$$

For example, the curvature correction of the two-prism spectrograph of the Dominion Astrophysical Observatory is +0.7 km/sec over the range of spectrum usually measured. This amount must be added to every velocity derived from spectrograms taken with the two-prism spectrograph in order to correct for the systematic displacement of the comparison arc with respect to the stellar spectrum due to the curvature of the images of the slit in different wave lengths. With grating spectrographs no such correction is necessary since the images of a long slit are straight lines.

In order to find the radial velocity of a star with respect to the sun, the gross radial velocity which is obtained from equation VIII-6 and the displacement $\Delta\lambda$ of the spectral lines from their zero-velocity positions must be corrected for the motion of

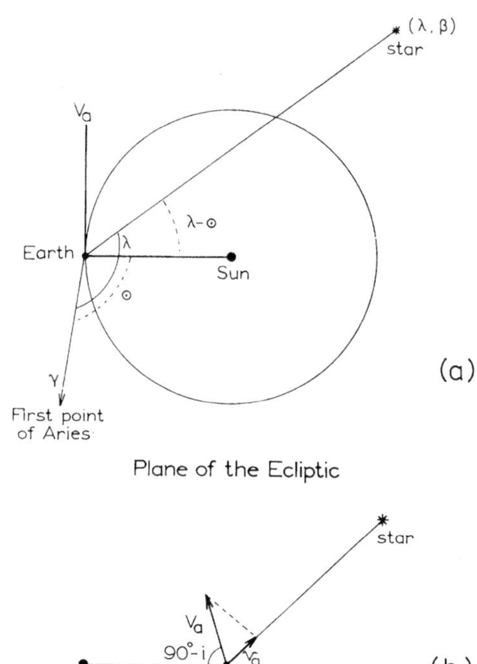

Fig. 30. The radial-velocity correction for the earth's motion about the sun.

the earth around the sun. When computing radial velocities a positive velocity is obtained if the displacement is towards longer wave lengths. This corresponds to motion in the line-of-sight away from the earth. Negative velocities, corresponding to motion towards the earth, are found from displacements towards shorter wave lengths.

The radial-velocity correction for the earth's motion about the sun may be found most easily by considering the problem in ecliptic coordinates. In Figure 30a the orbit of the earth about the sun is shown on the plane of the ecliptic, and a line is drawn from the earth to the projected position of a star which has ecliptic coordinates (λ, β). The longitude of the sun from the first point of Aries is $\odot$. Let the orbital velocity of the earth be V_a, and let i be the departure from 90° of the direction of the earth's motion from the line joining the earth and the sun, see Figure 30b. The desired radial-velocity correction, v_a, is the projection of V_a on the line joining the earth to the star. From Figures 30a and 30b it may be easily seen that

$$v_a = V_a \sin(\lambda - \odot + i) \cos\beta . \qquad \text{(VIII-7)}$$

The quantities $\odot$, the longitude of the sun, and i may be found from the *Nautical Almanac*, given the time of the observation. The ecliptic coordinates (λ, β) of the star may be calculated from the equatorial coordinates (α, δ) using the well known transformation between the two systems of coordinates. Note that $\odot$ and λ must be referred to the same mean equinox.

The quantity V_a can be found from the following equation:

$$V_a = \frac{a}{(1 - e^2)^{\frac{1}{2}}} \cdot \frac{2\pi}{T} \{1 + e\cos(\odot - \Gamma)\} \sec i . \qquad \text{(VIII-8)}$$

Here a is the semi-major axis of the earth's orbit in km, e is the eccentricity of the orbit, T is the number of mean sideral seconds in a solar year, and Γ is the longitude of the sun at perigee. These four constants can be found from the *Nautical Almanac*, thus V_a can be found as a function of $\odot$ and i.

SCHLESINGER (1899) has put the equations into a convenient form for the rapid calculation of the radial-velocity correction for the earth's orbital motion, v_a. He introduced the abbreviation

$$V = \frac{a}{(1 - e^2)^{\frac{1}{2}}} \cdot \frac{2\pi}{T} \qquad \text{(VIII-9)}$$

and then combined equations (VII-7) and (VIII-8) with the result known from spherical astronomy,

$$\tan i = \frac{e\sin(\odot - \Gamma)}{1 - e\cos(\odot - \Gamma)}, \qquad \text{(VIII-10)}$$

to obtain the equation

$$v_a = V\cos\beta\sin(\odot - \lambda) + Ve\sin(\Gamma - \lambda)\cos\beta . \qquad \text{(VIII-11)}$$

For any star λ and β are constant except for changes due to precession, thus one may define the parameters

$$b = V \cos \beta \tag{VIII-12}$$

and

$$c = Ve \sin(\Gamma - \lambda) \cos \beta . \tag{VIII-13}$$

These parameters will change very little in a century due to changes in Γ and e. It follows that one can write

$$v_a = b \sin(\odot - \lambda) + c . \tag{VIII-14}$$

The constants λ, b, and c have been computed by Schlesinger (1899) for all the bright stars and they are listed by him. Schlesinger's method will give v_a correct to 0.001 km/sec for periods of time greater than a century and it is most useful when many observations are made for one star, for one need only find $\odot$ for each observation and use the constants λ, b, and c which it is necessary to calculate only once for each star.

Herrick (1935) has given an alternative method for finding v_a. He noted that one can evaluate v_a by the following equation:

$$v_a = lX' + mY' + nZ' . \tag{VIII-15}$$

Here X', Y', and Z' are the heliocentric velocities of the earth in any rectangular system and l, m, and n are the direction cosines. If one uses the system of ecliptic coordinates, Z' is zero, and

$$l = \cos \alpha \cos \delta , \tag{VIII-16}$$

$$m = \sin \alpha \cos \delta \cos \varepsilon + \sin \alpha \sin \varepsilon . \tag{VIII-17}$$

Here ε is the inclination of the ecliptic, which is about 23°27′. Herrick has tabulated the coefficients X' and Y' as functions of calendar date from January 0, 1900 to December 31, 1996.

The observed gross radial velocity of a star must also be corrected for the component in the line of sight of the earth's motion due to the rotation of the earth. This quantity is

$$v_d = V_d \sin t \cos \delta \cos \phi . \tag{VIII-18}$$

Here V_d is the velocity of rotation of the earth at the equator which is 0.47 km/sec, t is the hour angle, taken to be positive for stars east of the meridian and negative for stars west of the meridian, and ϕ is the latitude of the observatory. The correction v_d varies from a few hundredths of a km/sec to a few tenths except at extreme hour angles.

In principle the gross radial velocity of a star should also be corrected for the motion of the earth in the earth-moon system. However, this correction can usually be neglected since it is of the order of ± 0.014 km/sec.

In summary, the heliocentric radial velocity of a star is

$$v = \Delta\lambda \cdot c/\lambda + v_a + v_d + \text{curvature correction}, \qquad \text{(VIII-19)}$$

where v_a is found from equation VIII-14 or VIII-15, v_d is found from equation VIII-18, and the curvature correction, if applicable, is derived empirically.

If one is merely measuring spectrograms in order to determine wave lengths in order to identify the various atomic spectra present in the star, it is not necessary to find specifically the correction for the earth's motion and for curvature. From a few lines of known identification, for instance the hydrogen lines and, in OB stars, the He I lines, one can find the average plate velocity, v_p, corresponding to the displacement of each well known line. Then

$$\lambda_{\text{true}} = \lambda_{\text{gross}} - v_p \cdot \frac{\lambda}{c}. \qquad \text{(VIII-20)}$$

5. Wave length and identification lists for O and B stars

In a stellar spectrum all possible lines from all atoms and ions that are present in the stellar atmosphere can appear in the stellar spectrum. Consequently most stellar spectra when examined at high dispersion and high resolution show very many lines. In the early type stellar spectra a few lines are strong; the remainder are very shallow and weak. Once the stronger lines of the spectrum from any atom or ion have been recognised, it is reasonable to expect all the weak lines also to be present, though many of them may not be detectable on low dispersion spectrograms. It is a statistical problem to decide whether the spectrum from a given ion or atom is present or not, particularly if many of the stronger lines are masked by lines from another ionic species.

Lists of wave lengths of lines seen in B type stars having sharp lines have been prepared by a number of investigators. A few of these studies will be mentioned here.

Adams and Dunham (1938) have given a list of the stronger lines in the extreme ultraviolet of B stars; Struve (1931, 1939) has given lists at longer wave lengths for several stars. Underhill (1948a, 1948b) has listed in considerable detail the lines visible in the spectrum of γ Pegasi. The most comprehensive list of lines occurring in early B or late O type spectra is that given by Underhill and de Groot (1965) for 10 Lacertae, O9V. A list of lines in the O7 star 9 Sagittae is given by Underhill (1958a). In the older lists the identifications are rather incomplete owing to the lack, at the time the lists were prepared, of comprehensive lists of lines in the second and third spectra of some atoms. Many of the unidentified lines in these lists are due to Fe III; very likely some at present unidentified lines are due to Ti III, Cr III and such spectra.

Although most of the lines in early type spectra have been identified, there remain a few moderately strong lines which are not identified. One of the most difficult regions in which to identify lines is that between 4800 Å and 6800 Å. The older laboratory lists of lines are often deficient here, probably owing to the lack of suitable photographic emulsions at the time the work was done. Furthermore, most of the lines in question arise as a result of transitions between high lying terms, and such transitions are not of great interest for the analysis of the term system of the atoms and ions and they are difficult to excite in laboratory sources.

6. Radial-velocity studies

One of the prime reasons for measuring wave lengths in stellar spectra is to determine the radial velocity of the star and to find differential motions in the atmosphere if they exist. In order to do this one must know what the wave length of the stellar feature would be if the star had zero velocity in the line of sight. Since many strong stellar lines are blends of several components, particularly on low dispersion spectra, it is not usually obvious what the zero-velocity wave length is. A special investigation must be made to set up a set of stable zero-velocity wave lengths. Clearly it is desirable to find a set of wave lengths that will be valid for a considerable range of spectral type at a given dispersion and that will be representative of the main part of the stellar atmosphere. Some strong spectral lines are formed in the superficial layers of early type stars. Their positions may reflect the peculiar motions of these superficial layers rather than the motion of the star as a whole.

Much radial-velocity work is done at about 50 Å/mm; higher dispersion, because of the larger scale on the spectrograms and the greater variety of useful spectral lines that is then available, is to be preferred if it can be used efficiently. The design of a spectrograph for radial-velocity work requires great care. The spectrum should be in good focus and definition for as long as possible a length, at least 500 Å. The optics must be fully illuminated in the same way whatever the attitude of the telescope. In particular, the light path of the comparison spectrum relative to that of the star light must not change. The optics must be rigid and suffer no flexure and it should be possible to regain a given setting of the optics and dispersing medium easily and securely. For the best performance, the spectrograph must be kept at a constant temperature.

The "new Mills spectrograph" of the Lick Observatory (CAMPBELL, 1928) was one of the most successful early spectrographs designed for radial-velocitiy work, and it has furnished some of the most precise radial velocities known. The one-prism and two-prism spectrographs of the Dominion Astrophysical Observatory (PLASKETT, 1919) have also proved to be very stable radial-velocity instruments. The use of large scale grating spectrographs at the coudé focus of large reflectors now permits very fine spectrograms of stars to be obtained and some very accurate radial-velocity work can be done.

6.1 *The selection of wave lengths for radial-velocity work*

Since blending is different at different dispersions, the same spectral lines cannot be used to determine radial velocities at all dispersions. Suitable lines must be selected for each spectrographic combination. One must also select good, well-defined comparison spectrum lines as wave length standards on each spectrogram. Not all the lines in the lists of Secondary Standards of Type B are useful with low and moderate dispersions because of blending and lack of resolution owing to the comparatively wide slit which is used with many stellar spectrograms.

The question of selecting appropriate wave lengths for radial-velocity work has been studied exhaustively by R. M. Petrie and his coworkers. The procedure is the following:

1. Establish a set of wave lengths for solar type stars, spectral types K5 to F4, using the motions of the solar system as a check. The radial velocities derived from the adopted wave lengths must reproduce the motions calculated from positional astronomy. Solar wave lengths are the starting point, for these are well known and on an internally consistent scale. By trial about 20 lines suitable for the selected range of spectral type have been found. Those suitable for use at high dispersion are listed by PETRIE (1946); those useful for low dispersion are given by McDONALD (1948). Each line is labelled according to its range of usefulness. The effective wave lengths of the stronger features change with spectral type at moderate dispersion because these lines are all blends and the relative strength of the components changes.

2. Using the wave-length system for spectral types K5 to F4, extend the wave-length system to types F and A by means of control stars for which the true radial velocity can be predicted. The control stars are either A and F stars belonging to clusters whose motion is known from positional work and from radial velocities of solar type stars, or A and F type stars which are one component of a binary, the other component being of type K5 to F4. Wide pairs are used in order to have separate images on the slit head. Consequently the relative orbital motion is very small. The lines suitable for use at high dispersion are given by PETRIE (1947b). Those suitable at low dispersion are given by PETRIE (1948).

3. Extend the system to the B type stars using as control stars those stars to which radial velocities may be assigned from observations other than wave-length measurements in B type spectra. Thus the control stars are chosen from visual binaries one component of which is of type B, the other of type A0 or later, and from members of clusters for which the radial velocity is known. For B type stars it has proved best to start from laboratory wave lengths and to use the control stars to test the suitability of each selected line. Each line is examined separately for stability and it is noted whether it is useful for main-sequence or for supergiant stars. The problem of determining wave lengths is made difficult by the diffuse character of the lines in most B type spectra and by the fact that at least half the known B stars are spectroscopic binaries or have atmospheres in which large scale motions occur.

The wave lengths suitable for determining the radial velocity of B type stars with spectrograms of 30 Å/mm or 50 Å/mm dispersion are listed by PETRIE (1953) and they are given in Table 20. The wave length of a few of the lines has been adjusted to yield the radial velocity given by the other lines. These cases are noted by the word "adj." in the third column of Table 20. The adjustment is necessary because several lines blend to give the observed feature. Only the dominant spectrum is listed. Most

TABLE 20

Wave lengths for determining the radial velocity of B stars with spectrograms of 30 Å/mm to 50 Å/mm

Spectrum	Wave length	Source	Sp. Type	Remarks
Ca II	3933.664	lab.	B0–B9	stellar and interstellar; Ca II H is too close to Hε to use.
He I	3964.727	lab.	B0–B6	strengthens in shell owing to dilution effects.
Hε	3970.074	lab.	B0–B9	use only if free from O II and Ca II H.
N II	3994.996	lab.	B0–B9	usually rather weak.
He I	4026.14	adj.	B0–B9	main-sequence stars.
He I	4026.189	lab.	B0–B9	supergiant stars.
O II	4069.766	lab.	B0–B3	other O II lines are unstable at low dispersion.
Hδ	4101.737	lab.	B0–B9	
He I	4120.837	lab.	B0–B7	O II and Fe III form possible blends.
Si II	4128.051	lab.	B3–B9	
Si II	4130.876	lab.	B3–B9	
C II	4267.167	lab.	B0–B9	blend of two closely equal components.
Hγ	4340.468	lab.	B0–B9	
He I	4387.928	lab.	B0–B8	Fe II is a possible blend at late B types.
He I	4471.325*	adj.	B0–B9	main-sequence stars.
He I	4471.398*	adj.	B0–B9	supergiant stars.
Mg II	4481.228	lab.	B3–B9	may blend with Al III at types earlier than B3.
Si III	4552.622	lab.	B0–B3	
Si III	4567.841	lab.	B0–B3	
Hβ	4861.332	lab.	B0–B9	only at dispersions near 30 Å/mm.
Hβ	4861.416	adj.	B0–B9	use at dispersions near 50 Å/mm.

* These wave lengths are for use with dispersions near 50 Å/mm; with dispersions near 30 Å/mm the wave lengths are 4471.477 and 4471.507 in main-sequence and supergiant stars respectively.

of the lines are due to He I or to hydrogen. The He I line at 4713 Å is not a stable line for radial-velocity work at moderate dispersion. A large unexplained scatter exists in the residuals from the mean plate velocity when this line is used. The strong lines due to Si IV at 4088 Å and 4116 Å are also unsuitable giving an unexplained large scatter in the results.

Petrie (1955) has found that his fundamental work on wave-length standards for B type stars, based on 16 stars, does not support the idea that the B stars in general will yield a positive K term of 1 to 2 km/sec due to relativity shift. Since the B type wave lengths given in Table 20 differ slightly from those used in older radial-velocity work, see for instance Campbell (1928) or Plaskett and Pearce (1930a), the older radial velocities must be corrected to the new system before being combined with the new results.

The problem of establishing standard wave lengths for the O type stars is difficult to solve because of the well nigh impossibility of finding O stars with constant radial velocity and sharp lines. The best thing to do appears to be to measure unblended lines from O III, N III, and other ions where possible, adopting laboratory wave lengths.

The strong lines of He I and H are formed in the superficial layers of the atmosphere and they may reflect atmospheric motions rather than the star's motion. Lines of the 4–*n* series of He II, when unblended with H, may give the stellar velocity for late O type stars, but at O6 and O5 they also are formed in the superficial layers of the star and their positions may reflect chiefly atmospheric motions.

Standard wave lengths for the Wolf-Rayet stars do not exist. Every distinct feature is a blend of several lines. It is possible to make internally consistent measurements on well defined emission features and from these to derive velocity changes such as those due to binary motion. It is quite another thing to place these velocities on an absolute scale for comparison with the velocities determined from absorption line spectra. Absorptions formed in expanding shells of gas may mutilate the shapes of the emission lines and the blending is usually different from star to star. Consequently it is difficult to find a standard wave length by means of which to derive the motion of Wolf-Rayet stars on the same scale as that used for the absorption line stars.

6.2 *Catalogues of radial velocities of O and B stars*

Much of the observing at the Dominion Astrophysical Observatory has been devoted to obtaining radial velocities of O and B type stars from spectra obtained with slit spectrographs. PLASKETT and PEARCE (1930b) have prepared a catalogue of the radial velocities of 1010 O and B stars, while PETRIE and PEARCE (1962) have given the radial velocities of 570 B stars most of which are north of +20° and have magnitudes between 7.5 and 8.6. SEYFERT and POPPER (1941) have published radial velocities for 118 B stars between the 9th and 11th magnitude, and POPPER (1944) has given radial velocities for 150 B type stars between the 8th and 12th magnitude. NEUBAUER (1943) has given radial velocities for 433 faint early type stars south of −23° declination. These radial velocities require a systematic correction of +11.4 km/sec (FEAST and THACKERAY, 1958). FEAST, THACKERAY, and WESSELINK (1955, 1957) have given the radial velocities of 277 southern B type stars. In these papers the radial-velocity spectrograph of the Radcliffe Observatory is described and the accuracy and reliability of the results is critically discussed. FEAST and THACKERAY (1963) have given the radial velocities of 122 more B stars, some stars being as faint as 11th magnitude. Radial velocities of a few southern B types stars are contained in papers by EVANS, MENZIES and STOY (1957, 1959) which give fundamental data on southern stars. Radial velocities of O and B type stars are being determined at the Haute Provence Observatory from objective-prism spectra by means of a technique developed by C. Fehrenbach. These results are reported in the *Publ. Haute Provence Obs.*, Vol. 3, and later issues. Data for areas in Cygnus and in Perseus have been published.

With an objective prism one obtains recordings of many spectra at one time and thus much more efficient use can be made of observing time than when a slit spectrograph is used. The radial velocities are determined by comparing the objective-prism spectra with standard spectra, using projection. In order to establish the plate constants there must be a number of stars with known radial velocities on each plate. If the method is to give accurate results it is essential that the field be good over a wide angle and that the spectra have good density and are of good quality. Fehrenbach and his colleagues have shown that one can obtain radial velocities for O and B stars with a probable error of ±5 km/sec when three plates are used.

RUBIN *et al.* (1962) have assembled a catalogue of the radial velocities, magnitudes, etc. for 1440 O and B stars within 3 kpc of the sun. This is a rather uncritical collection of data from all available sources, but it serves as a useful base from which to discuss the kinematics of the OB stars.

6.3 *The Probable Errors of Stellar Radial Velocities*

When radial velocities have been determined for a large number of stars using a homogeneous wave-length system, and the spectrograph and the measuring techniques have been kept unchanged, it is possible to deduce information about the intrinsic observational error and the proportion of stars varying in velocity by studying the distribution of the probable errors of the individual radial velocities. PETRIE (1960) has made such a study. His material consists of the radial velocities of 489 stars of types O9 to B5 determined from single-prism spectrograms measured by himself, of the radial velocities of 510 A type stars determined from single-prism spectrograms measured by HARPER (1937) at the Dominion Astrophysical Observatory, and of the radial velocities of 753 stars of types F5 to M (CAMPBELL, 1928) determined from three-prism and two-prism spectrograms at the Lick Observatory. Known binary stars are excluded from the analysis.

Petrie plots the frequency of the probable error in each group, B stars, A stars, and F5–M stars, against the size of the probable error and finds that the distribution curves rise rather sharply and then tail off slowly as the size of the probable error increases. His plots are shown in Figure 31. The first part of each curve may be fitted well with an error curve, shown by the dotted curve. The results for the B stars suggest an average probable error per plate of ± 4.5 km/sec. The long tail of each distribution curve is believed to be caused by including in each sample binary stars which have small range in radial velocity. Petrie concludes that about 51 per cent of the stars in his B type sample have variable radial velocity. He finds that 54 percent of his A type sample and that 51 procent of his F5–M type sample have variable velocity. Whether the variable velocity is due to binary motion or whether it is due to instability of the atmosphere cannot be determined from the present material.

7. Motions of the OB stars

The velocities of the nearby O and B stars can be used with the proper motions of the OB stars to demonstrate whether a group of stars has community of motion or not. The results for some important groups of O and B stars have already been mentioned in Chapter VI. EGGEN (1961) has discussed the space motions and distribution of the apparently bright B type stars. He has assembled a catalogue of 280 B type stars brighter than visual magnitude 5.0, giving magnitudes in three colours, spectral types, luminosity classes, proper motions, radial velocities, and space motions. He uses JOHNSON and IRIARTE's (1958) calibration of the MK luminosity classes in order to estimate the luminosities of his stars. Eggen concludes that the nearer B type stars do not form several isolated aggregates. Rather they form what might be called a "local association" which is some 1000 parsecs in diameter but less than 100 parsecs thick, and which shows a velocity gradient of 40 km/sec per kpc in the radial direction from

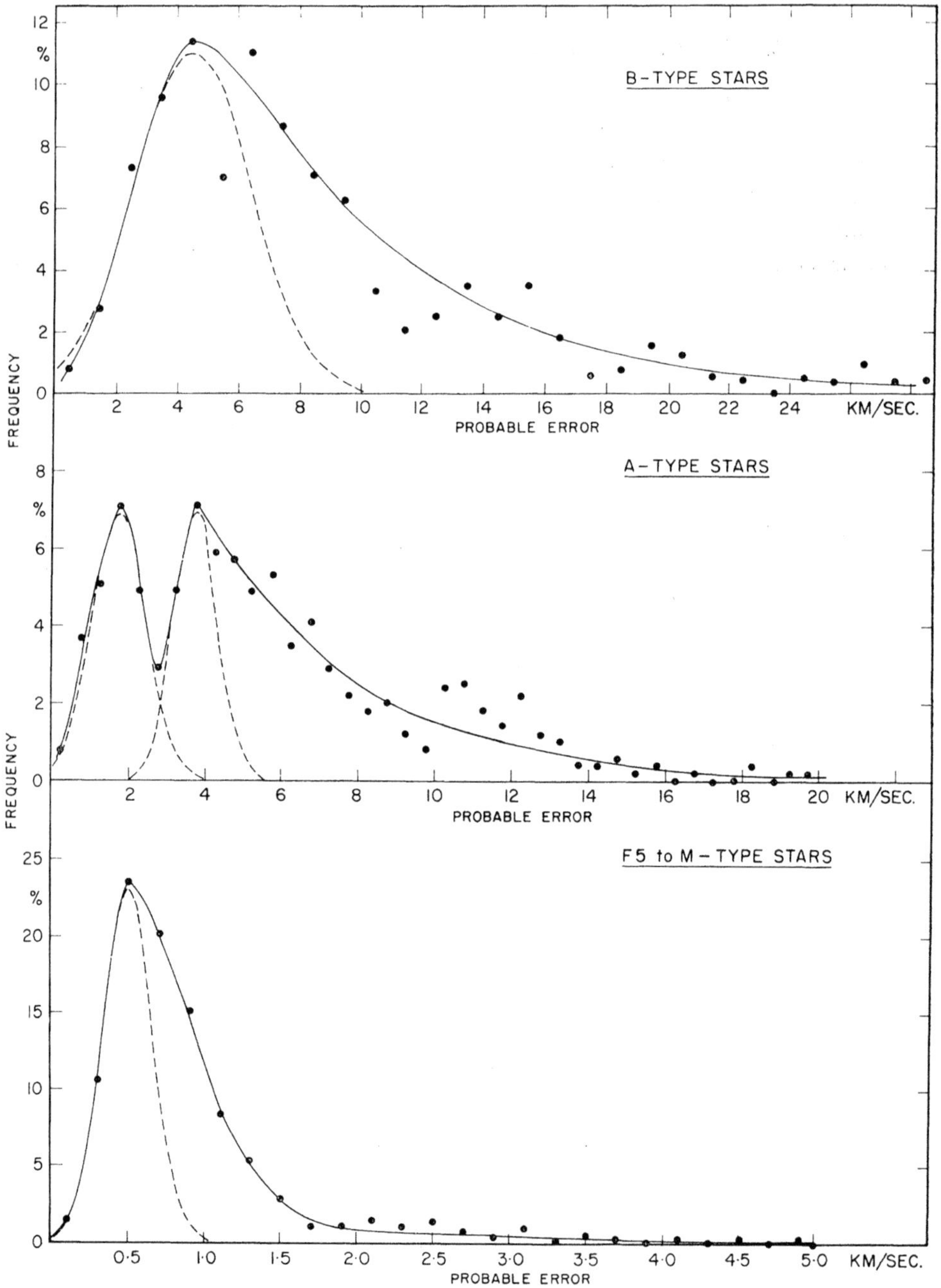

Fig. 31. The distribution of probable errors in B, A, and F type radial velocities.

the galactic centre. This means that relative to the sun the motion is 4 km/sec away from the galactic centre for stars near the centre of the proposed association and that the motion is some 24 km/sec away from the galactic centre for stars on the edge most distant from us.

The radial velocities of the more distant OB stars can be used to investigate the general motion of stars in the galaxy. PLASKETT and PEARCE (1934) demonstrated that the radial velocities of the OB stars confirm the hypothesis of galactic rotation. They used the information available then about the luminosities of the OB stars (from mean parallaxes derived from proper motions) to evaluate Oort's constants A and B and to estimate the distance to the centre of the galaxy from the sun, R_0. In view of later developments it is interesting to note that Plaskett and Pearce concluded that the best value of A was $+15.5 \pm 0.9$ km sec^{-1} kpc^{-1} and that R_0 was 10 kpc. They were aware of the obscuration in the plane of the galaxy caused by the interstellar dust, and they strongly suspected that the galaxy was a spiral nebula like the Andromeda Nebula, though they could not define actual spiral arms from their material.

The observed radial velocity of a star can be considered to consist of the following terms:

1. a term due to the motion of the sun with respect to the general standard of rest,

$$V = - u_0 \cos l \cos b - v_0 \sin l \cos b - w_0 \sin b \,, \qquad \text{(VIII-21)}$$

2. a term due to differential galactic rotation between the sun and the star,

$$V_{\text{rot}} = R_{\text{o}} (\omega - \omega_{\text{o}}) \sin l \cos b \,, \qquad \text{(VIII-22)}$$

3. terms describing a possible galactic dilation,
4. a K-term which may represent systematic motions of large stellar groupings or which may represent systematic errors in the radial velocities due to gravitational red shift, motions within stellar atmospheres, and erroneous wave-length systems, and
5. a term describing the star's peculiar motion.

This description assumes that the system of OB stars is cylindrically symmetric. In equations VIII-21 and VIII-22, l is the galactic longitude, b is the galactic latitude, u_0, v_0, and w_0 are the components of solar motion directed respectively towards the galactic centre along a radius, at right angles to a radius and in the direction of rotation of the galaxy, and perpendicular to the galactic plane, ω_0 is the angular velocity of rotation at the position of the sun.

For small values of $R - R_0$ one can write

$$\omega = \omega_{\text{o}} + (R - R_{\text{o}}) (d\omega/dR)_{\text{o}} \,. \qquad \text{(VIII-23)}$$

It follows that the component of stellar radial velocity due to differential galactic rotation can be written as

$$V_{\text{rot}} = - 2A (R - R_{\text{o}}) \sin l \cos b \,, \qquad \text{(VIII-24)}$$

where A is Oort's constant and we have

$$A = -\tfrac{1}{2} R_0 \left(\frac{d\omega}{dR}\right)_0 . \tag{VIII-25}$$

As more and more material about the radial velocities of distant objects has appeared, attempts have been made to solve for the constants describing the radial velocity of various groups of stars in the galaxy, namely u_0, v_0, w_0, R_0, ω_0, and A or $(d\omega/dR)_0$. It is beyond the scope of this book to discuss this subject in detail. For the more recent results reference may be made to the papers by WEAVER (1955); PETRIE, CUTTLE, and ANDREWS (1956); STIBBS (1956); GASCOIGNE and EGGEN (1957); RUBIN and BURLEY (1964); FEAST and SHUTTLEWORTH (1965), and PETRIE (1965).

The results of Feast and Shuttleworth are based on the most comprehensive and most critically selected body of data for the OB stars. They are confirmed by studies such as those made by Rubin and Burley and by Petrie from somewhat different material. Feast and Shuttleworth find that the K-term for the B stars varies with distance from the sun. This result is probably due to the peculiar motions of large stellar groupings and it tends to conceal a possible dependence on spectral type and luminosity class except for the O stars. Here a significantly positive K-term is found. It is probable that this is due to a gravitational red shift. The values of the Oort constant A in various subgroups of OB spectral types are in good agreement, indicating that the adopted absolute magnitude system (that of JOHNSON and IRIARTE, 1958) is self-consistent. It is found that the best values of A and R_0 are $+14.3 \pm 0.8$ km sec^{-1} kpc^{-1} and 9.9 ± 0.9 kpc respectively. The solar motion for B stars nearer than 0.25 kpc is appreciably different from that obtained from the more distant stars. This result is in agreement with the conclusions of Eggen that the motions of these stars are dominated by local group motions. The velocity dispersion of the peculiar motions of the B stars is found to be large, about ± 11.5 km/sec. It may be demonstrated that although the radial velocities of the OB stars do show the double sine-wave pattern to be expected for galactic rotation, the dispersion about this curve is so great that in any one part of the sky one cannot expect galactic rotation to give a sufficient description of the apparent motion of the stars (PETRIE, 1965). The rather large local group motions of the OB stars seriously hamper attempts to study the dynamical structure of the galaxy by means of galactic rotation. Petrie has discovered that the dispersion in the peculiar motions of OB stars fainter than -3.5 mag is ± 10.6 km/sec whereas the dispersion for OB stars brighter than -3.5 mag is ± 15.9 km/sec. Whether this result is to be interpreted in terms of a different distribution of kinetic energy among the two groups of stars, or whether it implies that the atmospheric motions in the intrinsically brighter stars are greater than those of the fainter stars cannot be decided at present.

CHAPTER IX

SPECTROSCOPIC BINARIES

1. Motion in an ellipse and the derivation of an orbit

Two bodies which are moving under a force which varies as r^{-2}, where r is the distance separating the bodies, will move along a conic section. Periodic motion is along an ellipse, with the centre of gravity at one focus. The motion of each star relative to the other is along a similar ellipse with the centre of gravity at one focus of the ellipse in which the star is moving. The geometry of the situation is illustrated in Figure 32. The XY plane is the plane of the sky and it is perpendicular to the line of sight which lies in the Z direction. The plane containing the orbit cuts the plane of the sky along the line of nodes. The ascending node, Ω, is the point A in the diagram. At the ascending node the star moves through the line of nodes with a positive radial velocity. The descending node, ℧, is at B. Here the radial velocity is negative. The orbit is inclined at an angle i

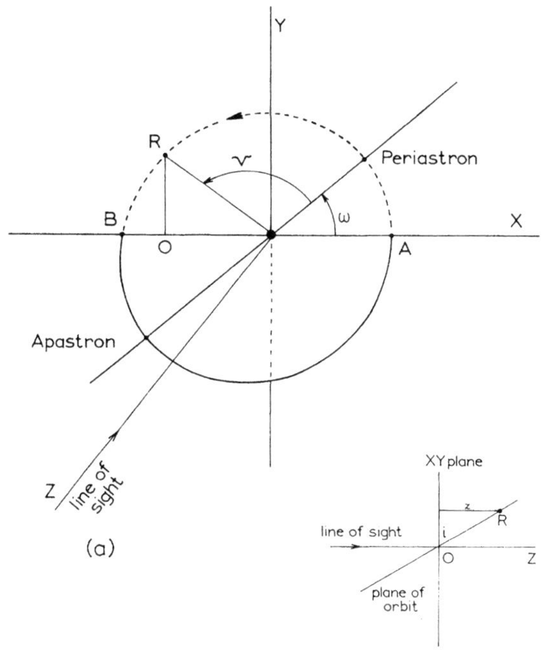

Fig. 32. The geometry of a radial-velocity orbit.

to the plane of the sky. This angle is measured as positive when at the ascending node the companion is moving away from the other star. (In the diagram i is negative since the companion is approaching periastron.) The angle from the line of nodes (ascending node) to periastron measured in the direction of motion is called the longitude of periastron and it is denoted by ω. The true anomaly, v, is the angle measured in the plane of the orbit in the direction of motion from periastron to the star. The distance of the star behind or in front of the plane of the sky is denoted by $+z$ or $-z$ respectively. The component of radial velocity of the star due to orbital motion is $\mathrm{d}z/\mathrm{d}t$.

Now, if OR is the distance to the star measured in the plane of the orbit in a direction perpendicular to the line of nodes, it follows that

$$z = \mathrm{OR} \sin i = r \sin(v + \omega) \sin i , \tag{IX-1}$$

and

$$\frac{\mathrm{d}z}{\mathrm{d}t} + r \sin i \cos(v + \omega) \frac{\mathrm{d}v}{\mathrm{d}t} + \frac{\mathrm{d}r}{\mathrm{d}t} \sin i \sin(v + \omega) . \tag{IX-2}$$

When the quantities $r\mathrm{d}v/\mathrm{d}t$ and $\mathrm{d}r/\mathrm{d}t$ are replaced in equation (IX-2) by the values which may be derived from the equation for an ellipse,

$$r = \frac{a(1 - e^2)}{1 + e \cos v} , \tag{IX-3}$$

and the condition that the areal velocity be constant (that is, no perturbing forces acting)

$$\tfrac{1}{2} r^2 \frac{\mathrm{d}v}{\mathrm{d}t} = \frac{\pi a^2 (1 - e^2)^{\frac{1}{2}}}{P} = \mathrm{const} , \tag{IX-4}$$

it is found that the radial velocity due to orbital motion can be expressed as

$$\frac{\mathrm{d}z}{\mathrm{d}t} = \frac{2\pi}{P} \frac{a \sin i}{(1 - e^2)^{\frac{1}{2}}} \left[\cos(v + \omega)(1 + e \cos \omega) + e \sin v \sin(v + \omega)\right] . \tag{IX-5}$$

Here a is the semi major axis and e is the eccentricity of the ellipse. After expanding terms and collecting, it is possible to write

$$\frac{\mathrm{d}z}{\mathrm{d}t} = K\left[e \cos \omega + \cos(v + \omega)\right] , \tag{IX-6}$$

where

$$K = \frac{2\pi}{P} \frac{a \sin i}{(1 - e^2)^{\frac{1}{2}}} . \tag{IX-7}$$

The quantity K is known as the semi-amplitude of the radial-velocity orbit. The radial velocity of the star with respect to the sun is

$$\frac{\mathrm{d}\zeta}{\mathrm{d}t} = \gamma + K\left[e \cos \omega + \cos(v + \omega)\right] . \tag{IX-8}$$

Here γ is the radial velocity of the centre of mass of the system with respect to the sun. The observational problem posed by spectroscopic binaries is to find the constants describing the motion in an ellipse which satisfy equation (IX-8), given a series of observations of the radial velocity $d\zeta/dt$.

When binary orbits are studied by means of the radial-velocity changes which are observed, it is customary to express the period in days. The average angular motion in the orbit is then

$$\mu = 2\pi/P \qquad \text{radians/day}. \tag{IX-9}$$

The mean anomaly is

$$M = \mu(t - T) \tag{IX-10}$$

where T is the date of periastron passage expressed in Julian days, and t is the date of the observation.

The mean anomaly, M, and the true anomaly, v, are related through the eccentric anomaly, E. Thus

$$M = E - e \sin E, \tag{IX-11}$$

where

$$\tan \tfrac{1}{2} E = (1 - e)^{\frac{1}{2}} (1 + e)^{-\frac{1}{2}} \tan \tfrac{1}{2} v. \tag{IX-12}$$

Methods of solving equations (IX-11) and (IX-12) to obtain v, knowing M, are well known, see for instance VAN DE KAMP (1960). In practice, if $e < 0.77$ one can use the *Allegheny Tables* (SCHLESINGER and UDICK, 1912). KENT, TAACK, and LARSON (1963) have prepared tables of the eccentric and true anomaly in elliptic orbits for values of the eccentricity between 0.01 and 1.00.

The projected semi-major axis of the orbit can be found from equation (IX-7) knowing e, P, and K. When P is expressed in days and K is in km/sec,

$$a \sin i = 1.375 \times 10^4 KP(1 - e^2)^{\frac{1}{2}} \,\text{km}. \tag{IX-13}$$

Kepler's third law expresses the relation

$$\mathfrak{M}_1 + \mathfrak{M}_2 = (a_1 + a_2)^3 P^{-2}, \tag{IX-14}$$

which is valid for two-body motion under gravitation. Here a_1 and a_2 are the semi-major axes of the relative orbits of star 1 and star 2 expressed in astronomical units. The stars have masses $\mathfrak{M}_1$ and $\mathfrak{M}_2$ in units of the solar mass, and the period is expressed in years. When two spectra can be measured, we can find the ratio

$$\frac{a_1}{a_2} = \frac{K_1}{K_2}. \tag{IX-15}$$

Here K_1 and K_2 are the semi-amplitudes of the two velocity curves. Consequently one can show that

$$(\mathfrak{M}_1 + \mathfrak{M}_2) \sin^3 i = \frac{a_1^3 \sin^3 i}{P^2} \left(\frac{\mathfrak{M}_1 + \mathfrak{M}_2}{\mathfrak{M}_2} \right)^3. \tag{IX-16}$$

For a single-spectrum binary, the quantity $a_1 \sin i$ can be found from the velocity curve (see equation IX-13) and one can find what is called the mass function, $f(\mathfrak{M})$. We have

$$f(\mathfrak{M}) = \frac{\mathfrak{M}_2^3}{(\mathfrak{M}_1 + \mathfrak{M}_2)^2} \sin^3 i = 1.035 \times 10^{-7} K_1^3 P (1 - e^2)^{3/2} \qquad \text{solar masses.} \tag{IX-17}$$

Here P is in days, and K is in km/sec.

When two spectra are visible, it is possible to find the reduced masses

$$\mathfrak{M}_2 \sin^3 i = 1.035 \times 10^{-7} K_1 (K_1 + K_2)^2 P (1 - e^2)^{3/2} \qquad \text{solar masses}, \tag{IX-18}$$

and

$$\mathfrak{M}_1 \sin^3 i = 1.035 \times 10^{-7} K_2 (K_1 + K_2)^2 P (1 - e^2)^{3/2} \qquad \text{solar masses}. \tag{IX-19}$$

One cannot solve for the inclination of the orbit, i, from a radial-velocity study alone. The inclination may be found either from the solution of a visual orbit or from an analysis of the light changes, should the binary star be an eclipsing pair.

In summary, the radial-velocity orbit of a binary star is defined by the six elements

P = period, which is expressed in days,
T = time of periastron passage, which is usually expressed as a Julian date,
e = eccentricity,
ω = longitude of periastron passage measured from the ascending node,
γ = radial velocity of the centre of mass of the system with respect to the sun,
K = semi-amplitude expressed in km/sec.

From these elements, one may find the projected semi-major axis of the orbit, $a \sin i$, and the mass function of the system, or the quantities $\mathfrak{M}_1 \sin^3 i$ and $\mathfrak{M}_2 \sin^3 i$ if the spectra of both stars can be seen and measured.

In solving a radial-velocity orbit, the first thing to do is to find the period as accurately as possible by successive trials of plotting the radial velocity against the date of observation. It has long been recognised that erroneous periods may affect materially all statistical studies of spectroscopic binaries. This subject has been reviewed by TANNER (1948). The problem is that at any one observatory the observations are obtained at intervals of one sideral day $\pm$ a small interval, α. Since most stars are observed for radial velocity when they are near the meridian, α is of the order of $0\overset{d}{.}1$.

A radial-velocity curve due to binary motion is a periodic function, thus one may write

$$V(\varphi) = V(\varphi \pm n). \tag{IX-20}$$

Here φ is the phase, which is equal to $(t-T)/P$, and n is an integer. Consider two possible periods P_1 and P_2. Since the velocity has the same value whether it is given a phase according to P_1 or to P_2, it follows that

$$\varphi_1 = \varphi_2 \pm n \tag{IX-21}$$

This implies that

$$\frac{t-T}{P_1}=\frac{t-T}{P_2}\pm n \tag{IX-22}$$

All times are measured from an arbitrary epoch, T. Let τ be the least possible interval of observation. Then

$$\tau=(1\pm\alpha)\,\text{sideral day}=0.9972\,(1\pm\alpha)\,\text{solar day}. \tag{IX-23}$$

Since it follows that

$$t-T=k\tau,$$

where k is close to an integer, we find the following relationship in terms of sideral days:

$$\frac{1}{P_1}=\frac{1}{P_2}\pm\frac{n}{k\tau}. \tag{IX-24}$$

In terms of solar days,

$$\frac{1}{P_1}=\frac{1}{P_2}\pm 1.0027\frac{n}{k}(1\mp\alpha). \tag{IX-25}$$

It is frequently very difficult to distinguish between periods which are related as

$$\frac{1}{P_1}=\frac{1}{P_2}\pm 1, \tag{IX-26}$$

the case which results when n and k have their smallest values, namely unity. When the true period is near an integral number of days, it is particularly difficult to distinguish between P_1 and P_2, for observations at any one observatory can be obtained at only a few phases in the orbit. Tanner has proposed a graphical method of determining the true period and he has reviewed a number of published orbits and detected spurious periods. One example is the star δ Capricorni. Tanner shows that the correct period is 1.022768 day instead of 39.50 days as announced in the original work.

In the case of double-lined binaries, it is particularly difficult to be sure of the correct period, because one is restricted to observations over only certain phases in the orbit, owing to blending of the two spectra when the radial-velocity separation is small.

Once the period has been found for a radial-velocity orbit, the next step is to determine preliminary elements, T, e, ω, γ, and K. One method of doing this is to compare the observed radial-velocity curve plotted on a predetermined scale with standard curves which have been computed for a series of values of e, ω, and T. This method has the advantage that one can mentally take into account different weights for the points and no use is made of free-hand curves which may contain points of inflection which do not correspond to motion in an ellipse.

Another method is the graphical method due to LEHMANN–FILHÉS (1894) which is described, for instance, by AITKEN (1935) and by PETRIE (1962). This method is useful when e is not too small. When e is small, one can adopt circular motion as the preliminary solution. Then $e=0$ and ω and T are not defined. In this case the origin of time taken to be T_0, the time of nodal passage at the ascending node. The preliminary elements then consist of P, T_0, K, and $e=0$.

The next step in determining the best set of elements is to correct the preliminary elements by the method of least squares. The method is described by LEHMAN-FILHÉS (1894). The predicted velocity which will be found by means of equation IX-8 is a function of six variables γ, K, e, ω, T, and P. We have

$$V = f(\gamma, K, e, \omega, T, P). \tag{IX-27}$$

The change in velocity produced by changes in each of the elements $\Delta\gamma$, ΔK, etc. is

$$\Delta V = \frac{\partial f}{\partial \gamma}\Delta\gamma + \frac{\partial f}{\partial K}\Delta K + \frac{\partial f}{\partial e}\Delta e + \frac{\partial f}{\partial \omega}\Delta\omega + \frac{\partial f}{\partial T}\Delta T + \frac{\partial f}{\partial P}\Delta P. \tag{IX-28}$$

If one assumes that all the corrections are of first order and independent, introduces the abbreviation

$$u = v + \omega,$$

and recalls the definition (equation (IX-9)) of mean daily motion, it follows that the velocity increment due to the use of improved elements is

$$\begin{aligned}\Delta V &= \Delta\gamma + (\cos u + e\cos\omega)\Delta K + K\left[\cos\omega - \sin u \sin v(2 + e\cos v)(1 - e^2)^{-1}\right]\Delta e \\ &\quad - K(\sin u + e\sin\omega)\Delta\omega + \sin u(1 + e\cos v)^2 K\mu(1 - e^2)^{-3/2}\Delta T \\ &\quad - \sin u(1 + e\cos v)^2 K(t - T)(1 - e^2)^{-3/2}\Delta\mu = O - C.\end{aligned} \tag{IX-29}$$

Here O represents the observed velocity at time t and C is the velocity predicted by the preliminary elements.

Equation (IX-29) is valid for each observation. The coefficients can be calculated from the preliminary elements. The problem reduces to solving the set of observational equations (IX-29) for the best set of corrections by the method of least squares. Usually several observations at about the same phase are combined into a normal point. In addition it is assumed that the period is exact, thus $\Delta\mu=0$. A solution is made for the five unknowns: $\Delta\gamma$, ΔK, $K\Delta e$, $-K\Delta\omega$, and $\mu(1-e^2)^{-3/2}\Delta T$.

For a double-lined binary (two spectra visible), equation IX-29 contains terms in K_1 and K_2 similar to those written and there are six unknowns, for there is ΔK_1 as well as ΔK_2.

When the solution of the set of observational equations is to be performed by hand computation, it is useful to use Gauss' method of elimination (see, for instance, WATSON, 1867) because this method contains checks against arithmetic errors and because one may at the same time find the probable errors of the correction terms. Weights are usually assigned to each observational equation.

STERNE (1941) has developed a method of correcting the preliminary elements which is particularly effective when the orbit is nearly circular, that is for $e \leqslant 0.10$.

The above methods of obtaining the best set of elements to represent radial-velocity variations believed to be due to the motion of two bodies under their mutual gravitation are designed for hand computation. Now that large electronic computers are available to perform the curve-fitting process, the outlook on the problem of solving binary orbits has changed. This subject has been discussed by several people, see for instance HEARD and MCRAE (1957). The scope of the problem can vary from using a computer only to obtain corrections to an adopted preliminary set of elements, to finding the preliminary elements and correcting them. The problem of finding preliminary elements from a set of radial velocities and times of observation is difficult because it involves programming judgement into the numerical process. The observed radial velocities are unavoidably uncertain, and furthermore it is not always certain that they represent only strict binary motion. An experienced observer when handling the observations in order to solve for the preliminary elements may become suspicious that perturbations of the radial velocity are occurring which do not represent binary motion. In this case he will modify his procedure in an appropriate manner, for instance by rejecting observations made at a certain phase in the orbit. Such judgement cannot readily be designed into a computer program.

The final stage of solving a radial-velocity binary orbit is to compute an ephemeris with the final elements and to see if the residuals from the new theoretical curve show any systematic trend. The sum of the squares of the residuals is examined to see if there has been a reduction when the final elements are used in place of the preliminary elements. It is a matter of some experience and judgement to decide whether one application of the method of differential corrections is sufficient to correct a given set of preliminary elements or whether another solution is required.

Unless extremely rough orbital elements are all that are required, any set of preliminary orbital elements should be improved by the method of least squares, for this method gives an impersonal adjustment of the five (six) elements to give the best fit to the observations and it yields an evaluation of the uncertainty in each correction term. By inspection of graphical solutions one cannot impartially adjust a set of elements because the effect of a change in one element cannot easily be distinguished from a change in another element.

The elements derived by PETRIE (1962) for AR Cassiopeiae are given here as one example of the use of the method of differential corrections. These corrections lead to a reduction in the weighted sum of the squares of the residuals by 28 per cent.

AR Cassiopeiae, B4

Preliminary elements	Corrections	Final elements
$P=6.06652\pm0.00006$ days		$P=6.06652\pm0.00006$ days
$e=0.25$	$\Delta e=-0.026$	$e=0.224\pm0.016$
$\omega=34°$	$\Delta\omega=2\overset{\circ}{.}3$	$\omega=36\overset{\circ}{.}3\pm3\overset{\circ}{.}8$
$T=$JD 2430648.050	$\Delta T=-0.058$ days	$T=$JD 2430648.044
$K=56.0$ km/sec	$\Delta K=-0.33$ km/sec	$K=55.66\pm0.93$ km/sec
$\gamma=-10.5$ km/sec	$\Delta\gamma=0.22$ km/sec	$\gamma=-10.27\pm0.63$ km/sec
$\Sigma p(O-C)^2=75.74$		$\Sigma p(O-C)^2=54.55$

This system is under study by BATTEN (1961) who has given a preliminary discussion of the peculiarities which are observed.

2. Catalogues of spectroscopic and eclipsing binaries

The study of double-lined spectroscopic binaries offers one of the best means for finding absolute stellar masses, particularly when coupled with studies of visual orbits or with analyses of the light curves of eclipsing binaries. J. H. Moore at the Lick Observatory prepared five catalogues of the orbital elements of spectroscopic binaries and he made an effort to separate the stars for which the radial-velocity variations are surely due to orbital motion from those in which the periodic variations are due to atmospheric motion, that is to pulsation. The final catalogue of this series is by MOORE and NEUBAUER (1948). Since that date Bouigue has published further lists of spectroscopic binaries in the *Ann. Obs. Toulouse*, volumes **21** to **24**.

PLAUT (1950, 1953) has given a catalogue, with critical discussion, of orbital and light elements for 130 binaries brighter at maximum light than 8.5 magnitude. Most of these stars are of types O and B. Plaut also gives a summary of the techniques which may be used to analyse the light curves. The eclipsing binaries have been discussed by GAPOSCHKIN (1948) also. He gives a somewhat uncritical catalogue of the absolute dimensions of eclipsing systems.

If the masses and luminosities of the stars are to be determined accurately, one must be very critical of what data are used in the analysis. This subject has been studied by POPPER (1957). Popper's object is to derive an accurate mass-luminosity relationship for the stars from observational data only. The fundamental quantities to be obtained from the observations are the mass $\mathfrak{M}$, the radius R, and the effective temperature, T_{eff}.

The observation of K_1 and K_2 is crucial for determining the masses of the two components; one also needs to know P, i, and e accurately. In order to determine radii, one must know accurately the quantities K_1, K_2, and k, which is the ratio of the radii as found from the solution of the light curve. In addition, one must know P, i, e, and r_1, where r_1 is the radius of star 1 relative to the distance between the centres of the two stars. The effective temperature is usually deduced from the B–V colour. One must also know the spectral types of both stars and the relative surface brightness, J_1/J_2. If the stars are reddened, it is necessary to have a light curve in a third colour, say U, as well as in B and V in order to estimate T_{eff} securely for both stars. Popper feels that if an observed mass-luminosity relationship is to be useful for guiding our choice between the several theoretical possibilities, the stellar masses must be known to within about 11 per cent ($\Delta \log \mathfrak{M} = 0.05$) and the visual absolute magnitudes must be known to better than 0.5 magnitude. These demands put a considerable stress on the observations and great efforts must be made to estimate accurately the parameters describing the secondary star. Frequently, owing to the faintness of the spectrum of the secondary star, the estimate of K_2 is quite uncertain and the spectral type of the secondary star is not known within even one major division.

3. The secular variation of orbital elements

Triple and multiple stars are rather common and some have been discovered from purely spectroscopic observations. The usual configuration is a close pair AB with a more distant companion C. The principle effect of the third body on the close pair AB is that the systemic velocity, γ, varies in a periodic manner. The stars β Persei (Algol), λ Tauri, and HD 100018 form three interesting triple systems.

The star Algol has been studied by EBBIGHAUSEN (1958) who has shown that the systemic velocity of the eclipsing pair AB varies with a period of 1.87 years as the eclipsing pair, which has a period of 2.8673 days, moves about the third star. The spectral type of Algol A is B8; that of component B is not known with any certainty. BORGMAN (1963) from a study of colour indices in his system of seven narrow-band colours has estimated that it is gG8. STRUVE and SAHADE (1957) during primary eclipse on 1956, November 23 obtained high-dispersion spectrograms of the Algol system and demonstrated the presence of component C. They estimate that the spectral type of component C is late A or early F, luminosity class IV or V. The semi-amplitude of the motion of the system AB about the centre of mass of the triple system is about 10 km/sec and that of component C is 33.6 km/sec. Consequently the mass of the pair AB is about 3.5 times that of component C. Not enough is known to solve for the mass ratio $\mathfrak{M}_1/\mathfrak{M}_2$.

The eclipsing pair λ Tauri has a period of 3.952952 days. It has been shown by EBBIGHAUSEN and STRUVE (1956) to move about a third companion with a period of 33.025 days. The spectral type of component A is B3. The spectrum of component B can be seen only when the stars are near the nodes; the spectral type is late B or early A. The spectrum of component C is never visible. The radial velocities from the hydrogen lines give some evidence of the presence of moving streams of gas around the eclipsing pair.

Three sets of absorption lines are seen on high-dispersion spectrograms of the system HD 100018 at phases when the lines are most widely separated, that is at times of nodal passage in the short-period orbit. This system has been described by PETRIE and LAIDLER (1952) and because the system is of considerable interest brief mention of it is made here though the stars are of spectral type F. The pair AB is a binary with a period of 7.39902 days while the combination AB–C forms a visual binary system, ADS 8189, which has a period of 84.734 years. The three stars are about the same magnitude, thus three sets of spectral lines are recorded.

Secular changes in the longitude of periastron, ω, have been observed for a few close binaries. The rotation of the line of apsides may result from perturbations due to the non-spherical form of the stars which are in close contact. Interest attaches to the problem of determining the rate of rotation of the line of apsides because from this numerical factor one can hope to deduce information about the central condensation of the stars. Theoretical discussions have been given by COWLING (1938) and by STERNE (1939).

The change in ω may be detected by studying light curves or from radial-velocity orbits determined at a number of epochs. The observational problem is difficult because the close binaries which are most likely to show the effect usually have low values of the eccentricity, e. When e is small, it is very difficult to determine ω accurately and to be sure of suspected changes in ω.

Apsidal motion has been suggested for a few well known eclipsing binaries, but none of the evidence is conclusive (PETRIE, 1962). According to REDMAN (1930) Y Cygni has a variable γ velocity, which indicates the presence of a third body in the system, as well as a variable ω. The variations in ω were first detected by DUGAN from the light curve. The eclipsing pair has a period of 2.996332 days and the two stars appear to be equally bright. The spectral type is O9.5 for each star and the lines are very diffuse and shallow, thus difficult to measure accurately for radial velocity. Redman had two orbits: one from observations in 1919–21 and one from 1928–29. STRUVE, SAHADE and ZEBERGS (1959) have added a third orbit confirming the variation in γ and ω.

Another example of a variable ω is the eclipsing pair AR Cassiopeiae which has spectral type B4. Changes in γ have also been detected. BATTEN (1961) has postulated that the changes in γ and ω are due to motion in a period of 0.71 years about the centre of gravity formed with a third body. However, this suggestion does not account for all the observed variations. The star AR Cassiopeiae undergoes annular eclipses. HUFFER and COLLINS (1962) have published extensive observations of the light variations in three colours, U, B, and V, and have given a solution for the elements describing the light curves. The results from this study of the light variations of AR Cassiopeiae are compatible with this being a triple system.

The best case of suspected rotation of the line of apsides is offered by the star HD 218407 (HR 8800) which is of spectral type B3. YOUNG (1920) determined the first orbit from observations obtained 1918–20. He found $\gamma = -15.1$ km/sec and $\omega = 125°.7 \pm 7°.1$. PETRIE (1959) determined a second orbit from observations obtained in the years 1957 and 1958. From this material it turns out that $\gamma = -14.3$ km/sec and $\omega = 215°.6 \pm 4°.6$. STRUVE, HUANG, and ZEBERGS (1959) observed HR 8800 for the four nights September 4–8, 1958 and derived an orbit. They found $\gamma = -16.5$ km/sec and $\omega = 202° \pm 3°$. Petrie remeasured Young's original spectrograms as well as studying a series of new spectrograms and was able to establish that a period of 3.33735 ± 0.00002 days would represent the velocity variations over 4213 cycles.

Petrie's results indicate that ω has changed by about 90° in 14061 days, and thus that the maximum period of apsidal motion (a change of 90 degrees, not 360+90 degrees, in the above time) is 154 ± 14 years. The observations of Struve, Huang and Zebergs suggest a period of 190 years for the apsidal motion. There is no indication of a variation of γ; thus a massive third body is not suspected in this system. The spectrum of the secondary component has been detected near nodal passage by Petrie and by Struve, Huang, and Zebergs but the lines are weak and difficult to measure. The difference in magnitude between the two components appears to be about 2 magnitudes; the secondary star is a B star. The observed dimensions of the system and the observed period of apsidal motion are not incompatible, but it is clear that further observations of this system at closer intervals than 40 years will be necessary if the problems posed by this binary star are to be solved in a satisfactory manner. Struve, Huang, and Zebergs have suggested that the structure observed in the $H\alpha$, profile indicates the presence of a gas stream in this system, but they conclude that the measured radial velocities (which depend upon lines in the blue-violet spectral region) have probably not been distorted by this gas stream.

4. The falsification of velocity curves

When a spectroscopic binary orbit is determined from a series of measurements of the changing radial velocity of the star, it is implicitly assumed that the displacement of the spectral lines is a true measure of the component of velocity of the star in the line of sight. This will certainly be true for stars which show uniformly illuminated discs and for those cases when there is no systematic motion of the stellar atmosphere with respect to the centre of gravity of the star. However, the displacement of the lines can be perturbed and a false radial-velocity orbit can result.

Three types of disturbance in the radial-velocity curve have been recognised, namely the rotation effect, the presence of gas streams in the system, and the reflection effect. The fact that the radial velocities do give a false representation of the line-of-sight motion of the star in its orbit frequently is discovered from the fact that the value for the eccentricity deduced from the radial-velocity curve is not the same as that deduced from the light curve. Unless there is a strong rotation effect in an eclipsing binary, it is not usually obvious from radial velocity observations alone that something is wrong.

The rotational disturbance in the radial velocity curve of an eclipsing star appears at the time of the eclipse. In the case of a star which is rotating rapidly as the disc of the star is covered by the companion, the lines are shifted in the direction in which the exposed limb of the star is moving. After totality, the lines appear shifted in the opposite direction as the other limb appears. As more and more of the disc of the star is exposed, the lines become symmetrical and lie once more at the wave length corresponding to the line-of-sight velocity of the star, for rotation of a star causes no displacement of the lines so long as the disc is fully exposed and uniformly illuminated.

The star U Cephei, HD 5769, spectral type B8, offers a good example of the distortion caused by the rotation effect and by gas streams. STRUVE and HUANG (1958) have discussed this interesting system and their radial-velocity curve is shown in Figure 33. The rotation effect is unusually large. During eclipse the secondary spectrum can be seen. Outside of eclipse the radial velocity curve is unsymmetrical leading to a spurious value of e. The system has been studied in detail by HARDIE (1950) who resolved the puzzle posed by the fact that photometric studies indicated that the eccentricity was less than 0.1 whereas spectroscopic studies suggested that e lay in the range 0.2 to 0.5. Hardie noticed that the profiles of the hydrogen lines were asymmetric, the core being displaced with respect to the wings. The rather sharp cores appear to come from a stream of gas in the system, whereas the underlying stellar H

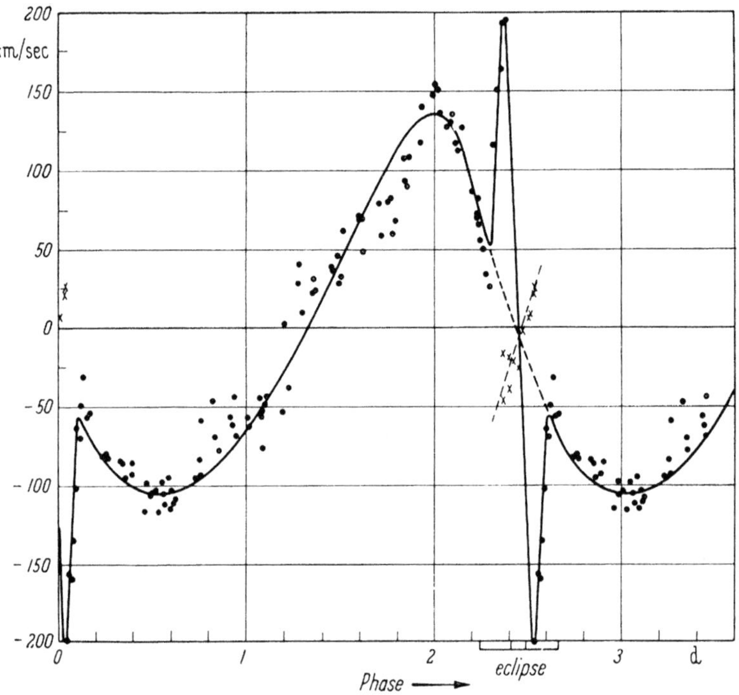

Fig. 33. The radial-velocity curve of U Cephei. The period is 2.5 days. The rotational distortion is exceptionally large. From STRUVE and HUANG (1958).

lines, whose positions can be determined from the wings of the composite profile, yield radial velocities in accord with the photometric orbit.

The reflection effect, like the rotation effect, arises because the stars are rotating, and because we do not observe a uniformly illuminated disc. In the reflection effect, the inner hemisphere of each star is illuminated by the other star, and it is surmised that the physical conditions in these inner hemispheres are changed slightly so that lines of a higher level of excitation arise there than the lines formed in the outer, cool hemispheres. Since the star is rotating, the centre of light of the inner hemisphere is not the centre of motion of the star. Thus it may be expected that in a star warmed by reflection effect a difference in velocity displacement will exist between lines from the hot side and those from the cool side. OVENDEN (1963) has investigated what displacements arise in the case of the binary star 57 Cygni, B3. He finds that there is indeed a spread in the measured displacements of lines of different origin. The N III lines λ4097 and λ4103, for instance, yield quite a differently shaped radial-velocity curve than does the Mg II λ4481 line. Ovenden concludes that it is not possible to utilise the velocity curves of a close binary system to deduce the masses of the components without due attention being paid to the reflection effect and to the physical behaviour of the particular lines used to define the radial velocities. He makes solutions for the orbital elements using individual lines and finds that a range of three to one occurs in the deduced mass function for 57 Cygni, depending upon which line is selected to indicate the motion of the star.

The presence of gas streams in a close binary system may obscure the secondary spectrum. In some cases it is now suspected that what was originally presumed to have been the spectrum of the secondary star is in reality only the spectrum of the gas moving in the close binary system. This type of disturbance can occur in any close binary system and it is particularly serious for the most massive early type stars. The radial velocities of HD 698, HD 1337 (AO Cass), HD 47129, and of β Lyrae are in each case strongly influenced by the presence of gas streams, and satisfactory radial-velocity orbits cannot be derived without taking these streams into account.

PEARCE (1932) found minimum masses $\mathfrak{M}_1 \sin^3 i = 113 \odot$, $\mathfrak{M}_2 \sin^3 i = 45 \odot$ for the components of the B8 star HD 698. These are unusually large values, particularly for a binary star in which the brighter star is B8, and they result from the relatively long period, 55.904 days, and large amplitudes, 85.5 km/sec and 215.5 km/sec found by Pearce. STRUVE and RUDKJØBING (1948) noted that such masses put the stars well off the mass-luminosity relation and they have suggested that the observed semi-amplitudes of the velocities are probably not real indications of binary motion. In particular the lines attributed to the secondary star may come from a stream of gas in the binary system or from a shell.

Early work (PEARCE, 1926) suggested that the system AO Cassiopeiae was very massive with the primary having a mass of 36, the secondary a mass of 34 solar masses. STRUVE and HORAK (1949) have re-examined the system. They find a primary velocity curve rather like that originally found by Pearce, but the velocities they attribute to the secondary star are quite different from those found by Pearce. The two stars are of type O and the lines attributed to the secondary star vary irregularly in intensity. The light curve of this system is peculiar (HILTNER, 1949b), the secondary minimum being wider than the primary. Almost certainly there are streams of gas in the system. In addition the secondary star may be surrounded by an envelope of electrons as has been suggested for the Wolf-Rayet stars CQ Cephei and V444 Cygni.

The star HD 47129, which was first described by J. S. PLASKETT (1922), is another O type star which appears to be buried in streams of gas. STRUVE, SAHADE, and HUANG (1958) have discussed the peculiarities of this system and have pointed out evidence for a stream moving at -700 km/sec as well as noting the periodic changes in the Hα emission.

Perhaps the most famous and most puzzling of all early type binary systems is β Lyrae. This is an eclipsing system in which the light changes are continuous around the cycle, thus the stars form what is called a contact binary. Attention was focussed on this star by Struve who did much to unravel the problems presented by the spectrum. The difficulty has been to determine which set of lines represents the primary star, which the secondary star, and which lines are due to streams of gas in the system. The appearance of the spectrum varies secularly and with phase in the orbit, and emission lines are seen. A summarising discussion has been given by HUANG (1962b) who gives reference to the extensive literature on this star. According to Huang, the primary star is a B8 III star with a mass between 3.5 and 8.0 solar masses; the secondary star has a mass in the range 12 to 20 solar masses but its spectral type is not surely known. It is possible that the variable velocity of the Hβ emission represents the radial velocity of the secondary star. The system β Lyrae is unusual in that the brighter star is the least massive of the two stars.

BELTON and WOOLF (1965) have obtained measurements of the light changes of β Lyrae in the six-colour system of STEBBINS and WHITFORD (1943) throughout the 12.9 day period of the eclipsing system. From this material they conclude that the spectral type of the primary star is B8.5II at maximum light and B9.5II at secondary minimum. They estimate that the spectral type of the secondary star must be about A7, and that not more than 25 per cent of the secondary star can be visible at secondary minimum. WOOLF (1965) has rediscussed the observational material about β Lyrae and he comes to the conclusion that HUANG's (1962, 1963) suggestion that the secondary star is imbedded in a disc is the model which best accounts for all the observed facts. Woolf concludes that the mass of the primary is 2.05 ± 0.30 solar masses, a value appropriate for a main-sequence star of type B9, and that the mass of the secondary is 11.74 ± 0.40 solar masses. Since the six-colour photometry and the spectroscopic evidence agree that the secondary star is less luminous than the primary star, it is clear that the secondary star must be a quite unusual object. Woolf concludes that the secondary star is about two magnitudes underluminous for its mass, yet the disc has a radius about twice that of the primary star.

5. Estimates of the masses of the early type stars

From the radial-velocity orbit of a double lined spectroscopic binary star one can derive the minimum masses $\mathfrak{M}_1 \sin^3 i$ and $\mathfrak{M}_2 \sin^3 i$. If the system is also eclipsing, the inclination, i, may be found and an estimate of the masses may be obtained. Although in many systems secondary effects, such as those due to gas streams or due to the presence of a third body, are observed, it is possible to obtain an idea of the masses of the early type stars. PEARCE (1957) has summarised the available material for main-sequence stars and his results are given here:

No. Stars	Spectral Type	Average Mass
5	O5–O7	32.7 ± 1.2 M$_\odot$
9	O8–O9	23.3 ± 2.0
14	B0–B2	13.6 ± 0.5
18	B3–B4	10.2 ± 1.1
12	B5–B7	4.7 ± 0.3

The quoted probable errors represent the scatter of the individual values about the mean value. It is to be noted that none of the main-sequence stars which are components of binary systems are more massive than 35 times the sun. We can only presume that the few early type main-sequence stars which appear to be single have similar masses. No information about the masses of the B type supergiants is available from the study of binary stars.

6. The observed mass-luminosity law

EDDINGTON (1926) pointed out that theory indicated that the luminosity of a star should be chiefly determined by its mass and only slightly by its radius, and he derived an appropriate theoretical expression. One can show for the standard model of a star (see, for instance, CHANDRASEKHAR, 1939) that the luminosity, L, of a star of mass $\mathfrak{M}$ and radius R (both in solar units), relative to the luminosity of the sun will be given by an equation of the following form

$$L/L_{\odot} = \text{const}\, \mathfrak{M}^{5+s} R^{-s}. \qquad \text{(IX-30)}$$

Here s is the absolute value of the exponent of the temperature in the expression adopted to give the opacity. It is expected that s will be close to 0.5. Now the quantity $2.5 \log(L/L_{\odot})$ is the difference in bolometric absolute magnitude between the sun and the star. Thus since the bolometric absolute magnitude of the sun is constant, theory suggests that a relation of the following form should exist between the bolometric absolute magnitude of a star, its radius in solar units, and its mass in solar units:

$$M_{\text{Bol}} = \text{const.} - 2.5\, s \log R + 2.5(5 + s) \log \mathfrak{M}. \qquad \text{(IX-31)}$$

The bolometric absolute magnitude of a star cannot usually be measured, because an important part of the radiation field from the star often lies in a spectral region that is inaccessible from the surface of the earth. However, the bolometric absolute magnitude can be estimated from the visual absolute magnitude if the bolometric correction appropriate to the spectral type is known. We have

$$M_{\text{Bol}} = M_V + B.C. \qquad \text{(IX-32)}$$

The bolometric corrections for early type stars may best be estimated by means of theoretical studies using model atmospheres. The value adopted for the bolometric correction at any spectral type is closely linked to the value adopted for the effective temperature at that type, and if the predictions of the theory of stellar atmospheres are to be used in a logical way, the relationship between spectral type and bolometric correction and that between spectral type and effective temperature which are used should come from the same set of calculations. Some values (UNDERHILL, 1966) are given in Table 21. They are adequate for finding the observed constants in equation (IX-31). Other, slightly different values have been suggested from other series of model atmospheres (MIHALAS, 1965; STROM, 1964). To test the validity of an equation of the

TABLE 21

Bolometric corrections and effective temperatures for early type stars from models without line blanketing

Spectral type	T_{eff}	B.C.	Spectral type	T_{eff}	B.C.
	degrees	mag		degrees	mag
B9 V	12690	− 0.92			
B6 V	15330	− 1.37	B7 V	14870	− 1.28
B3 V	19210	− 1.91			
B2 V	25670	− 2.59	B2 IV	25890	− 2.61
B1.5 V	27820	− 2.78	B1 IV	27780	− 2.74
B0.5 V	30030	− 2.93	B0 IV	29190	− 2.81
B0 V	31020	− 2.96	O9.5 IV	30360	− 2.87
O9	33950	− 3.14	O8	33410	− 3.07

form of equation (IX-31), we need to know not only the mass of the star, but we should also have an estimate of the radius. The radius can be estimated if the star is an eclipsing binary. Another possibility is to use the relationship

$$\frac{L}{L_{\odot}} = R^2 \left[\frac{T_{\text{eff}}}{T_{\text{eff}}(\odot)}\right]^4 \qquad \text{(IX-33)}$$

and to put the equation for the mass-luminosity relation into the form

$$M_V + B.C. - \frac{10s}{2-s} \log \frac{T_{\text{eff}}}{T_{\text{eff}}(\odot)} = \text{const} + \frac{5(5+s)}{2-s} \log \mathfrak{M}. \qquad \text{(IX-34)}$$

Since the combination of the second and third terms on the left varies very little between types B9 and O9, whatever the details of the adopted model atmospheres (Underhill, 1966) and model of the interior of the star, we are essentially looking for a relation of the form

$$M_V = A + B \log \mathfrak{M}, \qquad \text{(IX-35)}$$

for stars of this limited spectral range. Over a wider spectral range we may expect the "constants" A and B to vary somewhat.

Harris, Strand, and Worley (1963) have discussed the data available from binary stars for determining the observed mass-luminosity relationship and they have made a selection of the most reliable data. Their material contains nine B stars for which the data are of the first quality and six for which the data are somewhat less reliable. These 15 stars together with their masses, radii, and bolometric absolute magnitudes are listed in Table 22. It is to be noted that these masses are about the same as the average values derived from less carefully selected data by Pearce. The radii of main-sequence B stars run from about 2 solar radii at type B8 to about 6 solar radii at type O9. It seems clear that the radii of main-sequence stars do not go through a large range. However, the estimates of radius are rather uncertain for stars of the earliest spectral types, because the radius is derived from an analysis of the light curve and any distortions caused by streams of gas and extended atmospheres around the stars will make the results uncertain. Such secondary effects are very frequent among stars of early spectral type.

TABLE 22

The mass-luminosity data on B type stars selected by Harris, Strand and Worley

Star	Type	Mass solar masses	Radius solar radii	M_{Bol}	Quality
Y Cyg A	O9.5	17.5	5.9	− 6.5	1
Y Cyg B	O9.5	17.2	5.9	− 6.5	1
μ^1 Sco A	B1.5 V	14.1	5.2	− 4.9	2
μ^1 Sco B	B3	9.3	5.8	− 4.9	2
V356 Sgr A	B3	12.0	4.9	− 3.9	2
31 Cyg B	B3 V	8.9	4.7	− 3.8	2
32 Cyg B	B3 V	8.1	3.9	− 3.4	2
Z Vul A	B3 V	5.4	4.7	− 3.6	1
U Oph A	B5	5.30	3.4	− 2.5	1
U Oph B	B6	4.65	3.1	− 2.0	1
ζ Phe A	B6 V	6.1	3.4	− 2.2	1
ζ Aur B	B7 V	5.6	–	–	2
AR Aur A	B8	2.55	1.8	− 0.2	1
AR Aur B	B9.5	2.30	1.8	+ 0.4	1
RX Her A	B9.5	2.75	2.4	− 0.2	1

Hanbury Brown, and Twiss (1958) have measured the diameter of α Canis Majoris, A1V, using an intensity interferometer and they find that the radius is 2.0 solar radii. Such a result is in good accord with what the eclipsing binaries suggest for the radii of main-sequence stars having types near A0. The results of Hanbury Brown, Hazard, Davis, and Allen (1964) for α Lyrae, A0V, however, are in conflict with the results from eclipsing binaries for main-sequence stars. Hanbury Brown at al find that the radius of α Lyrae is 3.2 solar radii. Such a radius is more appropriate for a B6 star or for a star which has moved a significant distance off the main sequence towards greater luminosity than for an A0V star.

Studies of the intensity distribution in the continuous spectrum of α Lyrae and of α Canis Majoris using model atmospheres have indicated that the effective temperature of each star is close to 9500 degrees and that the shapes of the hydrogen lines can be reproduced well with models in which the acceleration of gravity is close to 10^4. Such a value for g is typical of a main-sequence star. Petrie (1965) has noted that the apparent magnitude and parallax of α Lyrae make the star about 0.5 magnitude brighter intrinsically than the luminosity suggested by the strength of the Hγ profile. He has suggested that possibly the star is a binary with such a value of the inclination that it is very difficult to detect radial-velocity variations. If this were the case, the rather large radius found by Hanbury Brown et al could be considered to be a weighted mean of the dimensions of the system rather than the radius of a single star. The star α Lyrae deserves further study.

Harris, Strand, and Worley have shown that the empirical mass-luminosity relation for stars brighter than bolometric absolute magnitude +7.5 is given by

$$M_{Bol} = 4.6 - 10.0 \log \mathfrak{M} . \qquad \text{(IX-36)}$$

The data from the early type eclipsing binaries falls on the same line as that from the visual binaries of later spectral type. For stars fainter than bolometric absolute magnitude +7.5, Harris, Strand, and Worley find

$$M_{\mathrm{Bol}} = 5.2 - 6.9 \log \mathfrak{M} . \tag{IX-37}$$

PETRIE (1950) has applied an indirect method first suggested by LUNDMARK and LUYTEN (1927) to estimate the mass-luminosity relationship. By studying Δm_V, the difference in visual magnitude between the components of a binary, and the mass ratio, Petrie is able to estimate the slope of the mass-luminosity relationship at a number of values of the luminosity. It is assumed that Δm_V, which is estimated from the relative depths and equivalent widths of the blended spectral features at various phases in the orbit, is equivalent to the difference in bolometric absolute magnitude between the stars. Instead of taking the bolometric absolute magnitude as the parameter in the mass-luminosity relationship, Petrie follows Eddington and uses

$$M_c = M_{\mathrm{Bol}} + 2 \log (T_{\mathrm{eff}}/5200) . \tag{IX-38}$$

He finds the mass-luminosity relationship

$$\log \mathfrak{M} = 0.548 - 0.144\, M_c + 0.002\, M_c^2 . \tag{IX-39}$$

The zero point of this relation is set by adopting the values

$$M_c = 1.27, \log \mathfrak{M} = 0.371$$

for Sirius. Petrie has used data from 93 spectroscopic binaries to establish equation (IX-39), and he considers that this equation will be valid for stars with M_c lying in the range +3 to −5, that is from type F to type O approximately. The visual binaries appear to fit quite well on to equation (IX-39) at the less bright end.

As with the direct approach to the mass-luminosity relation, which is exemplified by the work of Harris, Strand, and Worley, the outstanding problem remains still to determine accurately the position of the bright end of the mass-luminosity relationship. For the B stars, Harris, Strand and Worley suggest that

$$L/L_\odot \sim \mathfrak{M}^{4.0} ;$$

Petrie suggests

$$L/L_\odot \sim \mathfrak{M}^{6.9} .$$

Theory suggests

$$L/L_\odot \sim \mathfrak{M}^{5+s} ,$$

with s of the order of +0.5. The observational data available at present are not yet reliable enough to permit one to say which of the empirical mass-luminosity relationships is to be preferred for the O and B stars.

CHAPTER X

SPECTROPHOTOMETRIC TECHNIQUES

1. Photographic spectrophotometry

Photographic spectrophotometry is a basic method of studying stellar spectra that is of continuing value. This method is most valuable for making measurements of the relative intensity distribution in the stellar spectrum over short wave-length intervals. The determination of absolute intensities by photographic techniques is a very difficult problem, for one must take account of the varying transmission and sensitivity of the instruments and of the sky. These factors will vary with time and with the wave length at which the observations are made. Absolute measurements of the light intensity in stellar spectra are rarely attempted. Furthermore, the theory of stellar spectra is used to predict *relative* intensities within a line profile for comparison with observed intensities, rather than absolute values, and all present methods of analysis are aimed at obtaining information from relative intensities.

The dominant factor in recording stellar spectra, particularly at high dispersion, is the low intensity per angstrom which is incident on the photosensitive surface. This means that fast photographic emulsions which have fairly large grain and low contrast must be used in most cases.

Reciprocity failure at stellar light levels is often serious. With the Kodak IIaO emulsion, when exposures are longer than about one hour, baked plates will give a greater blackening per unit time at low light levels than unbaked plates, yet the photographic grain will not appear worse to the eye. However, the gain for spectrophotometry is not great since the effect of baking is to flatten the calibration curve. Thus small fluctuations in density of a given size will correspond to larger changes in intensity than with unbaked plates. This has the result that the noise on a tracing of the spectrum will be worsened, particularly in the case of lightly exposed spectra. In spite of this, it is worthwhile to use baked plates for all exposures longer than about one hour.

Experience has shown that good results are obtained when IIaO plates are baked for 72 hours at a temperature of 50°. To avoid undue background on the plates the time of baking should not greatly exceed 72 hours and the temperature should be kept very constant, that is $\Delta T \approx 0°.5$ C. The air in the oven should be dry.

Other techniques such as bathing in distilled water or ammonia followed by rapid drying are used for sensitising infrared plates. Particular attention to detail is required in order to obtain clean, adequately sensitized plates.

1.1 *Intensity calibrations*

Since the blackening on a photographic plate is not linearly related over a wide range of light intensities to the amount of light falling on the plate, the relationship between I_ν, the intensity of the light of frequency ν falling on the plate, and the blackening, which is measured by the deflection, d, of a microphotometer must be found empirically. The shape of the calibration curve will change somewhat depending upon the sensitivity of the emulsion, the wave length of the incident light, the length of the exposure to light, and the development process which has been used. Each manufacturer of photographic plates for spectroscopic use usually recommends a development process which will give the optimum results for the selected emulsion.

Each star spectrogram should be accompanied by a calibration exposure, the calibration spectrum being preferably on the same plate, or at least on a plate out of the same box as the star spectrogram and the star spectrogram and intensity calibration should be developed together. It is advisable to adjust the light level in the calibration system so that the calibration exposure is at least one third the length of the stellar exposure and to use a filter with an appropriate source of continuous light so that the distribution of intensity with wave length in the calibration spectrum is comparable to that in the star. Most stellar exposures are intermittent at a slow rate, for the image of the star often is shorter than the length of slit being used, and the star must be drifted up and down the slit in order to obtain a spectrogram of convenient width. Consequently any point in the slit is illuminated for less than the full exposure time. Stellar exposures with high dispersion may be a few minutes to several hours long.

A known gradation of intensity in the calibration exposure may be obtained by means of a rapidly rotating step sector or continuous spiral sector placed before the entrance slit of the calibrating spectrograph, or by means of step slits, a linear, or a wedge slit. If the sector is rotating rapidly, about 1000 rpm, it has been found that there will be no errors caused by intermittency effect. Since latent images change most rapidly within the first 24 hours after exposure, the calibration exposure and the stellar spectrum exposure should be made at about the same time and both should be developed together. If it is not possible to expose both at about the same time, it is better to delay the development for 24 hours or so, in order that both latent images can come to about the same stage of aging.

Since the shape of the calibration curve does depend somewhat upon wave length, the calibration device should disperse the light and form a spectrum. Then one can derive calibration curves at various wave lengths from one calibration exposure. An auxiliary spectrograph giving moderate dispersion may be used or the stellar spectrograph itself may be used to form the calibration spectrum. In any case, the calibration spectrograph should be well baffled and diaphragmed in order to eliminate all sources of stray light. Prism spectrographs are to be recommended because it is relatively easy with these to obtain a low level of scattered light. If a grating is used as the dispersing element, care must be taken to ensure that light from unwanted orders is not scattered into the main light beam. Also it is necessary to ensure that the Rowland and other ghosts due to irregularities in the rulings ar sufficiently weak to cause no trouble. Particular attention must be paid to obtaining uniform illumination over the full length of entrance slit used for the calibration exposure and to keep any narrow openings clean and free from dirt.

The accuracy of any calibration system should be tested by using step weakeners such as sputtered platinum on quartz which have little or no wave length dependence. Photoelectric recording is useful for making such tests because of the linear response, but difficulties may be encountered when the tests are run at the light levels usually used with stellar spectrograms because these light levels are very low.

1.2 *Methods for reducing microphotometer tracings*

The first thing to do is to establish empirically the relationship between the light intensity incident on the photographic plate and the amount of light transmitted by

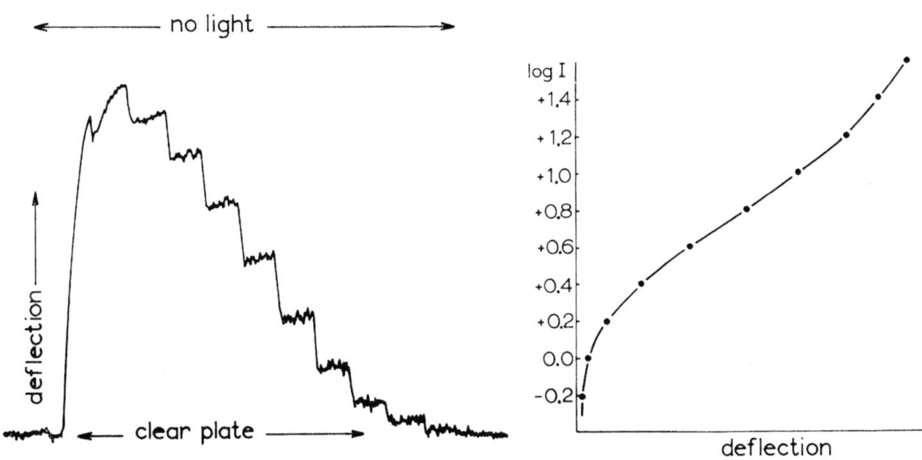

Fig. 34. A microphotometer tracing across a series of intensity steps and the resulting calibration curve.

the resulting blackening. This is done by tracing across an intensity calibration with the microphotometer using similar width of slits, illumination, etc., as will be used for the stellar spectrogram.

A typical microphotometer tracing across a pattern of intensity steps is shown in Figure 34 together with the resulting calibration curve. In this case the relative intensities of the steps are so arranged that $\Delta \log I = 0.2$ between any two successive steps. A rapidly rotating sector was used to produce the intensity pattern. The deflections of the microphotometer correspond to the transmission of the plate at each level of blackening and they are read from the line indicating full transmission or clear plate.

With the resulting calibration curve one may transform the tracing produced by the microphotometer, which gives an analogue of transmission *vs* position on the plate, into a curve giving log I *vs* position on the plate. A similar process can be followed to obtain a curve of I *vs* position on the plate. In each case one value of the transmission is arbitrarily selected to correspond to an intensity 100, and all the other values are derived relative to this point.

The transformation of the transmission-position-on-the-plate curve by means of the calibration curve into an intensity-position-on-the-plate curve can be done (1) by using point by point, graphical interpolation, (2) by a suitable transformation, using electronic techniques, into deflection of a pen of the voltages corresponding to transmission which are created by the microphotometer, or (3) by running the transmission-position-on-the-plate tracing through a semi-automatic mechanical system for interpolation such as the Intensitometer which has been developed at the Dominion Astrophysical Observatory (see WRIGHT (1962) for a description of this instrument), or (4) by a simple computer program. In the latter case the input is first a set of pieces of information (intensity, transmission) which define the calibration curve, and a set of data defining the translation of information giving the position-along-the-plate into wave length. Then the computer is given a set of information in the form (transmission, position on the plate), and it interpolates to find a set of points (intensity, wave length). This information may be output in digital or analogue form as desired and as is available with the computer. Further data handling of the input data can be devised as required.

All four methods of finding the relation between intensity and wave length along a stellar spectrogram have their use. It is generally most useful to have the output in analogue form, that is as a tracing, for then one may readily see irregularities and flaws due to accidental occurrences at some part in the process of transforming the incident light into a curve in which both coordinates, intensity and wave length, are known with an accuracy of a few percent.

So far attention has been directed to obtaining an intensity wave-length tracing such as that shown in the upper part of Figure 35. This, however, is not the most convenient way of presenting the available information, for on each spectrogram an arbitrary point is set to intensity 100 and the intensities of all other points relative to this arbitrary reference point vary not only in the manner produced by the star, but also according to the distortions introduced by interstellar absorption, by absorption and refraction in the earth's atmosphere, and by the spectral sensitivity of the complete observing equipment. One cannot readily compare the intensity tracings of two spectra of the same star, or of different stars by superimposing the intensity-wave length curves.

It is more convenient to take out the distortions caused by interstellar space, the earth's atmosphere, and by the equipment, by studying the intensity at any point in the spectrum relative to the intensity given by a datum line which is usually considered to represent the continuous spectrum of the star. This continuum can be readily sketched on tracings of spectra which are not too full of lines. When the lines overlap

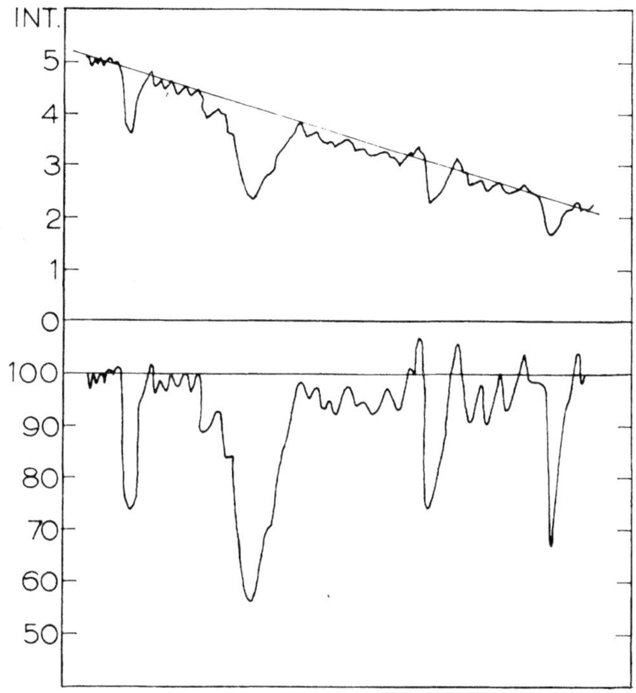

Fig. 35. An intensity *vs.* wave-length tracing and a rectified intensity tracing of a section of a stellar spectrogram.

to such an extent that no pieces of undisturbed continuous spectrum can be found, the datum line can be defined by stating which apparent high points of intensity it goes through.

Let the intensity at any wave length λ in the datum line be $I_0(\lambda)$, and let the actual intensity at this wave length be $I(\lambda)$. Then a rectified intensity tracing is one in which the ratio $I(\lambda)/I_0(\lambda)$ is given as a function of wave length. Such a tracing is shown in the lower part of Figure 35. Astrophysical theory is concerned with the interpretation of $I(\lambda)/I_0(\lambda)$ in the case where the datum line, $I_0(\lambda)$, represents the intensity distribution which would result if no lines were present. When the spectral lines are not too crowded, it is possible to estimate the apparent variation of $I_0(\lambda)$ quite well.

Rectified intensity tracings from different spectrograms can readily be intercompared by superposition when care has been taken to make all the tracings on the same scale. It is valuable to make such an intercomparison in order to minimize the random errors due to photographic grain and to accidental markings on the spectrograms. Also the resulting line profiles and equivalent widths are directly comparable to the data predicted by astrophysical theory.

Rectified tracings may be obtained by an arithmetic transformation of intensity-wave length tracings, given the position of $I_0(\lambda)$; that is $I(\lambda)$ is multiplied by $1/I_0(\lambda)$.

This transformation may be performed manually, point by point, or it may be done by introducing a further piece of information, namely the deflection corresponding to $I_0(\lambda)$ into the data received by the computer, and adding the necessary manipulation to the program, or it may be done by a mechanical process using the Intensitometer. The possibility of making a transformation of this sort by a mechanical trick is one of the great advantages of the Intensitometer. Indeed, $I_0(\lambda)$ does not have to be the continuum of the star. It may be the spectrum of another star which is blended with the star under consideration. WRIGHT (1954) in his study of *Capella* has demonstrated the usefulness of the Intensitometer in separating composite spectra.

To obtain a rectified intensity tracing by means of the Intensitometer is a two stage process. First a tracing giving log I *vs.* wave length is produced. Then a line representing the datum line $I_0(\lambda)$ is drawn on the log $I(\lambda)$ tracing. Finally this modified tracing is run through the Intensitometer and the difference $\log I(\lambda) - \log I_0(\lambda)$ is read and translated by means of an anti-log curve into the desired ratio $I(\lambda)/I_0(\lambda)$. The original Intensitometer determined the difference $\log I - \log I_0$ and made the translation by mechanical means. It is clearly also possible to make this translation by an appropriate electronic manipulation, the difference $\log I - \log I_0$ on the tracing being represented as a voltage. The two curves $\log I(\lambda)$ and $\log I_0(\lambda)$ are followed with different pointers by the operator. If the function $\log I_0(\lambda)$ is complicated, a second operator must assist because one person cannot accurately follow two rapidly varying curves at one time.

When the rectified tracing has been obtained, a line profile may be drawn and the area of the profile found. This measurement gives the equivalent width W_λ of a line. We have

$$W_\lambda = \int_a^b \frac{I_0(\lambda) - I(\lambda)}{I_0(\lambda)} \, \mathrm{d}\lambda , \qquad \text{(X-1)}$$

where a and b are the limits between which the line profile extends. Equivalent widths are expressed in angstroms or in milliangstroms. Occasionally the unit for equivalent width is given as km/sec. In this case a transformation has been made:

$$W_{\mathrm{km/sec}} = W_\lambda \, c/\lambda . \qquad \text{(X-2)}$$

When spectra that are full of lines are being analysed it is often convenient to approximate the line profile by one of a series of standard line profiles which all have the same shape but which have been scaled to have different areas. If *rectified* intensity tracings on a selected scale are used, it is very easy to estimate equivalent widths of lines by comparing the observed profiles with the set of standard profiles. This method is particularly useful when only one side of a profile is available, the other side being seriously distorted by a blend. This method cannot be used efficiently when only $I(\lambda)$-wave length tracings are available because its use depends upon the fact that all the tracings are made on the same scale in both coordinates at all wave lengths.

For some types of study it is adequate to estimate equivalent widths by representing the line profile as a triangle and computing the area from measurements of the depth and total width.

1.3 *Standard equivalent widths and line profiles*

The problem of establishing a set of standard equivalent widths and line profiles in a few stellar spectra has been studied by a small group under the leadership of K. O. Wright. The purpose of this work has been to obtain an estimate of the accuracy attainable by photographic spectrophotometry of stellar spectra and to set up a set of standard equivalent widths which any observer may use to check the behaviour of his spectrograph and system of spectrophotometry. The results have been published by WRIGHT *et al.* (1963).

The following twelve stars were selected as standard stars:

Type B ρ Leonis, γ Pegasi, ι Herculis
Type A γ Geminorum, θ Leonis, 68 Tauri, 15 Vulpeculae.
Type F σ Bootis, α Canis Majoris, 110 Herculis,
Type G λ Serpentis, μ Herculis.

Equivalent widths were measured in the wave-length region 3900 Å to 4520 Å and profiles have been determined for Hγ. Altogether 173 lines were measured in the B type spectra, 511 lines in the A type spectra, 552 lines in the F type spectra and 361 lines in the G type spectra. Careful intercomparisons have been made between the results from the various high-dispersion spectrograms which were available, and mean equivalent widths were derived.

When several spectrograms are available for one object, the average percentage deviation from the mean equivalent width of any line is five percent both for Victoria and for Mount Wilson spectra. The Victoria equivalent widths tend to be slightly larger than the Mount Wilson equivalent widths for the same object, but it is felt that this is rather the result of different interpretations of the continuous spectrum and line profile than the result of systematic errors in the spectrophotometric procedures followed at either Observatory.

The results from the high-dispersion spectrograms were compared with those from spectra having a dispersion of 30 Å/mm or less. In this case a distinct systematic trend was found, the equivalent widths from the low-dispersion spectra being as much as 25 per cent greater than those from the high-dispersion spectra. This systematic error is most marked for weak lines.

The *spectral purity* of a spectrum is given by the projected slit width in angstroms on the plate. The purity is usually considerable less than the theoretical resolving power of the spectrograph because of the necessity of using a relatively wide slit in order to obtain enough light in the spectrograph. Frequently with moderate- or low-dispersion spectrograms the spectral purity is of the order of 1 Å; it may be 2 Å. In such cases accurate measurements of equivalent width can only be made for strong lines.

2. Photoelectric scanning of spectral lines

When a stellar spectrum is to be scanned at high resolution in order to derive line profiles and equivalent widths, a narrow slit must be used in order to maintain adequate spectral purity. In this case compensation for the spectral intensity changes caused by seeing is a serious difficulty. A high-resolution scanner has been developed at the Mount Wilson Observatory and a few results for three early type stars, τ Scorpii B0V, 10 Lacertae O9V, and 9 Sagittae O7f, have been reported by GREENSTEIN and OKE (1961). MIHALAS (1964) has given more scanner measurements of equivalent widths for six O stars.

The instrument scans a limited spectral region at one time by rotating a prism in front of the exit slit in order to move the spectrum past the photocell. The dispersing grating and the recording photocell do not move. The effects of seeing are compensated by monitoring bands of spectrum 100 Å wide adjacent to the section being analysed. The initial dispersion is 2.8 Å/mm and the effective half-width of the analysing system is 0.3 to 0.4 Å.

The equivalent widths obtained by this scanner in the case of 10 Lacertae are the same as those obtained from photographic spectrophotometry of high-dispersion, high-resolution spectrograms for lines with $W_\lambda \geqslant 50$ mÅ (UNDERHILL and DE GROOT, 1965), but for weaker lines the scanner equivalent widths tend to be larger than the photographically determined equivalent widths. It seems probable that the result is due to the limited spectral purity of the scanner, *cf.* the systematic effects with spectrograms of moderate purity reported in Section 1.3.

Photoelectric scanning of spectral lines at present is most useful for determining accurately the shapes of wide, shallow features such as the wings of H and He II lines in O type spectra. This type of measurement is very difficult on photographic plates because of the disturbing effects of photographic grain. However, the greater spectral purity of the best high-dispersion spectrograms make such spectrograms a better means by which to measure the equivalent widths of weak lines than a scanner. Furthermore, because one normally records 500 Å or more at one time with a high-dispersion photographic spectrograph, the use of photography is still one of the most efficient methods of using telescope time for spectroscopic studies.

3. Correction for the instrumental profile

The intensity distribution that is recorded in a spectral line is not the true profile of the line, but is the true profile broadened by the instrumental profile. Thus

$$R(\lambda) = \int_0^\infty T(l)\, A(\lambda - l)\, \mathrm{d}l\,. \tag{X-3}$$

Here $R(\lambda)$ as the recorded line intensity at wave length λ, $T(l)$ is the true profile of the line and $A(\lambda - l)$ is the apparatus function. That is, the function $A(\lambda - l)$ gives the shape which an infinitely narrow spectral line will have after passing through the spectrograph. The problem is to invert equation (X-3) to obtain the function $T(l)$, knowing $R(\lambda)$ and the apparatus function, $A(\lambda)$. The equation has been written as though the integration were taken over an infinite range on either side of the position λ in the line profile. In practice finite limits are imposed.

Various techniques for solving equation (X-3) have been proposed, and mention of a few possibilities will be made here. The methods depend upon being able to evaluate sufficiently accurately the various moments of the recorded profile $R(\lambda)$ and of the apparatus profile $A(\lambda)$. The nth moment of a function $f(x)$ is

$$\mu_n = \int_0^\infty x^n f(x)\,\mathrm{d}x. \tag{X-4}$$

If the higher moments are to be evaluated accurately, one must know accurately the shape of the extreme wings of the profile. This raises practical difficulties, because it is difficult if not impossible to determine the wings of $R(\lambda)$ and of $A(\lambda)$ accurately. The apparatus profile may be found by evaluating the shape of a single spectral line from the ^{198}Hg spectrum or from the ^{86}Kr spectrum. If it is not possible to make a special study of the shape of the apparatus profile using a source of light which gives very narrow lines, an estimate of the shape may be obtained from the shapes of lines in the Fe arc comparison spectrum or by studying the shapes of absorption lines due to the O_2 in the earth's atmosphere. Such estimates will in general be wider than the true apparatus profile.

Burger and van Cittert (1932, 1933) and Shane (1932) independently suggested an iterative process for solving equation (X-3) for the true profile $T(\lambda)$. Shane observed that the extreme wings of the apparatus function contribute an effectively constant term to the distorted profile, and he wrote equation (X-3) in the following form

$$R(\lambda) = S_0 + \int_{\lambda-\varepsilon}^{\lambda+\varepsilon} T(l)\,A(\lambda - l)\,\mathrm{d}l. \tag{X-5}$$

Here the term S_0 is called the scattered light and it is given by

$$S_0 = \int_0^{\lambda-\varepsilon} T(l)\,A(\lambda - l)\,\mathrm{d}l + \int_{\lambda+\varepsilon}^{\infty} T(l)\,A(\lambda - l)\,\mathrm{d}l. \tag{X-6}$$

With a well designed spectrograph and modern gratings, S_0 can be very small, but it is not always negligible. The profiles $R(\lambda)$, $T(\lambda)$, and $A(\lambda)$ may be asymmetric.

De Jager and Neven (1964, 1966) have experimented with the method of Burger and van Cittert and the method of Fourier transforms for inverting equation X-3 in order to find the true profiles of lines observed with the solar spectrograph at the Jungfraujoch Observatory. They corrected for the scattered light in the instrument and they determined the apparatus profile accurately using a line from a ^{198}Hg isotope lamp. They find that the numerical methods will not converge unless the half width of the recorded profile exceeds the half width of the apparatus profile by at least a factor 2, and they note that numerical methods must be used if the asymmetries of the profiles are to be preserved. The apparatus profile of a grating spectrograph is usually asymmetric owing to the irregularly placed ghosts that may occur; also many stellar and solar lines are asymmetric.

The results of correcting a Ti I line near $\lambda 8400$ Å in the solar spectrum are displayed in Figure 36 which is due to de Jager and Neven. The observed profile is

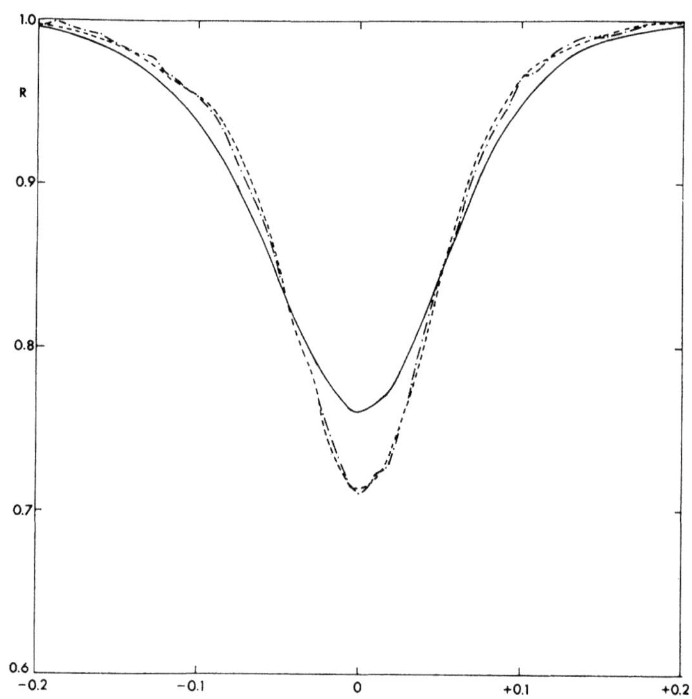

Fig. 36. The result of correcting the solar line Ti $\lambda 8412$ for the instrumental broadening of the Jungfraujoch solar spectrograph. The solid line is the uncorrected, observed line profile. The dash-dot line gives the corrected profile by the method of Fourier transforms; the broken line gives the corrected profile using the method of Burger and van Cittert. The abscissa is measured in angstroms.

shown by the solid line. The corrected profile found by the method of Fourier transforms is shown by a dash-dot line. That found by the following iterative method is shown by a broken line. Equation X-3 is solved by repeatedly evaluating the equation

$$T_n(\lambda) = T_{n-1}(\lambda) + R(\lambda) - \int_0^\infty T_{n-1}(l)\, A(\lambda - l)\, \mathrm{d}l\,. \qquad \text{(X-7)}$$

The initial step is to set

$$T_1(\lambda) = R(\lambda), \tag{X-8}$$

and the iteration is continued until

$$T_n(\lambda) - T_{n-1}(\lambda) \leqslant \Delta, \tag{X-9}$$

where Δ is a pre-assigned tolerance which depends upon the quality of the observations. In practice, it is usually not useful to iterate more than three or four times. If the tolerance Δ is made too small, the method will diverge owing to the incompatibilities which are present between the observed functions $R(\lambda)$ and $A(\lambda)$ because of the normal uncertainties of measurement.

The method of Fourier transforms and this iterative method yield essentially the same result. Both methods can only be carried out efficiently with the help of a high-speed computer, and both require accurate observed profiles from which to start.

Analytical methods have been proposed for solving equation X-3, one of the most comprehensive discussions of the problem having been given by VAN DE HULST (1946). Analytical methods represent the profiles $R(\lambda)$, $T(\lambda)$, and $A(\lambda)$ by functions of known form and symmetry. A convenient method in the case that the lines and apparatus function are symmetrical is to use Voigt functions. This method was proposed by VAN DE HULST and REESINCK (1947) who gave auxiliary tables for carrying out the work. In this case each profile is represented by a function which depends upon two parameters, β_1, and β_2. Thus if $\beta_1(R)$ and $\beta_2(R)$ are the parameters for the recorded profile and $\beta_1(A)$ and $\beta_2(A)$ are the parameters for the apparatus profile, one determines the parameters $\beta_1(T)$ and $\beta_2(T)$ for the true profile from the following relations

$$\beta_1(T) = \beta_1(R) - \beta_1(A) \tag{X-10}$$

and

$$\beta_2^2(T) = \beta_2^2(R) - \beta_2^2(A). \tag{X-11}$$

Using the auxiliary tables which are given, one can readily find β_1 and β_2^2 for the recorded profile and for the apparatus profile from measurements of the total half widths, $h(R)$ and $h(A)$, and of the total widths at one tenth the central depths, $b_{0.1}(R)$ and $b_{0.1}(A)$. The accuracy of this method depends upon how well the adopted symmetrical Voigt functions fit the profiles $R(\lambda)$, $T(\lambda)$, and $A(\lambda)$, and how well the selected measurements serve to determine a representative set of parameters β_1 and β_2.

ELSTE (1953) rediscussed the problem and gave more extensive and homogeneous tables of the Voigt function and the necessary parameters than was given by van de Hulst and Reesinck. He suggested that a more reliable determination of β_1 and β_2 might be found from measurements of the total width at 0.2 and 0.4 times the central depth rather than at one tenth and one half the central depth as suggested by van de Hulst and Reesinck. A further tabulation of the Voigt function has been given by POSENER (1959), who corrects some errors in the tables of van de Hulst and Reesinck which were drawn from various sources. With these auxiliary tables one may readily correct observed line profiles for instrumental broadening.

De Jager and Neven, however, have shown that the results obtained by using Voigt functions do not always agree well with those found by direct numerical solutions of equation (X-3) particularly when the line and apparatus profiles are asymmetrical. One difficulty is that the observed profiles frequently do not have the shape of Voigt functions. Another difficulty is finding truly representative values of the parameters β_1 and β_2. In the case of the Ti line studied by de Jager and Neven, the derived corrected central intensity of the line varies from 0.713 to 0.642 depending upon the method used to correct for instrumental broadening. The observed central intensity was 0.760. In the case of stellar spectra where the resolution of the spectrograph is not so high, and the observed profiles are more inexact owing to the difficulties of photographic spectrophotometry it is rarely worthwhile to attempt to correct line profiles for instrumental broadening. Some information can be derived for strong lines, particularly among the O and B type stars, for it is found that even the apparently sharpest lines are not very sharp. If the apparent width at half intensity of the line is at least twice the width at half intensity of the instrumental profile, some reliance may be placed on the results of a correction procedure.

CHAPTER XI

THE INTERPRETATION OF NORMAL, MAIN-SEQUENCE, EARLY TYPE SPECTRA

1. The problem

The problem facing astrophysicists is to build up an exact physical picture of the conditions in a stellar atmosphere by analysis of the part of the radiation field from the star which is accessible to observation. In practice the most readily obtained suitable observational material has been high-dispersion spectrograms covering the spectral range 3000 Å to 6800 Å. In fact most studies have been done using a restricted spectral range from about 3500 Å to 5000 Å. The observations yield relative intensities within the spectrum over short wave-length regions, that is line profiles, and the relative strengths of the lines, that is equivalent widths. Each observational fact is expressed relative to the intensity of an adopted datum line which is called the continuum of the star. Little or no use is made in the analysis of the relative intensity of the continuum at different wave lengths because such information, on an absolute scale, cannot be easily obtained from normal stellar spectrograms. The profiles and equivalent widths of the spectral lines are obtained by the methods described in Chapter X. Most methods of analysis are concerned with equivalent widths only, since reliable line profiles can only be obtained for a few strong lines.

Useful data for astrophysical analysis can be obtained for early type stars from spectrograms having a dispersion of 10 Å/mm or better. The spectra of early type stars contain a few strong lines and many weak shallow lines that are at the limit of detection. It is usually easy to define the continuum in a unique manner except at wave lengths between about 3650 Å and about 3900 Å where the wings of the Balmer lines of hydrogen overlap. The continuum is intended to represent the intensity distribution which would result from a stellar atmosphere having the same temperature and pressure structure as the real stellar atmosphere but in which there is no line absorption, only absorption in continua.

The methods of analysis which have been developed aim at determining the temperature, pressure, and state of motion in the stellar atmosphere at each point in the atmosphere and at finding the relative abundances of the elements. The method used is to compare observed line profiles where available and equivalent widths with those predicted from a model chosen to represent the stellar atmosphere. The model is characterised by a few parameters and the values of these parameters give, in more or less detail, the desired information about temperatures, pressures, state of motion, and abundances. The ideas behind the methods which are used are summarised in the Appendix.

It is beyond the scope of this book to review in detail the methods of analysis which have been developed, but references to the major papers giving these methods and the results from their application will be given. In these papers are found practically all existing lists of equivalent widths and line profiles that have been measured for early type stars. The observational studies have been limited to a handful of bright O and B stars which have sharp spectral lines. Sharp-lined stars are selected for study because then line profiles and equivalent widths may be determined more accurately, and also it is hoped in this way to select stars for which mass motion of the atmosphere either by irregular pulsation or by rotation is not important. Such mass motions add complexity to the radiation transfer problem. The usual theories ignore such complications.

All models designed to represent the atmospheres of normal main-sequence stars of early spectral type consist of plane parallel layers of gas in hydrostatic equilibrium under the gravitational attraction of the star. The atmospheres contain a mass fraction X of hydrogen, Y of helium, and Z of all the other elements. Characteristic values of X, Y, and Z are about 0.70, 0.28, and 0.02. One major problem is to determine accurately the values of X, Y, and Z, in particular the relative abundances of the elements C, N, O, Si, etc. which give the lines observed in O and B type spectra. Since hydrogen is almost completely ionised in O and B type atmospheres, and the degree of ionization of helium is also large, it is only necessary to consider X and Y when calculating the relative values of the electron and gas pressure.

The simplest model is a single layer of gas at one temperature and one pressure in which the interaction between radiation and matter takes place as though the atmosphere were in thermodynamic equilibrium at the temperature T. More detailed models allow the pressure and temperature to increase inwards with depth, the temperature distribution being found by making the total radiation field (calculated for the case of no lines) obey the constraint radiative equilibrium. In such detailed models, account is taken of the fact that the opacity comes from electron scattering as well as from absorption in the continua due to H, H^-, He I and He II, and some account is taken of the radiation-pressure gradient in the atmosphere. Even more sophisticated models in which account is taken of the opacity due to the lines are at present being constructed.

Once a model, simple or complex, has been constructed, it is necessary to predict the line spectrum of the model for comparison with the observed equivalent widths and line profiles. This step requires a theory of the formation of absorption lines in stellar spectra as well as an adequate knowledge of the physical constants defining the interaction between atoms, ions, and radiation. The theory which has been applied for early type stars assumes that, so far as lines are concerned, it is adequate to represent the interactions as though they occurred in thermal equilibrium at the local temperature. This point of view can be justified for many of the lines observed in early type spectra. However, in the case of strong lines more realistic representations of the actual physical situation are needed. Attempts to solve the realistic problem are now being made for some representative cases.

The curve of growth is used to show in graphical form how the equivalent width of a spectral line increases with increasing abundance of the element. The shape of the curve of growth changes in a characteristic way when the temperature and pressure in the atmosphere change. The methods of analysis which are used for the quantitative study of O and B type spectra result in a set of parameters for the curve of growth, and these more or less internally consistent parameters are translated by well defined formulas into the desired information about temperature, pressure, state of motion, and abundance of the elements. The Doppler displacements due to the motion of the atoms resulting from the kinetic energy the atoms possess at a given temperature enter into the analysis. This motion causes the thermal Doppler broadening of spectral lines. In some cases the results of the analysis suggest that the atoms have a Maxwellian distribution of velocities corresponding to a higher temperature than expected. It is then said that microturbulence is present in the stellar gas. In the case of early type stars, microturbulence cannot be unequivocally determined from the analysis of equivalent widths only, because usually a suitable selection of weak and strong lines from the same ionic spectrum is not present in the accessible spectral range. However, when line profiles are also studied using models, it is seen that microturbulence is an important factor in producing the observed spectrum. The concept of microturbulence is an arbitrary mathematical device which was introduced to obtain better agreement between theory and observation. Modern studies of line formation are aimed at finding out the true physical implication of this fitting factor.

In reading the following sections of this chapter it must be continually borne in mind that the observational material available for making the analysis is in no case sufficient for a complete, internally consistent analysis, and that each theoretical representation is a simplification of the complex situation that really exists in an early type stellar atmosphere. It is a measure of the dediction and skill of the few workers in this field that so much has been found out about the atmospheres of normal main-sequence stars of types O and B. However, all values of the resulting parameters should not necessarily be taken at their face value.

2. Methods of analysis developed at Kiel and the results

The methods used have been developed chiefly by Unsöld. Throughout these analyses, from the simplest to the most complex representation, it is assumed that the concept of local thermodynamic equilibrium gives an adequate representation of the formation of lines in O and B type spectra. The first method used, *coarse analysis*, represents the stellar atmosphere by a single layer of gas at a characteristic temperature and pressure. After an orientation has been obtained in this way, a multi-layer representation may be used. This makes it possible to apply the techniques of *fine analysis*.

2.1 *Coarse analysis*

The methods were developed by UNSÖLD (1942a) for a study of the spectrum of the B0V star τ Scorpii. The stellar atmosphere is represented by a single representative layer at a characteristic temperature and pressure. In this paper Unsöld assembles much information about the f-values of the atomic transitions which are observed, and shows how these data can be estimated when no experimental or quantum mechanical results are available. (Since that time much of the necessary information has appeared; see, for instance, GLENNON and WIESE (1962).)

Unsöld began his analysis by estimating the number of H atoms in the second quantum state from the equivalent widths of the higher members of the Balmer series, making use of the hypothesis that these lines are formed in a thin layer. He then estimated the electron pressure from the Teller-Inglis formula (INGLIS and TELLER, 1939) which states that

$$\log N_e = 23.26 - 7.5 \log n_m . \tag{XI-1}$$

Here N_e is the number of electrons per cubic centimeter in the atmosphere and n_m is the upper quantum number of the last visible line in the Balmer series. This formula is derived from a consideration of the depression of the ionization limit of hydrogen due to Stark effect.

A second estimate of the electron density is obtained by assuming that the observed profiles of the wings of Hβ, Hγ, and Hδ can be represented by the Minnaert interpolation formula

$$\frac{1}{R_\nu} = \frac{1}{R_c} + \frac{1}{l_\nu} . \tag{XI-2}$$

Here R_ν is the depth of the line at frequency ν, R_c is the central depth of the line, and l_ν is the value of the line-absorption coefficient at frequency ν.

In main-sequence stars, the Balmer lines of hydrogen are broadened by Stark effect and according to a simple theory (which has been greatly improved by later theoretical developments) one can write that

$$l_\nu = C \Delta\lambda^{-5/2} \tag{XI-3}$$

in the wings of the line. The constant of proportionality, C, depends upon the electron density. Unsöld shows how to use formulas (XI-2) and (XI-3) and the observed equivalent widths of Hβ, Hγ, and Hδ to obtain an estimate of N_e.

Unsöld uses a "thin-layer" theory, similar to that used for the Balmer lines, to obtain an estimate of the number of neutral helium atoms in the 2^3P level from the measured equivalent widths of the He I lines. Then, using his final estimates of T and P_e, he is able to find the relative abundance of helium to hydrogen making use of the Boltzmann law to calculate the relative populations of the excited levels of helium and the Saha law to estimate the degree of ionization of helium. He checks his result by noting that the derived abundance, T, and N_e give a satisfactory prediction of the equivalent width of He II λ4686.

The curve of growth for a single layer of gas is used to estimate relative abundances from the measured equivalent widths of lines in the following spectra: C II, C III, N II, N III, O II, O III, Ne II, Mg II, Al III, Si III and Si IV. Unsöld's formulation of the curve of growth is based on Minnaert's interpolation formula, equation (XI-2). It is found that the equivalent widths of lines in the spectrum of τ Scorpii can best be represented by a layer of gas having a temperature of 28150 ± 750 degrees, that log P_e is 3.07 ± 0.25, and log N_e is 14.48 ± 0.25. Finally Unsöld is able to estimate the relative abundances of the elements.

This was the first such analysis for an early type star. The results are shown in Table 23 together with the results of other somewhat similar analyses. The relative abundances are by number, the value of log N being entered opposite the symbol for the element. These numbers are relative to an abundance log $N = 12.00$ for hydrogen.

It must be stressed that in the method of coarse analysis the following assumptions are made in order to obtain a description of the physics of line formation:

1. That the same, single representative layer is suitable for interpreting the equivalent widths of all lines in the stellar spectrum.
2. That the process of line formation for all lines studied can be represented by the empirical formula given in equation (XI-2).
3. That the degree of excitation and of ionization in each spectrum are those which would occur in thermodynamic equilibrium at the adopted temperature and electron pressure.
4. That the motions of the atoms in the atmosphere are completely represented by the Maxwellian distribution of velocities corresponding to the temperature T.

These assumptions suffice for a first survey of the situation. Later work, particularly that concerned with the interpretation of line profiles as well as equivalent widths has shown that more sophisticated methods of analysis should be applied and that large errors can occur in the estimated relative abundances if improvements are not made in the basic physical picture used to describe the process of line formation in early type stellar atmospheres (see Section 4).

In later papers, UNSÖLD (1942b, 1944a) developed further approximate techniques for analysing the equivalent widths of lines broadened by Stark effect and he applied his methods (UNSÖLD, 1944b) to the data of E. G. WILLIAMS (1936a) and of others to obtain an estimate of the H/He ratio in stars of types O8 to B3. The average value which he found was H:He = 85:15 by number.

2.2 *Fine analysis*

The techniques developed at Kiel and known as "fine analysis" were first demonstrated by TRAVING (1955) in a further study of the spectrum of τ Scorpii and by HUNGER (1955) in a study of the spectrum of α Lyrae, A0V. The methods of fine analysis relieve assumption 1 of coarse analysis by representing the stellar atmosphere by a model atmosphere consisting of many layers. The temperature law is found by requiring that the condition of radiative equilibrium be satisfied. The pressure in the atmosphere is found by requiring that the atmosphere be in hydrostatic equilibrium under the adopted value of gravity. Assumption 2 is also relieved, the method of weighting functions being used to calculate the equivalent widths of the lines. It is assumed that the source function in the line is identical with the Planck function.

TABLE 23

Results of coarse analysis of early type spectra

Star	Type	Author	Year	T	$\log P_e$	$\log g$	H	He	C	N	O	Ne	Mg	Al	Si	P	S
10 Lac	O9 V	ALLER	1946	29600°	2.80	4.44	12.00	10.93	8.01	8.05	8.69	8.64	7.51		7.62		
τ Sco	B0 V	UNSÖLD	1942a	28150°	3.07	4.3	12.00	11.19	8.24	8.58	8.99	9.05	7.76	6.56	7.80		
γ Peg	B2 IV	MICZAIKA	1948	20000°	2.63		12.00	11.15	7.85	8.64	8.54				8.08		
γ Peg	B2 IV	ALLER	1949	20000°	2.60		12.00	10.84	7.53	7.22	7.82		7.35	6.01	6.6:	4.2:	6.6:
δ Cet	B2 IV	ALLER	1949	20400°	2.46		12.00	10.56	7.16	7.95	8.34						6.7
ι Her	B3 IV	ALLER	1949	18600°	2.39		12.00	10.69	7.74	7.43	7.98		8.02	6.09	6.58		7.6
3 Cen A	B4 IV p	JSG*	1961	14700°	2.27		12.00	10.4	8.5	8.8	7.9		8.6	7.3		7.5	7.4
ε CMa	B2 II	ALLER	1956a	19800°	2.00		12.00	11.23	8.04	8.30	8.50		8.03	6.46	7.48		
55 Cyg	B3 Ia	VOIGT	1952	16250°	1.55	2.87	12.00	11.39	8.14	8.54	9.18			7.68	6.79	7.55	
55 Cyg	B3 Ia	ALLER**	1956b	17500°	2.30	2.90	12.00	11.18	8.41	8.63	8.98				7.46		
+ 33° 2642	B2	TRAVING	1962	18000°	1.72	2.34	12.00	11.05	7.86	7.79	8.04			6.81	7.17		

* JUGAKU, SARGENT and GREENSTEIN.

** This is a simple model atmosphere analysis; the T and $\log P_e$ values correspond to the level $\tau = 0.6$ in the model.

TABLE 24

Results of fine analysis of early type spectra

Star	Type	Author	Year	T_{eff}	log g	H	He	C	N	O	Ne	Mg	Al	Si	P	S
10 Lac	O9 V	TRAVING	1957	37450°	4.45	12.00	11.23	8.37	8.37	8.77	8.72	8.22	7.07	7.75		
τ Sco	B0 V	TRAVING	1955	32800°	4.8	12.00	11.23	8.37	8.57	9.12	8.72	7.73	6.58	7.95		
τ Sco	B0 V	AEJ*	1957	32800°	4.45	12.00	11.08	7.70	8.26	8.63	8.86	8.3	6.4	7.63		
γ Peg	B2 IV	AJ**	1959	24000°	3.8	12.00	11.17	8.58	8.01	8.63	8.73	7.95	5.76	7.03	5.50	7.80
3 Cen A	B4 IV p	HARDOP	1966	19350°	3.87	12.00	9.99	7.38	7.94	8.36	8.54	6.80		7.27	7.54	
ζ Per	B1 Ib	CAYREL	1958	25900°	3.6	12.00	11.31	8.26	8.31	9.03	8.61	7.76	6.78	7.96	7.48	

* ALLER, ELSTE, and JUGAKU.

** ALLER and JUGAKU.

Assumptions 3 and 4 of coarse analysis are continued. That is Saha's and Boltzmann's laws are used to calculate the ionization and excitation at each depth in the model and the lines are assumed to be broadened by thermal Doppler broadening and by damping. Thus the representation of the actual physical situation encountered in line formation is made considerably more realistic than in the case of coarse analysis, though it is still not physically exact. The assumptions which are made in this theory of line formation are a reasonable approximation for the formation of weak subordinate lines from high-lying levels which are close to the continuum. Many of the lines in B type spectra fall into this category.

The most serious limitation of the method of fine analysis is that because the method makes use of equivalent widths only, it is unable in the case of B stars to demonstrate clearly whether motions of the atoms are important in causing the observed equivalent widths. In principle it should be possible to separate the effects on the line strengths of abundance and of motion, but in practice so few lines are available in any one ionic spectrum and the range of strength which is encountered is so small that the separation cannot be made. Thus, although a set of internally consistent "abundances" can be determined by applying the methods of fine analysis, it is by no means certain that these numbers represent true relative abundances of the elements. Motions of the gas in the stellar atmosphere exceeding the thermal motions by a factor 2 to 6 can increase significantly the equivalent widths of most lines. This point has been demonstrated by ABHYANKAR (1964) who has calculated the curve of growth for an atmosphere in which the velocity changes linearly with depth and the absorption coefficient has a Gaussian shape.

Some results from the methods of fine analysis developed at Kiel and at other places are given in Table 24. Little can be said about these numbers. Like the results in Table 23, they give an internally consistent description of the stellar spectrum in terms of the theory which is used. What remains to be proved is that the theory has not omitted a factor which has a significant effect on the line strengths and which will lead to significantly different values of the parameters describing the stellar atmosphere.

3. Methods of analysis developed at Michigan and the results

Similar methods of fine analysis to those developed at Kiel were developed independently at the University of Michigan under the direction of L. H. Aller. The idea was to improve the representation of a stellar atmosphere by using a detailed model atmosphere and to develop rapid methods of computing the equivalent widths and curve of growth for lines formed in the model atmosphere. The techniques are based on the concept of the saturation function and they are described by ALLER (1960). These techniques, like those known as the method of weighting functions, do not permit one to handle blended lines. Careful attention is paid to making use of the most accurate available representation of the line absorption coefficients and of the partition functions. The degree of ionisation and excitation in the atmosphere is computed as though local thermodynamic equilibrium exists and the source function in frequencies within the lines is usually assumed to be identical with the Planck function, although the actual source function in the continuous spectrum as found from model atmosphere computations is also used in some cases.

This work culminated in the detailed analysis of the spectra of two sharp-lined B stars, τ Scorpii, B0V, (ALLER, ELSTE, and JUGAKU, 1957) and γ Pegasi, B2IV (ALLER and JUGAKU, 1959). New observational data were obtained in each case. The results are summarised in Table 24. The model for τ Scorpii is a revision of the model by TRAVING (1955), changes being made so as to obtain a better representation of the profile of Hγ and of the strengths of the strong lines of Si III and Si IV. A model for γ Pegasi was developed at the University of Michigan, and also the models of UNDERHILL (1956b) were used.

These studies confirm a discovery made by UNDERHILL (1956b) that lines from the spectra of all ions of the same element cannot be equally well represented by the same model atmosphere. There is difficulty in representing strong and weak lines at the same time. ALLER and JUGAKU (1959) discuss this problem in some detail for Si II, Si III, and Si IV in the case of γ Pegasi, and they note that other strong lines such as Mg II λ4481 are observed to be stronger than can be predicted by a model which predicts the weak lines reasonably well. They also discuss the problems raised by the lines of the He I spectrum (see Section 5).

4. Methods of analysis developed at Victoria and the results

The first significant results in the detailed study of early type spectra by Underhill were published in 1948 (UNDERHILL, 1948b). Here profiles and equivalent widths were given for some of the stronger lines in the spectra of six supergiant B stars (ρ Leo, χ^2 Ori, o^2 C Ma, χ Aur, η CMa, β Ori), one giant star (ε CMa) and three main-sequence stars (γ Peg, τ Her, ζ Dra), and an attempt was made to interpret these profiles using a very simple model atmosphere. It was shown that the simple concept of a single layer at one representative temperature and pressure and the use of the Minnaert interpolation formula were inadequate for representing line formation in B type atmospheres.

This work was followed by the development of techniques to compute detailed model atmospheres for O and B stars and to obtain numerical solutions of the Milne-Eddington transfer equation, so that the source function and radiation field could be calculated. Because the Milne-Eddington transfer equation includes the effects of coherent, isotropic scattering (by electrons) as well as absorption and re-emission as though the material were in thermodynamic equilibrium at the local temperature and pressure, an iterative process is required to solve for the source function. This section of the numerical work demands much computation and until high-speed electronic computers became available it essentially limited the number of models that could be investigated. Six model atmospheres were found by hand computing methods and the equivalent types of the models were estimated from the relative strength of a few spectral lines (UNDERHILL, 1950, 1951a, 1956b). In each case an analytical expression for the variation of temperature with depth was assumed, and the radiation field in the continuous spectrum was computed in detail in order to check that the models were close to a state of radiative equilibrium. It is assumed that the appropriate temperature law is that which will yield a state of radiative equilibrium in the model atmosphere.

The first model has a type near O9.5V according to the computed relative intensities of lines of C II and C III. It was used to predict the strength of the He II line λ4541, and an estimate was made of the helium abundance, see Section 5. The same He II line profile was also computed for the second model (UNDERHILL, 1951a) which has a type near O5. It is well known that the He II lines are observed to increase in strength from type O9 to O5, but the calculations showed that the predicted He II line in the O5 model was past maximum strength. The meaning of this result was not clear at the time it was obtained, for it is true that neither model is in strict radiative equilibrium and one could not say decisively how much this lack would affect the predicted line profile. However, in view of later results, this discrepancy between theory and observation is significant and it can be explained (as was noted in 1951) by postulating that the observed strong stellar line is formed chiefly in a halo or outer atmosphere which is not represented by the model atmosphere. Model atmospheres are generally constructed to represent well the layers in which the continuous spectrum of O and B stars and weak features such as the Stark broadened wings of Hγ or Hδ are formed.

Line profiles of Si II λ4130, Si III λ4552, and Si IV λ4088 were predicted by means of the four B type models (UNDERHILL, 1956b) and it was discovered that all three stages of ionization could not be represented equally well by one silicon abundance. If the abundance of Si was adjusted to give a reasonable representation of the Si II and Si IV lines, then the Si III line was predicted to be too weak. It seems likely that much of this difficulty is caused by dilution effects which lead to an extra strengthening of Si III λ4552 in spectral types where this line is strong (see Chapter VII). The computations clearly show that at spectral types near B2 and B1, the strong lines of Si III are formed in an outer halo which is not well represented by the models.

Since by hand computation it is not easy to iterate the models in order to obtain strict radiative equilibrium, nor to handle blended lines (the selected Si II and Si IV lines fall in the wing of Hδ), attention was next turned to the problem of developing computing programs for making detailed model atmospheres in radiative equilibrium and for computing theoretical spectra allowing for blends and for moderately complex forms of the line absorption coefficient. The radiation transfer problem both in the continuous spectrum and in the lines is represented by the Milne-Eddington equation. That is, radiation is permitted to interact with material by coherent, isotropic scattering and by absorption and re-emission as though thermodynamic equilibrium exists at the local temperature and pressure. Line profiles are computed point by point, and the equivalent widths of the lines are found by plotting the profiles and measuring their area with a planimeter just as is done with observed spectra. One can readily handle problems of line blending by adding all contributions to the monochromatic absorption coefficient in contrast to the case when the method of weighting functions or the saturation function is used to find the equivalent widths. Furthermore, the program can be modified to introduce an extra Doppler broadening equivalent to the effect which is known as microturbulence. The methods have been described in detail by UNDERHILL (1962a, 1962b) and some models and theoretical line spectra have been computed.

The first results showed that a series of model atmospheres which represent main-sequence B stars well so far as the continuous spectrum is concerned, also produce a hydrogen line spectrum like that which is observed. No detailed comparison with particular stars is possible because a suitable series of accurate observed H line profiles does not exist at this time. The equivalent spectral types of the models are estimated from the computed Balmer jump (no line absorption) using the empirical relationship between spectral type and D of CHALONGE and DIVAN (1952). The effective temperatures of these models and their bolometric corrections can be computed from the standard definitions. The models are listed in Table 25. The temperature at a level near $\tau=0.1$ is also listed, for this temperature is characteristic of the level where weak and moderate strength lines are formed. A similar, but more extensive series of models has been published by MIHALAS (1965); STROM (1964) has also made a series of B type models.

All the models in Table 25 except Model 73 are close to a state of radiative equilibrium throughout the levels important for line formation. The departures from strict radiative equilibrium are a little larger than desirable in Model 73. Each model was forced to a fixed boundary temperature. Revised, unpublished models are available in which the boundary condition has been relaxed. If the temperature law is to

TABLE 25
A series of model atmospheres

Model No.	log g	T_{eff} degrees	B.C. mag.	Sp. Type	$T(\tau \sim 0.1)$ degrees	Reference*
73	4.00	12690	− 0.92	B9	10000	*P.D.A.O.* **11**, No. 23
66	4.00	15333	− 1.37	B6	12000	*P.D.A.O.* **11**, No. 23
55	4.00	19215	− 1.91	B3	15000	*P.D.A.O.* **11**, No. 23
63	4.00	25673	− 2.59	B2	20000	*P.D.A.O.* **11**, No. 23
89	4.00	27820	− 2.78	B1.5	22000	*B.A.N.* **17**, 161
97	4.00	30027	− 2.93	B0.5	24000	*B.A.N.* **17**, 161
64	4.00	31023	− 2.96	B0	25200	*P.D.A.O.* **11**, No. 23
98	4.00	33947	− 3.14	O9	30000	*B.A.N.* **17**, 161
119	3.70	14870	− 1.28	B7	12000	unpublished
148	3.70	25897	− 2.61	B2	20000	unpublished
143	3.70	27777	− 2.74	B1	22000	unpublished
146	3.70	29194	− 2.81	B0	24000	unpublished
145	3.70	30355	− 2.87	O9.5	25200	unpublished
140	3.70	33411	− 3.07	O8	30000	unpublished

* The abbreviations in this column must be read as follows:
P.D.A.O. *Publ. Dominion Astrophys. Obs.*
B.A.N. *Bull. Astron. Inst. Neth.*

be determined at very small values of τ ($\tau<0.01$) by forcing the condition of radiative equilibrium, the calculation of the radiation field must be carried out to great accuracy. Mihalas and Strom have made use of elegant computing techniques which do this.

The models in Table 25 are composed of hydrogen and helium, the relative abundances by weight being $X=0.68$ and $Y=0.32$. The only sources of opacity that are considered are absorption in the continua of H, He I, He II, and H^- and electron scattering. No account is taken of the blanketing effect by the lines, in particular in the spectral region between 911.6 Å and 1500 Å. Line blanketing has the effect to reducing the effective temperature by some $2500° \pm 500°$ and of reducing the bolometric correction by approximately 0.4 mag from the values for an unblanketed model which has about the same temperature near $\tau=0.1$. Investigations are at present underway to determine whether line-blanketed models produce a theoretical spectrum in the usually observed spectral region which differs significantly from the spectrum of an unblanketed model.

The problem of the relative strengths of the Si II, Si III, and Si IV lines was re-examined with the series of models given in Table 25, and it was confirmed (UNDERHILL, 1963a) that with a reasonable value of the silicon abundance, the models would predict line strengths like those observed in B stars when the lines are weak, but at the spectral types where the lines (Si II, Si III, or Si IV) approach maximum strength the predicted lines are too weak. Since the lines lie on the flat part of the curve of growth when they are strong, it is impossible to obtain the observed strengths by a reasonable adjustment in the abundance. The numerical work indicates that when the lines are strong, they are formed chiefly in the outermost part of the atmosphere, in a part which is not really well defined by putting the constraint radiative equilibrium on the continuous spectrum as is done in all methods of model computation. It is an open question how best to define the physical conditions in this part of the model atmosphere.

The problem of line formation was further investigated by comparing profiles of lines in the spectrum of 10 Lacertae with those predicted from Model 98 (UNDERHILL and DE GROOT, 1964; DE GROOT and UNDERHILL, 1963) and it was discovered that the observed line profiles and equivalent widths of medium and strong lines can only be represented by introducing microturbulence. It was found that the most probable velocity which gives the best agreement with observation is four times the thermal most probable velocity. Thus microturbulence is present in the atmosphere of 10 Lacertae. Since most of the lines which are used to estimate abundances in early type stars fall on the flat part of the curve of growth, the introduction of turbulence causes a significant reduction in the estimated abundances.

A sample of the results of this analysis is shown in Figure 37. The observed profiles of six lines of multiplet 2 of Ne II in the spectrum of 10 Lacertae are compared with the computed profiles from Model 98 in the case that the lines are broadened by microturbulence. The observed line at 3311.30 Å is blended with a line of Ar III. The theoretical curves of growth for this Ne II multiplet are shown in Figure 38. The considerable increase in equivalent width which results from the introduction of

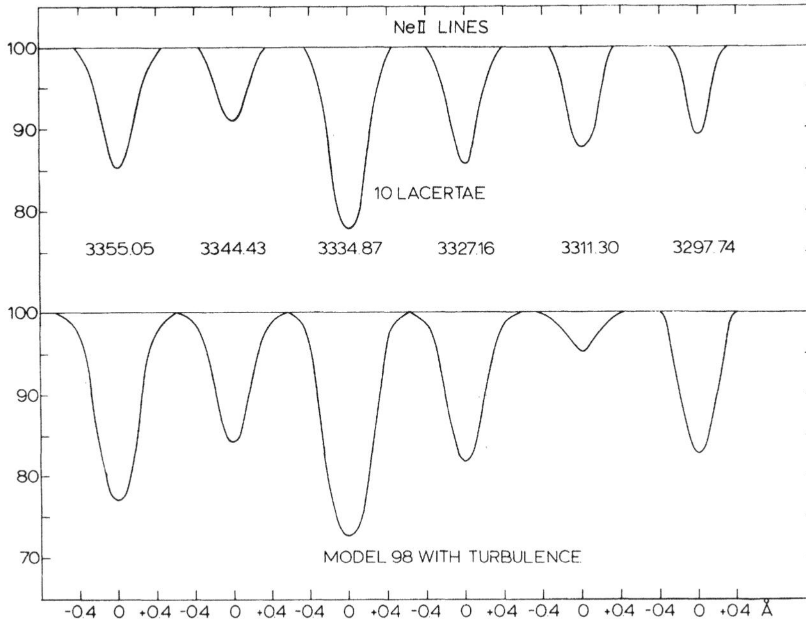

Fig. 37. Observed Ne II line profiles in 10 Lacertae compared with computed profiles.

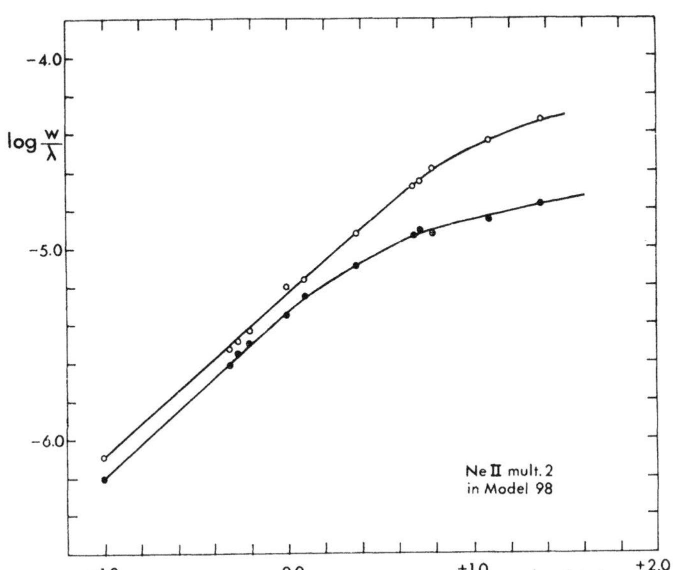

Fig. 38. The theoretical curve of growth for multiplet 2 of Ne II in Model 98. The lower curve is for thermal Doppler broadening plus damping; the upper curve is for the case of broadening by microturbulence.

microturbulence leads to a significant raising of the flat part of the curve of growth. Thereby the neon abundance is reduced from what would have been estimated, were no allowance made for the presence of microturbulence.

The following relative abundances have been found for 10 Lacertae:

Element	Fractional abundance by weight	log N (scale log H = 12.00)
O	$<9.7 \times 10^{-3}$	<8.78
Ne	3.4×10^{-3}	8.23
Mg	$>3.3 \times 10^{-4}$	>7.15
Si	1.21×10^{-3}	7.64

These values tend to be smaller than those found by Traving, as may be expected from his neglect of microturbulence.

It is quite clear from this work that the state of motion in the outermost layers of O and B stars has much to do with the apparent strength of the readily observed lines. This factor should not be left out of the analysis. The cores of the stronger lines of γ Pegasi are also not really sharp, not so sharp as thermal motions alone would produce, and it is suspected that in this case also the abundance estimates may be in error due to the neglect of turbulent broadening.

The results of comparisons between the observed spectra of sharp-lined main-sequence O and B type stars with detailed theoretical spectra computed from model atmospheres have indicated that the present models represent well weak features such as the continuous spectrum, weak lines of all elements, and the Stark broadened wings of H and He II lines, but that the theoretical representations fail when attempts are made to predict the strong lines. The strong lines are observed to be stronger than predicted. The observations can be reconciled with theory by introducing the concept of microturbulence. This procedure is, however, only a mathematical device. One of the chief problems facing astrophysicists at this time is to discover the origin of the field of motion which appears to control the strength of most of the lines which are used for classifying early type spectra and to find its effect on the temperature, pressure, and state of ionization and excitation of the outermost layers of O and B stars.

Qualitatively it is easy to see that the interpretation of O type spectra in particular is not a simple thing, for almost all of the lines which are observed are strong lines, and consequently are formed in the turbulent outer atmosphere. Weak lines, in the theoretical sense of $l_\nu/\kappa_\nu < 1$ over all parts of the line profile, are not readily observed because the line is then too shallow for the profile to be determined accurately by the usual techniques. One can expect great variety in O type spectra, for thermodynamic and hydrostatic equilibrium do not give a good first estimate of the physical state of the atmosphere. Rather account must be taken of the individual events which occur in the process of line formation in each case. The fact that many O stars show Of emission (N III λ4640 and He II λ4686) is supporting evidence for this conclusion.

5. The helium abundance

The characteristic feature of B type spectra is the presence of lines of He I. In O type stars lines of He II appear as well as lines of He I. Thus it may be expected that a quantitative analysis of the strengths of the He I and He II lines in O and B type spectra will yield information about the relative abundance of helium in the atmospheric layers of O and B stars. Conclusions about the total content of helium in stars can only be drawn by combining the results from spectrum analysis with deductions concerning the energy generation of stars, the internal structure of stars, and the amount of mixing that may take place between the surface layers of a star and the interior. Here we shall attempt to summarise the information about the helium abundance which is available from the study of O and B type spectra.

In normal stars helium is the second most abundant element. Consequently the helium lines are strong lines. This raises serious problems. If it is desired to carry through an exact analysis of a stellar spectrum and to determine accurately the helium abundance from the observed strengths of lines due to helium, one must have reliable models of the atmospheric levels where the lines are formed. However, the presently available models of B and O type atmospheres are not very useful, for they represent reliably only those layers of the stellar atmosphere where weak features such as the continuous spectrum are formed. In particular it appears (see Section 4) that the state of motion of the gas in the outermost layers of O and B type stars has much to do with determining the observed strengths and profiles of strong lines. Very little is known about these motions at this time.

A second problem arises because the He I and the He II lines are sensitive to Stark effect. In particular it is a difficult problem in theoretical physics to determine accurately the shape of the line-absorption coefficient of the He I lines as a function of the temperature and electron pressure in the stellar atmosphere. GRIEM *et al.* (1962) have given solutions for certain isolated He I lines. UNSÖLD (1944a) has indicated how to treat many of the lines in a more approximate manner. There is no satisfactory theory completely worked out for some of the strongest He I lines such as λ4471 and λ4026 which are used for classification purposes, although in principle the theory could be developed. One is therefore restricted to the analysis of the weaker lines of He I. Adequate observations of line shape and equivalent width of such lines exist only for one or two stars. Since the He II spectrum is hydrogen-like, the theory of the Stark broadened line absorption coefficient of He II is, in principle, more tractable. One is restricted to the analysis of lines of the 4–n series, with n odd, since it is known that the 3–4 line, λ4686, frequently appears in emission owing to a monochromatic fluorescent effect, and the 4–n lines with n even blend with the Balmer lines of hydrogen.

A number of estimates of the helium abundance from the observed strengths of the He I lines have been made by the method of coarse analysis which was developed by Unsöld. These results have been summarised by UNDERHILL (1953a). The derived relative abundance by number of hydrogen to helium runs from 3.2 to 27.5, the smaller ratios being found from analysis of the spectra of supergiant stars. These results are only useful in confirming the conclusion that in normal O and B spectra helium is the second most abundant element. With the method of coarse analysis it is impossible to take account of the fact that with strong lines the cores of the line profiles are formed in parts of the atmosphere where the temperature and pressure and state of motion are entirely different from the parts where the wings are formed. Furthermore, these analyses make use of equivalent widths only and little or no attempt is made to check whether the adopted model and theory of line formation does produce lines shaped like those which are observed. The derived abundance is a fitting factor to make the (often uncertain) observed equivalent widths agree with the predicted equivalent widths. These methods have been applied to stars of type B3 and earlier.

With the development of methods of fine analysis at Kiel and at Michigan, the problem of the helium abundance was once more investigated, and again attempts were made to find that helium abundance which would give the best representation of the observed equivalent widths of the He I lines. TRAVING (1955, 1957) in his analyses of the spectra of τ Scorpii and of 10 Lacertae found he could justify the helium abundance originally found by UNSÖLD (1942a).

ALLER and JUGAKU (1959) in their analysis of the spectrum of γ Pegasi discussed the problem presented by the He I lines in considerable detail and they showed that the theory and models in use at that time were quite unsatisfactory for interpreting many of the He I lines and they drew attention to the fact that the strongest and best known He I lines are largely formed in the outermost layers of the stars. Furthermore, they noted that the same line observed in different stars tends to give a similar abundance, but that the resulting abundances within a given star derived from a group of lines differ among themselves more than should be the case.

JUGAKU (1959) has made a particular study of the ratio of helium to hydrogen in the spectrum of τ Scorpii. In addition to providing a list of equivalent widths of He I lines lying between 3354.55 Å and 6678.15 Å and profiles for many of these lines, he gives equivalent widths for He II λ5411, λ4686, λ4541, and λ4200 and the profile of He II λ4686. Jugaku uses a model atmosphere derived from that of TRAVING (1955) and he uses the method of the saturation function in a manner similar to that developed by ALLER, ELSTE, and JUGAKU (1957).

Jugaku finds that in τ Scorpii the He I lines are formed high in the atmosphere and that since at such levels helium is almost completely singly ionised at the relevant temperatures and pressures, the He I lines cannot be used for temperature determination in early B type stars. He discusses the stronger He I lines individually and concludes that the differences in the helium abundance found from the various lines are due to an inadequate understanding of the mechanism of line formation rather than to the uncertainty of the observed intensities. The possibility that motions in the atmosphere (microturbulence) may affect the observed line profiles is not taken into account in Jugaku's study. From his study of the He I lines Jugaku concludes that the ratio $N(\mathrm{H})/N(\mathrm{He})$ is about 9 in τ Scorpii.

Finally Jugaku calculates the profile of He II λ4686 using the hydrogen-like theory of Stark broadening and he notes that with this line the uncertainties due to the lack of exact knowledge of the temperature at great depths in the model make it impossible to decide between a H/He ratio by number of 6:1 or 20:1.

The first attempt to determine the helium abundance from the strength of the He II lines was made by UNDERHILL (1951b). Here the Stark broadening of λ4541 was estimated by the Holtzmark-Verwey theory and the computed line profiles from models representing O9.5 and O5 stars were compared with observed profiles. It was concluded that the ratio $N(\mathrm{H})/N(\mathrm{He})$ could not be close to 6:1 as suggested by UNSÖLD (1942a), but that it might be close to 20:1. In this work no account was taken of possible broadening of the lines by motions due to thermal Doppler effect or to microturbulence. Such broadening would lead to a smaller helium abundance. This work has been criticised by TRAVING (1955, 1957) in particular on the grounds that the temperature law in the adopted models was not well enough determined and that the computed equivalent width of the line λ4541 is so sensitive to the structure of the model, particularly in the deeper layers where the Stark broadened wings are formed, that the derived small helium abundance was meaningless. JUGAKU (1959) has also drawn attention to this weakness of that first attempt to estimate the helium abundance from the observed strength of the He II lines in O type spectra.

Although these criticisms have a certain foundation, they miss the essential point which has become abundantly clear as a result of the analysis (Section 4) of the spectrum of 10 Lacertae by Underhill and de Groot using a series of well determined model atmospheres. The point is that, as JUGAKU (1959) remarked, even in stars as cool as τ Scorpii (B0V), helium is essentially entirely singly ionised in the outer part of the atmosphere. Thus there is a great abundance of ions to absorb the lines of the 4–n series of He II. Even the high excitation potential of 50 volts of the $n=4$ level does not reduce the population of excited ions too severely. Thus the cores of the lines of the 4–n series of He II are formed in the outer, extended atmospheres of the O stars. Here considerable motions, which may be described as microturbulence, exist. The result is that the observed He II lines have rather strong and comparatively deep cores. These are shown in Figure 39, left side, for the case of 10 Lacertae. Somewhat similar profiles with strong wide cores occur in λ Orionis, O8, and in AE Aurigae, O9 V (UNDERHILL, 1961). The computed profiles for Model 98, which is a reasonable match for the deeper layers of 10 Lacertae, are shown on the right side of Figure 39. These lines were computed with the best available theory of the Stark broadening of hydrogen-like lines, but no broadening due to thermal Doppler effect or to microturbulence is included. The match between theory and observation in the wings of the profiles is reasonably good. One cannot match the cores. Increasing the helium abundance over the adopted ratio of H/He equal to 8.5:1 by number would not solve the difficulty. Clearly, in order to represent the profiles of the He II lines in O stars one should make a model with an extended halo or shell in which considerable turbulence occurs. At the present time it seems best to conclude that in normal O and B stars the relative abundance by number of hydrogen to helium lies between 10:1 and 20:1.

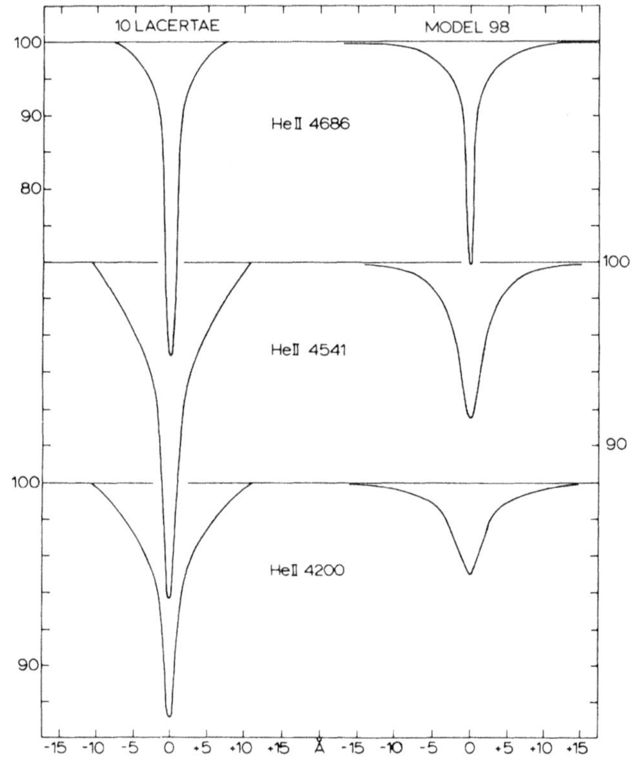

Fig. 39. Observed and computed He II lines in the spectrum of 10 Lacertae.

MIHALAS (1964) has attempted to estimate the helium abundance by fitting equivalent widths of He II lines observed in O stars to the values computed from a series of reliable model atmospheres, and he concludes that the ratio $N(\mathrm{H})/N(\mathrm{He})$ lies in the range 5:1 to 7:1. By using equivalent widths *only* he must arrive at a large helium abundance, for he has insufficient information to make clear the fact that the He II lines in O stars have cores which are dominantly broadened by microturbulence and that these cores account for a large part of the measured equivalent width. Mihalas does examine this point for 10 Lacertae and confirms the suggestion of Underhill and de Groot that microturbulence exists, but he does not go further and clarify the situation as regards the estimation of the helium abundance using the equivalent widths of He II lines in O stars.

At this moment the situation with regard to obtaining an accurate estimate of the abundance of helium from the strength of the He I and He II lines in O and B type spectra is unsatisfactory. It is essential to be able to define accurately the conditions in the outermost parts of the stellar atmospheres and to take into account the broadening effects of motion on the line strengths as well as departures from equilibrium populations in the He I levels in particular (dilution effects) before comparing observed and predicted line profiles and equivalent widths and deducing from the fit an abundance. It is more probable that the helium abundance will be reduced from the estimates that are given in Tables 23 and 24 than that it will be increased. It might be advantageous to study the He I lines in late B and early A type stars, for here the lines are weak. However, a good theory of the Stark broadening of the most easily observed lines such as λ4471 and λ4026 would be required, and one would need reliable model atmospheres in which the possible effects of convection were taken into account.

6. Other model atmosphere studies of early type stars

The studies reviewed in the foregoing sections contain the chief part of the information that is available from attempts to interpret in detail the spectra of normal O and B type stars. RUDKJØBING (1947) constructed the first model atmosphere for an early type star and he emphasised the great importance of taking account of the large variations in the value of the absorption coefficient which occur at different wave lengths. PECKER (1950) developed the first models corresponding to about type B1 in which the condition of radiative equilibrium was rigorously enforced in the outer layers of the star in order to establish securely the appropriate temperature law. He demonstrated that in the early B type stars the boundary temperature was depressed greatly with respect to the effective temperature from what was expected according to the analytical theory of radiative transfer in a grey atmosphere.

Further model atmospheres for B type stars, constructed by hand-computing techniques, according to the classical principles, have been obtained by MCDONALD (1953) and by SAITO (1954, 1956). The advent of the high-speed computer has relegated all the hand-computed models to a place in history. The comprehensive series of models by MIHALAS (1965) and by STROM (1964) are superior in that they meet the selected restraints on the problem more accurately than was possible when the computing was done by hand and they provide a truly homogeneous set of model atmospheres. These newer series of models, however, introduce few new physical ideas; rather, they greatly improve the techniques of numerical analysis.

NEVEN and DE JAGER (1954) attempted to find the appropriate temperature law in the outer regions of B type atmospheres by analysing the apparent central intensities of the hydrogen lines in the stars τ Scorpii, B0V, δ Ceti, B2IV, γ Pegasi, B2IV, and ι Herculis, B3V. By doing this they found a temperature law for values of $\tau<0.4$. At greater depths they continued their models by assuming the simple relation

$$T^4(\tau) = \tfrac{1}{2} T_{\mathrm{eff}}^4 (1 + \tfrac{3}{2}\tau). \tag{XI-4}$$

The pressure at each layer was found by requiring that the model be in hydrostatic equilibrium. With these models, Neven and de Jager used the method of weighting functions to interpret the observed strengths of the He I lines and to find the helium abundance. Because the temperatures which they adopt for the outer layers of their models are lower than the values which are found by the usual methods, Neven and de Jager arrived at a lower helium abundance than was found by the Kiel group, for instance. (At the spectral types under consideration, helium is nearly completely once-ionized in the outer part of the stellar atmosphere. Consequently, lowering the temperature makes more neutral helium atoms available for absorbing the lines of the He I spectrum.) The average result of Neven and de Jager is $N(\mathrm{H})/N(\mathrm{He})=18$. This result demonstrates how important it is to determine accurately the exact physical conditions in the outermost layers of stars of type B3 and earlier if one is to find an accurate value for the helium abundance. Neven and de Jager made no attempt to see if their models, and the models they developed later (NEVEN and DE JAGER, 1957), satisfy the condition of radiative equilibrium.

CAYREL (1958) has studied the spectrum of the supergiant ζ Persei, B1Ib, in detail and he has constructed a model to represent this stellar atmosphere. The model is in radiative and mechanical equilibrium (turbulent pressure being included) and it is shown that the model reproduces some of the observed profiles of the hydrogen lines. Cayrel then calculated line profiles and equivalent widths for lines of He I, C II, O II, N II, Ne II, Mg II, Al III, Si II, Si III, Si IV, and S III, using a somewhat simplified theory of line formation, and derived a set of abundances similar to that found by Traving for τ Scorpii (see Table 24).

The model atmospheres which are available to represent the atmospheres of O and B stars can be used to predict the line spectrum in the ultraviolet spectral region which is not accessible from the surface of the earth. MORTON (1965) has shown that at wave lengths shorter than about 1500 Å many very deep, very broad absorption lines can be expected in the spectrum of a B2 star. Most of the lines falling in this spectral region are resonance lines or lines from low lying levels. Thus, it is to be expected that these lines will be formed very high in the stellar atmosphere. In fact, the theory of radiative transfer and line formation which has been used in this first survey is not a good means of representing the formation of these lines. A more complex theory taking account of non-coherent scattering and of departures from local thermodynamic equilibrium would be more appropriate. Nevertheless, Morton's work clearly demonstrates that very strong absorption lines can be expected in the far ultraviolet spectrum of O and B stars. GUILLAUME, VAN RENSBERGEN, and UNDERHILL (1965) have surveyed the type of line which may be expected in the spectral region between 2000 Å and 3000 Å. Here the theory of line formation which is used can be better justified. It is found that lines from the third spectra of the metals (Cr III, Ti III, etc.) will be strong and deep, and that lines from the second spectra of the metals will be weaker, though not negligible. There will also be a few very strong lines of He I, C II, C III, and Si III in this spectral range. If microturbulence is present in O and B type atmospheres, as seems likely from the analysis reported in Section 4, the lines in the region 2000 Å to 3000 Å will be wide and strong. In fact, the spectrum of a B star between 2000 Å and 3000 Å should resemble that of an F star in the ordinary spectral range, consisting of a few strong, deep lines and many lines of moderate strength.

A good many model atmospheres for B type stars now exist. It is extremely difficult to make a choice between them for representing any single star when only spectral features observed between 3050 Å and 6700 Å are used to relate the model to the star. It is to be hoped that detailed observations of the line spectrum of O and B stars at wave lengths shorter than 3000 Å will help to solve this problem. The preliminary calculations of the ultraviolet spectra of B stars indicate that it will be very difficult to interpret broad or narrow band colour observations in terms of the physical structure of the stellar atmosphere because the amount of line blanketing, or possibly line emission, in any spectral range will depend sensitively upon the adopted temperatures, pressures, and mode of line formation used to derive the model. The existing theories and observations at this time do not permit a sufficiently precise separation among the many possibilities that exist for constructing model atmospheres and predicting theoretical spectra. It is very probable that O and B stars which have been given the same spectral type from a study of the spectrum between 3900 Å and 5000 Å will have rather different ultraviolet spectra.

CHAPTER XII

APPARENTLY UNUSUAL ABUNDANCES IN EARLY TYPE STARS

1. The helium stars

There are five early type stars known in which the hydrogen lines are very weak or absent and in which the helium lines are strong. All these stars are apparently faint and three are at unfavourable declinations for observation from the northern hemisphere. They are listed in Table 26 together with some general information about

TABLE 26
The Helium stars

Star	Approx. sp. type	α (1900)	δ (1900)	m_V	b^{II}	Discovery paper
BD + 10° 2179	B3p	$10^h\ 33^m.7$	+ 10° 34′	10.0	+ 54°	KLEMOLA (1961)
HD 124448	B3p	14 08.6	− 45 49	10.1	+ 14°	POPPER (1942)
BD + 13° 3224	B1p	16 43.5	+ 13 27	10.5	+ 33°	BERGER and GREENSTEIN (1963)
HD 160641	Op	17 36.0	− 17 51	9.8	+ 6°	BIDELMAN (1952)
HD 168478	B5p	18 14.7	− 70 36	8.9	− 23°	THACKERAY and WESSELINK (1952)

each. A detailed analysis of the spectrum of HD 168476, of HD 124448, and of BD+10°2179 has been made by HILL (1964, 1965) who used high-dispersion spectrograms obtained at the Radcliffe Observatory and the equivalent widths measured by KLEMOLA (1961) from Lick coudé spectrograms. A little quantitative information from high-dispersion spectrograms has been given by BERGER and GREENSTEIN (1963) about BD +13°3224. ALLER (1954) has given some information about HD 160641.

The four helium stars HD 124448, HD 160641, HD 168476, and BD +10°2179 are high velocity stars, their radial velocities being respectively -66 ± 2 km/sec, about 100 km/sec, -165.0 ± 0.8 km/sec, and $+155 \pm 9$ km/sec. The radial velocity of BD +13°3224 is, however, about +15 km/sec. These stars do not appear to be particularly luminous and the forbidden He I lines near λ4471 and λ4026 may be seen on some spectra. However, the forbidden He I lines are not conspicuous, and lines of the $2P^\circ - nP^\circ$ series have not been reported. These lines are visible in normal B type main-sequence stars such as γ Pegasi. BERGER and GREENSTEIN (1963) show that the profile of He I λ4471 in BD +13°3224 is quite symmetrical with no obvious blend due to [He I] λ4469. They have suggested that BD +13°3224 is a subdwarf.

HILL (1965) has estimated that log N_e is 14.06±0.10 in HD 124448, 14.04±0.17 in BD +10°2179, and 13.32±0.30 in HD 168476. The observations of Hill and of THACKERAY (1954) of line intensities in the spectrum of HD 168476 can best be interpreted by concluding that this star is surrounded by a shell which produces lines from the second spectra of the metals of greater strength than is normally associated with a B5 star. Because of the rather high galactic latitudes of most of the helium stars, the equivalent widths of the interstellar Ca II lines are not a secure indication of the distance of each star. Nevertheless, a minimum distance may be estimated. It seems unlikely that the helium stars (except BD +13°3224) are so close by that one must deduce that they are significantly less luminous than normal main-sequence B stars of about the same spectral type. The colours of the helium stars are similar to those of normal, unreddened B stars of a type equivalent to the line spectrum, excluding the hydrogen lines.

The hydrogen lines are exceedingly weak in all of these stars and no hydrogen emission lines are seen. BIDELMAN (1952) has noted that in HD 160641 the wave lengths of the lines visible near the positions of the Balmer lines of hydrogen correspond to the wave lengths of lines of the 4–n series of He II. Thus he concludes that the hydrogen lines are very weak or absent in HD 160641. POPPER (1946) has estimated that the equivalent widths of Hγ and Hδ in the spectrum of HD 124448 cannot exceed 0.2 Å. In a following paper (POPPER, 1947) he demonstrated that the limit of the He I $2^3P°–n^3D$ series is visible at 3420 Å. KLEMOLA (1961) has noted that Hγ and Hδ are present as very weak lines in the spectrum of BD +10°2179. The line Hε is seriously blended with Ca II λ3968; H8 and H9 may be present but lost in blends with other lines. Lines H10 and higher and the Balmer jump cannot be detected. THACKERAY (1954) was unable to detect any hydrogen lines in the spectrum of HD 168478. He therefore concluded that the equivalent widths of the Balmer lines are probably less than 0.15 Å. HILL (1964) also was unable to detect hydrogen lines in the spectrum of HD 168478 and in the spectrum of HD 124448, thus sharpening the conclusion of Popper. BERGER and GREENSTEIN (1963) find that Hγ and Hδ are stronger in BD +13°3224, the equivalent width being about 0.6 Å.

The absorption lines from the light elements in the spectra of the helium stars are about twice as strong as in normal B type stars of approximately the same spectral type. This strengthening probably occurs because, owing to the absence of hydrogen, the continuous opacity in the atmospheres of the helium stars is very small. The apparent strength of an absorption line depends upon the value of the ratio l_ν/κ_ν where l_ν is the line absorption coefficient and κ_ν is the continuous absorption coefficient at the frequency ν in the spectrum. If κ_ν is small, a stronger line will result for a given value of l_ν. The lines are not very sharp. HILL (1965) has shown that the turbulent velocity is near 14 km/sec in HD 160641 and in HD 168476, but near 7 km/sec in HD 124448 and in BD +10°2179.

In the helium stars which resemble B type stars, lines of C II, N II, Ne I, Si II, Si III, Ca II and Fe III are seen. HILL (1965) finds evidence that carbon and possible neon and nitrogen are overabundant. The lines of O II are unusually weak. However, O II lines in normal B type spectra only begin to be moderately strong at spectral types B2 and earlier. The usually observed O II lines have lower excitation potentials of 22 to 24 volts, values rather higher than the lower excitation potentials for the other lines observed in the B type helium stars. It is possible that the abundance of oxygen is low in helium stars, but it is also possible that the rate of excitation to the O II levels is reduced from that in normal B type stars owing to extra strong He I lines of the 1^1S–$n^1P°$ series of He I which coalesce to the primary He I ionisation limit at 504 Å, or 24.58 volts. Neither C II nor C III lines are visible in the spectrum of BD +13°3224.

BIDELMAN (1952) has given a description of the spectrum of HD 160641. This star is a hydrogen-deficient O star. Weak lines of O II and O III are present as well as strong lines of He I, Ne II, C III, N III, and Si IV. Bidelman notes that C III λ4650 is particularly strong in absorption. It is possible that this strengthening is a dilution effect (see Chapter VII) occurring in an extended atmosphere around the star. However, no emission features of the Of type are seen. The absence of Of emissions indicates that He II λ303 is not strongly in emission. Furthermore no strong dilution effects have been reported in the He I spectrum (ALLER, 1954).

Attention has been directed to HD 37479, σ Orionis E (BERGER, 1956; GREENSTEIN and WALLERSTEIN, 1958; WALLERSTEIN, 1959; CHADEAU, 1960) and to HD 96446 (COWLEY and ALLER, 1962; BUSCOMBE, 1965) as early type stars which are helium rich. However, these stars are not similar to the stars given in Table 26, for the hydrogen lines are rather strong in both stars. Certainly the hydrogen lines are not absent or nearly absent. Buscombe suggests that the spectral type of HD 96446 is B1V, but the lines which he lists are very like what are found in the spectrum of 10 Lacertae (UNDERHILL and DE GROOT, 1965). It seems probable that this star is a late O type star with a shell spectrum. The radial velccity is $+6 \pm 0.6$ km/sec. The spectrum of HD 37479 deserves further analysis and a detailed radial-velocity study over a period of some months. This star has rather diffuse lines, presumably due to rapid rotation, and it will be difficult to make a detailed analysis such as has been done for the true helium stars, for it will be very difficult to detect and measure many of the lines from the light elements and the metals. A few stars of spectral type A and later appear also to have unusually large helium abundances.

2. Stars with strong lines of P II and P III

The discovery by BIDELMAN (1960a) that the spectrum of the B type star 3 Centauri A is quite unusual in that it has very sharp lines and the lines of P II and P III are unusually strong, has focussed attention on a group of B type stars with rather strange spectra. These stars have sharp lines and many show strong lines of P II as well as other lines from the metals that are not usually seen in B type spectra. BIDELMAN (1960b) has drawn attention also to the strong phosphorus lines in the spectrum of κ Cancri, B8p, which in addition shows unusually strong lines of Mn II. A short section of spectrum from these two stars is shown in Figure 40. This illustration is due to Bidelman. The extreme sharpness of the absorption lines is evident, for the Mg II

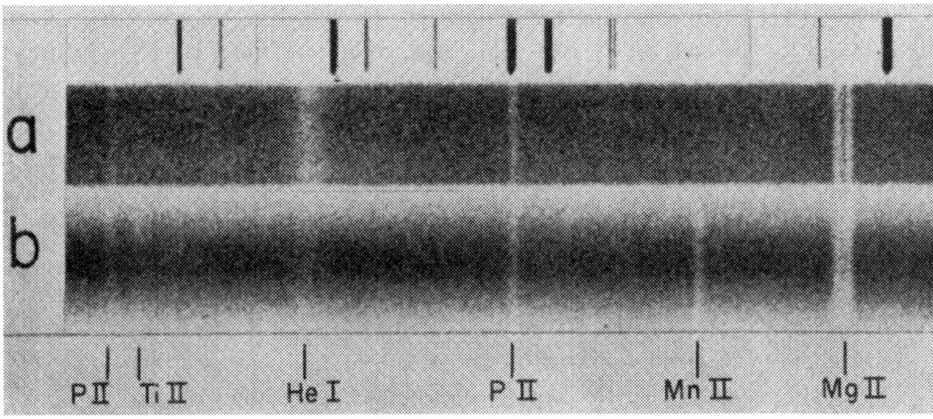

Fig. 40. Spectra of 3 Centauri and of κ Cancri in the neighbourhood of He I λ4471 and of Mg II λ 4481; (a) is 3 Centauri A, (b) is κ Cancri.

doublet is resolved although the separation of the lines is only 0.2 Å. The P II lines are strong in both stars and the Mn II line is unusually strong in κ Cancri for type B8. According to Bidelman, Babcock (unpublished) has been unable to detect a magnetic field for 3 Centauri A; however, κ Cancri is known to have a magnetic field.

There are a number of strong lines in the spectra of 3 Centauri A, κ Cancri, and of 112 Herculis, B9p, which BIDELMAN and CORLISS (1962) have identified as due to Ga II. Bidelman and Corliss suggested that all stars which have abnormally strong P II lines also have unusually strong lines due to Ga II. JASCHEK, JASCHEK, and LAVAGNINO (1963) have examined the spectra of nine B8p and B9p stars of the so-called Manganese class at high dispersion and they find that Ga II lines are found in the spectra of eight of the stars. They suggest that the ninth star is not truly a "manganese star". They have also looked for lines of P II in these spectra and they find that there is no correlation between the intensity of the Ga II lines and the intensity of the P II lines. Some stars exist with strong Ga II but weak or absent P II lines, while in other stars both Ga II and P II are strong. The Ga II lines are found to be absent from the A0p stars in which lines of silicon are unusually strong.

JUGAKU, SARGENT, and GREENSTEIN (1961) have made a detailed study of the spectrum of 3 Centauri A from six Palomar coudé spectrograms having dispersions between 4.5 Å/mm and 13.5 Å/mm. They measured the equivalent widths of 200 lines between 3700 Å and 8600 Å and made a curve of growth comparison with the spectrum of γ Pegasi, B2IV. It is unfortunate that there is not a curve of growth analysis of a B5 star available at this time for comparison with 3 Centauri A, for the atmospheric temperatures and pressures in a B2 star differ rather considerably from those in a B5 star. Jugaku, Sargent, and Greenstein adopt an effective temperature of 17000° for 3 Centauri A and find a compatible ionisation temperature and electron pressure from the Saha equation for Si II/Si III and for P II/P III. The final values are $T_{\rm ion} = 14800°$ and $\log P_e = 2.27$.

Lines from the following spectra are detected: H, He I, C II, N II, O I, O II (?), Ne I, Mg II, Si II, Si III, P II, P III, S II (?), Ar II, Ca II, Ni II, Fe II, Fe III, Ga II and Kr II. SARGENT and JUGAKU (1961) have reported that the He I lines are shifted in such a way as to suggest the presence of a large quantity of He^3. They suggest that the ratio He^3/He^4 lies between 5:1 and 16:1 and they discuss various possible ways of forming such a large abundance of He^3 in the atmosphere of 3 Centauri A. Sargent and Jugaku have also studied the profile of Hγ in an attempt to detect deuterium. They find an upper limit of $D/H \leqslant 0.01$.

The results from the comparison with the spectrum of γ Pegasi can be summarised as follows:

1. Helium is deficient by a factor of about 6.
2. The abundances of C, Ne, Mg, Si, Ca, and Ar are about normal.
3. Oxygen is deficient by a factor about 6.
4. Nitrogen is overabundant by a factor about 5.
5. Phosphorus is overabundant by a factor about 100.
6. Iron is overabundant by a factor about 4.
7. Gallium is overabundant by a factor about 8000.
8. Krypton is overabundant by a factor about 1300.
9. Sulphur is deficient by a factor of the order of or greater than 10.
10. Aluminium and chlorine are not observed, which is consistent with these elements having normal abundances.

It is to be noted that when each stellar atmosphere is represented by a single layer of gas at a characteristic temperature and pressure and a comparative curve of growth study is made between two stars, the only free parameters that can be adjusted in order to make the two sets of observations consistent are the relative abundances of the elements. Any differences in equivalent width due to a differing mode of line formation or to the presence in one of the stars of an extended atmosphere or shell in which considerable motions may occur, are interpreted as due to abundance differences. Consequently it is very difficult to assess the true meaning of the "abundance differences" given above. The star γ Pegasi is significantly earlier in spectral type than 3 Centauri A, in as far as a normal spectral type can be assigned to as unusual a star as 3 Centauri A, and it has wider lines. Measurements by Underhill (unpublished) on the intensity tracings of the spectrum of γ Pegasi used by WRIGHT, LEE, JACOBSEN, and GREENSTEIN (1963) to obtain standard equivalent widths, clearly indicate that the total width at half-intensity of lines of moderate strength in the spectrum of γ Pegasi is about 0.35 Å. This value is twice the instrumental width and also twice the width expected according to thermal Doppler effect. The lines of 3 Centauri are sharp. The Mg II doublet which has a separation of 0.2 Å is separated (see Figure 40). This doublet cannot be resolved in the spectrum of γ Pegasi.

Jugaku, Sargent, and Greenstein conjecture that the anomalous abundances reported for 3 Centauri A have been produced by the acceleration of particles on the surface of 3 Centauri A.

3. The star α Sculptoris

This star is classified as B5 in the *Henry Draper Catalogue* and it is one of the few B stars known to have extremely sharp lines. The pattern of line strengths is not normal, thus it is difficult to assign a spectral type to this star. JUGAKU and SARGENT (1961) suggest the type B4V. They have studied five high-dispersion spectrograms obtained at the Mount Wilson Observatory covering the spectral range 3600 Å to 8600 Å and they have made a preliminary curve of growth comparison with the spectrum of γ Pegasi using two lines of Si II and three lines of Si III to establish the characteristic temperature. They can detect the Balmer lines to $n=17$ and thus find that $\log N_e$ is 13.8. The relative strength of the Si II and Si III lines and this value of $\log N_e$ can be reconciled by adopting an ionisation temperature of 14100° and $\log P_e=2.3$. These data are consistent with the suggested spectral type B4V.

Jugaku and Sargent also give equivalent widths of eight He I lines, three C II lines, Mg II λ4481, and the three O I lines near λ7774. They conclude that in comparison to normal stars (that is to γ Pegasi), oxygen and helium are underabundant by about a factor five and that carbon is overabundant by a factor two while silicon has a normal abundance. These estimates must be treated with great reserve, for the spectrum of α Sculptoris is quite dissimilar to that of the normal stars of type B4 to B5. The following survey shows what is visible in the spectrum.

Element	Comments
H	Strong Balmer lines; visible to $n=17$.
He I	Many lines visible, but they are weak for type B4; the strengths of the He I lines are suitable for type B7.
C II	The lines are a little strong for B4.
N II	One multiplet observed (No. 12); normal strength for B4.
O I	Slightly weak for B4.
Na I	Weak lines; probably interstellar.
Mg II	The line λ4481 has normal strength for B4 or B5.
Si II	Rather strong; appropriate for B4.
Si III	Weak, but definitely present, appropriate for B4.
S II	Weak lines of one multiplet are found.
Ca II	The H and K lines are moderately strong.
Ti II	Many moderately strong lines are seen.
V II	May possibly be present as weak lines.
Cr II	Many moderately strong lines are seen.
Fe II	Many moderately strong lines are seen.
Fe III	A few weak lines appear to be present.
Sr II	The resonance multiplet appears weakly.
Nd II	Possibly present, but questionable.

The presence of many moderately strong lines of Ti II, Cr II, Fe II, and weak lines of Ca II, V II, Sr II, and possibly Nd II strongly suggests that α Sculptoris is a shell star. However, this explanation of some of the spectral peculiarities does not account for the extraordinarily weak He I lines.

The He I lines have strengths appropriate for about B7. It is possible perhaps to reconcile the strengths of the He I lines, and those of Mg II, Ca II, Ti II, V II, Cr II, Fe II, Sr II, and Nd II (if present) with a type near B7. Then one would have to account for the rather great strength of the lines from relatively highly excited levels in the spectra of C II, N II, Si II, Si III, and Fe III. One could postulate a hot chromosphere around a cooler B star (B7) in place of a cool shell around a hotter B star (B4), which has unusually weak He I lines. The profiles and intensities of the hydrogen and He I lines would be of great importance in making a decision between the two models. Unfortunately Jugaku and Sargent give no quantitative information about the spectrum of hydrogen in α Sculptoris.

Another interesting point about this unusual star is that Jugaku and Sargent find lines neither of Mn II nor of P II, in contrast to the situation with other peculiar B stars such as 3 Centauri A and κ Cancri where these lines are strong.

4. B stars in the galactic halo

Stars on or above the horizontal branch of Population II in the Hertzsprung-Russell diagram are thought to be very old objects. Some of these stars have spectra that may be classified as B type. TRAVING (1962) has made a preliminary analysis of the spectrum of two of these objects, namely BD +33°2642 for which the visual magnitude is 10.7 and Barnard 29 for which the visual magnitude is 12.9. Both stars have UBV colours which are similar to those of a B2 star. The star BD +33°2642 is at galactic latitude +49° and it is thought to be a member of the halo population; Barnard 29 is a member of M13.

Traving studied spectrograms obtained at the Palomar Observatory by Greenstein and Münch covering the region 3700 Å to 4900 Å at 18 Å/mm for BD +33°2642 and at 40 Å/mm for Barnard 29. He was able to measure equivalent widths of H, He, and the light elements on the spectrograms of BD +33°2642, but only the lines of H and He on the spectrograms of Barnard 29. Traving applied the methods of coarse analysis which have been developed at Kiel (see Chapter XI).

Since no element appears in two stages of ionisation, it is difficult to estimate the ionisation temperature. Since lines of Si III are observed in BD +33°2642 but not lines of Si II and of Si IV, Traving is able to estimate that the ionisation temperature lies in the range 17100° to 19020° for both stars. Such a temperature is consistent with the observed colours of the stars. He estimates the electron density from the number of Balmer lines visible (to $n=21$ for BD +33°2642 and to $n=18$ for Barnard 29), and from the Stark broadening of the H and He I lines. The average result is $\log N_e = 13.32$ for BD +33°2642 and 13.60 for Barnard 29. These values are somewhat lower than what is found for normal B2 main-sequence stars.

When he has established the characteristic temperature and pressure for each atmosphere, Traving is able to estimate the abundances of the light elements using simple curve of growth theory. He finds that the H/He ratio in BD +33°2642 is about the same as in τ Scorpii, but that it is about two times larger in Barnard 29. The elements C, N, O, Mg, and Si appear to be less abundant than in τ Scorpii by a factor between 5 and 10. Without further information about line profiles one cannot say whether this difference corresponds to a real abundance difference, or whether it is produced by neglecting microturbulence as a broadening agent in τ Scorpii.

From his abundances and the gas pressure, Traving is able to estimate log g_{eff} for both stars. After correcting for the radiation pressure gradient term in the equation of hydrostatic equilibrium, Traving finds that log g is 2.34 for BD +33°2642 and 2.66 for Barnard 29. These are unusually small values for B type stars. Since Barnard 29 is a member of Messier 13, one knows its distance and can thus estimate the visual absolute magnitude. The result is $M_V(\mathrm{B29}) = -1.36 \pm 0.3$, a value which is reasonable for a B2 main-sequence star. Since both log g and M_V are known, it is possible to make an estimate of the mass of the star. It turns out from this chain of arguments that the mass of Barnard 29 is 0.2 $m_\odot$. When all sources of error are considered, one is brought to the conclusion that the mass of Barnard 29 is less than or equal to one solar mass. This is a very low value for a main-sequence B star.

A problem thus is posed: How can a star of such low mass have such a high luminosity, i.e. $M_V = -1.36$? Another way of phrasing the dilemma is to ask how a B2 star of normal luminosity for the main sequence can have such a low value of g. It is not possible to estimate the mass of BD +33°2642 by this method because no estimate of M_V is available.

So far as the hydrogen-helium ratio (by number) in stars is concerned, we now have the following results:

1. In normal main-sequence B stars this ratio lies between 10 and 20.

2. A few stars exist in which hydrogen is completely undetected. The ratio of helium to the other elements appear to be as in normal stars. These are the "helium stars".

3. A few stars exist in which the helium lines are remarkably weak, considering the level of excitation and ionization shown by other lines in the stellar spectrum. We can say that helium is deficient by a factor of the order of five; thus the hydrogen to helium ratio lies between 50 and 100. These are stars such as 3 Centauri A or κ Cancri. It is not obvious that it is correct to interpret the unusual line strengths seen in these stars as abundance differences.

4. The halo population B stars appear to be composed of hydrogen and helium at about the normal ratio. However, the abundances of the light elements may depart from the adopted normal composition.

CHAPTER XIII

THE WOLF-RAYET STARS

1. Introduction

Wolf-Rayet stars are few in number and their spectra are easily recognised, for wide emission lines dominate the spectrum. The Wolf-Rayet stars form one of the few groups of stars which are named after astronomers. Wolf and Rayet (1867) at the Paris Observatory used a visual spectrometer to examine the spectra of stars and they noted that the spectra of the stars now known as HD 191765, HD 192103, and HD 192641 consist of broad bands of emission superposed on a weak continuous spectrum rather than of relatively narrow absorption lines crossing a continuous spectrum as is usual with stars. The emission features are sometimes referred to as bands. This word is used only to signify that the emission features are rather broad with steep sides. These emission features are due to spectral lines from several ionised atoms, and often two or more spectral lines blend together. Molecular bands are not present in the spectra of Wolf-Rayet stars.

The width of the emission lines differs from star to star; it is believed to be due to the motions of the emitting ions. Most emission features that can be distinguished in Wolf-Rayet spectra are blends of several lines. Only a few absorption lines appear in Wolf-Rayet spectra. Sometimes the absorption lines cause merely a reduction of the intensity of part of an emission feature rather than a depression of the continuous spectrum.

1.1 *General characteristics of Wolf-Rayet stars*

Wolf-Rayet stars are believed to be among the hottest stars that exist. Since the part of the radiation field from these stars that is observed, that is from about 3100 Å to 12000 Å, does not closely approximate a blackbody radiation field, care must be taken to make clear what is meant by temperature. Quite generally one may say that the level of excitation and ionisation that is observed in Wolf-Rayet spectra corresponds to electron temperatures between about 30000 °K and 50000 °K. By interpreting the observed line intensities and continuous spectrum one would like to be able to deduce the effective temperature which characterizes the total radiation field from the star. In order to do this it must be possible to relate the observed line intensities uniquely to the energy emerging from the star as radiation.

The methods of spectral classification which have been applied to normal absorption-line stellar spectra have been quite successful in arranging most of the stars in a sequence of increasing effective temperature. However, attempts to arrange the

Wolf-Rayet stars in a single temperature sequence which represents the total radiative flux emitted by these stars have long met defeat. Failure has resulted because the prominent features of Wolf-Rayet spectra come from different parts of these complex atmospheres and at present there is no adequate theory for relating local conditions uniquely to the total radiation field or effective temperature. That each Wolf-Rayet spectrum is a collection of spectra from sources of rather widely different character was long ago pointed out by PAYNE-GAPOSCHKIN (1935).

Normal stellar spectra also are collections of features arising from a number of different sources, that is from the different layers in the atmosphere, but often the physical characteristics of the discernable layers are sufficiently alike that one characteristic temperature and one characteristic pressure can be assigned to represent the ensemble. Then a model obeying simple constraints which define the temperature and pressure variations in terms of assigned parameters such as the effective temperature and the acceleration of gravity at the surface of the star can be computed (see Chapter XI). In these cases, the spectral details produced by the atmosphere can be understood in terms of two parameters and some simple physical relationships.

One should like to be able to carry through a similar type of analysis for Wolf-Rayet stars and to assign representative temperatures and pressures to the observable layers of Wolf-Rayet stars. The first thing to do is to determine which spectral phenomena are representative of the effective temperature and the acceleration of gravity of the underlying star and which are characteristic of the exterior layers of the atmosphere. (The outer layers may be only loosely bound to the star.) Once this separation has been made, the problem of arranging the Wolf-Rayet stars in a single sequence becomes simplified and comparable to that encountered in classifying the spectra of normal stars.

Wolf-Rayet spectra are similar in some respects to the spectra from Be stars, shell stars, and Of stars (see Chapter XV). Analogy suggests that a suitable model for a Wolf-Rayet star consists of a star with a fairly compact atmosphere which is surrounded by one or more extended, low density outer atmospheres or shells. This picture is confirmed by the analysis by KRON and GORDON (1950) of light curves in three colours of the eclipsing system HD 193576 (V444 Cygni).

A list of the Wolf-Rayet stars brighter than magnitude 9.5 is given in Table 27. About half of these stars are members of binary systems in which either a second spectrum is seen or the system is known to be eclipsing. Thus the Wolf-Rayet stars are like the O and B stars in that they frequently occur as members of a binary system.

1.2 *The spectral classification of Wolf-Rayet stars*

The purpose of a spectral classification system is to permit one to segregate stars of similar physical characteristics by examining the stellar spectrum. Thus as classification criteria one should choose not only spectral features which may be readily detected on low dispersion spectra, but also spectral features which change in appearance in a monotonic manner as the basic parameters of the star change. The

TABLE 27
The Wolf-Rayet stars brighter than magnitude 9.5

HD Number	α (1900)	δ (1900)	Mag.	Spectral type
50896	$6^h50^m.0$	− 23°48′	6.6	WN5
66811	8 00.1	− 39 43	2.3	WN7 + O5
68273	8 06.5	− 47 03	2.2	WC7 + O7
76536	8 51.6	− 47 13	9.0	WC6
86161	9 51.6	− 57 15	8.5	WN8
92740	10 37.4	− 59 09	6.5	WN7
92809	10 37.8	− 58 15	9.0	WC6
93131	10 40.1	− 59 36	6.7	WN7
93162	10 40.3	− 59 12	8.4	WN7 + O7
96548	11 02.3	− 64 58	7.8	WN8
97152	11 05.8	− 60 26	8.1	WC6 + B0
113904	13 01.7	− 64 46	5.6	WC6 + O
115473	13 12.2	− 57 37	9.3	WC6
136488	15 15.8	− 62 19	9.4	WC8
143414	15 55.0	− 62 24	9.5	WN6
151932	16 45.3	− 41 41	6.6	WN7
152270	16 47.3	− 41 40	6.7	WC7 + O8
156327	17 11.8	− 34 18	9.4	WC7 + B0
156385	17 12.1	− 45 32	7.2	WC7
164270	17 55.2	− 32 43	8.8	WC8
165763	18 02.5	− 21 16	7.8	WC6
168206	18 13.5	− 11 40	8.9	WC7 + O
190918	20 02.2	+ 35 31	7.0	WN5 + B0
191765	20 06.5	+ 35 53	7.8	WN6
192103	20 08.1	+ 35 54	7.9	WC7
192163	20 08.4	+ 38 03	7.4	WN6
192641	20 10.8	+ 36 21	7.9	WC7 + Be
193077	20 13.3	+ 37 07	8.0	WN6
228766	20 13.8	+ 37 00	9.1	WN7
193576	20 15.8	+ 38 25	8.0	WN6 + B1
193793	20 17.1	+ 43 32	6.8	WC6
193928	20 17.8	+ 36 36	9.4	WN6
211853	22 15.0	+ 55 37	8.9	WN6 + B
214419	22 32.9	+ 56 23	8.9	WN6 + O
219460	23 10.8	+ 59 55	9.0	WN5 + B0

basic parameters are the effective temperature and the stellar radius. The adopted classification criteria for Wolf-Rayet spectra permit one to segregate stars into groups with somewhat similar spectra. However, it is not possible to say with certainty whether the adopted spectral classes correspond to a monotonic sequence in effective temperature and radius or not. In order to make the latter correlation it is necessary to have a satisfactory theory relating the strength of the empirically selected classification criteria to the basic theoretical parameters, effective temperature and radius, in a unique way. At present such a theory does not exist.

Wolf-Rayet stars may be readily separated into two groups. Those stars in which the strongest emission features, other than those from helium and hydrogen, are from the ions of nitrogen are called WN stars, while those stars in which emission features from the ions of carbon are prominent, the lines of nitrogen ions being weak or absent, are called WC stars. The emission lines from the ions of oxygen are usually strong in WC stars but weak or absent in WN stars. It is very difficult to define representative sub-classes for Wolf-Rayet stars because, as SMITH (1955) has emphasised, each Wolf-Rayet star, when examined in detail, appears to be different from every other Wolf-Rayet star.

The basic classification system is that adopted by the International Astronomical Union and it has been described by BEALS (1938). The Wolf-Rayet stars are arranged in two sequences as follows using the relative intensities (estimated by visual inspection of spectrograms) of selected line groups.

The WN sequence:

Class WN5. Typical stars are HD 187282 and HD 21156. The criteria are that (λ4605 to λ4622)/λ4686 = 0.2, λ5875/λ5411 = 0.1, and λ4945 is present.

Class WN6. Typical stars are HD 191765 and HD 192163. The criteria are that λ5875/λ5411 = 0.5, the band λ4600 to λ4660 is strong, and a feature at λ4938 is present.

Class WN8. Typical stars are HD 177230 and HD 96548. The criteria are that λ4640/λ4686 = 1.5 and λ5875/λ5411 = 5.0.

The WC sequence:

Class WC6. Typical stars are HD 16523 and HD 165763. The criteria are that λ5696/λ5812 = 0.3, λ5696/λ5592 = 1.2, λ4267 is not present, λ4650 and λ4686 are not resolved, and λ5812 and λ5875 are not resolved. The emission features have a width of about 70 Å.

Class WC7. Typical stars are HD 192103 and HD 119078. The criteria are that λ5696/λ5812 = 0.7, λ5696/λ5592 = 8.0, λ5875/λ5411 = 1.5, λ4650/λ4686 = 4.0, λ4267/λ4786 = 1.0, and λ4650 and λ4686 are just resolved. The emission features have a width of about 35 Å.

Class WC8. Typical stars are HD 184738 and HD 164270. The criteria are that λ5696/λ5812 = 3.0, λ5875/λ5411 = 5.0, λ4650/λ4686 = 9.0, λ4267/λ4786 = 2.0. The width of the emission features is about 10 Å.

This classification system for Wolf-Rayet stars serves as a basic guide in selecting somewhat similar stars. However, it is not possible to place every Wolf-Rayet star uniquely into a spectral class using only the above criteria. BERTOLA (1963) has suggested using two new criteria for WN stars, namely the intensity ratio N III λ4634–40/N V λ4603–19 and the ratio C IV λ5806–12/He I λ5875. Bertola confirms Beals' observation that there does not appear to be a correlation between spectral type and line width for the WN stars though there is a clear correlation for WC stars. In fact this correlation is used as a subsidary classification criterion.

1.3 *The number of Wolf-Rayet stars known and their distribution*

Wolf-Rayet stars are comparatively rare objects, about 100 brighter than the 11th magnitude being known. ROBERTS (1962) has prepared a catalogue of the known Wolf-Rayet stars. This catalogue is probably very nearly complete because Wolf-

Rayet spectra can be detected readily on objective prism survey plates. The Wolf-Rayet stars known to be nuclei of planetary nebulae are omitted from Roberts' list. Thus he does not list HD 184738 which is "Campbell's hydrogen star" and a standard WC8 object. WESTERLUND and SMITH (1964) have given a catalogue of 58 Wolf-Rayet stars in the Large Magellanic Cloud and finding charts for 53 of these stars together with photometric data in blue, visual, and red pass bands.

The Wolf-Rayet stars are associated with the O and B stars and they are closely confined to the plane of the galaxy. They are usually considered to be Population I objects. Some of the central stars of planetary nebulae have broad emission lines in their spectra like those observed in Wolf-Rayet spectra. However, these stars are generally considered to be members of Population II and they should not be considered statistically with the normal Wolf-Rayet stars. The similarity in spectrum is only an indication that the physical conditions in the extended atmosphere are rather alike in the two cases.

ROBERTS (1958) has investigated whether Wolf-Rayet stars occur preferentially in open clusters. He finds that at least half the Wolf-Rayet stars occur in clusters or associations, and that it is more probable to find a WN star in a cluster than a WC star in spite of the fact that when the list of Wolf-Rayet stars is examined, it is found that the number of WN stars is about the same as the number of WC stars. Since clusters and associations are considered to be young groups of stars, Roberts concludes that the Wolf-Rayet stars in clusters and associations are young stars. Roberts suggests the fact that a large number of WC stars are not in clusters indicates that WN stars evolve into WC stars and that the field WC stars have come from clusters which have broken up. It is remarkable that no Wolf-Rayet stars are known to occur in the quadrant centered on the anti-centre of the galaxy. One explanation of this observation could be that this section of the galaxy is sufficiently old that the Wolf-Rayet stars have evolved past the stage of development which gives the characteristic Wolf-Rayet spectrum and that the stars which were Wolf-Rayet stars are now on the main sequence and appear at other spectral types.

2. The properties of the Wolf-Rayet stars

Wolf-Rayet stars are a rather inhomogeneous group of objects. The spectra of very few of them have been studied at moderate or high dispersion owing to the faintness of the objects which are accessible from the northern hemisphere. Thus it is really not possible to make a statistically sound evaluation of the physical characteristics of these stars. Nevertheless, a summary of what is known or suspected about the physical characteristics of these stars will now be given.

2.1 *Absolute magnitudes and colours*

It is very difficult to determine the visual absolute magnitudes of the Wolf-Rayet stars because the chief available criterion for distance is the strength of the interstellar absorption lines and it is known that this is not a very reliable criterion for distance.

Results obtained in this way have statistical value only and little weight should be placed on the values obtained for single stars. SANFORD and WILSON (1939) estimated from the strength of the interstellar lines that for WC stars the average visual absolute magnitude is -2.8 and for WN stars it is -2.1. It is now known, however, that the interstellar extinction is greater than Sanford and Wilson estimated. If one adds a magnitude in order to allow more correctly for the interstellar absorption, it is found that the average visual absolute magnitude is -3.8 for WC stars and -3.1 for WN stars.

HOUCK (1956) has given photometric and spectroscopic data for two Wolf-Rayet stars (HD 151932, WN7 and HD 152270, WC7+O8) in the apparent neighbourhood of the cluster NGC 6231. From his photometry and upon the assumption that these Wolf-Rayet stars are members of the cluster NGC 6231 (which contains many O and B stars) Houck finds that the Wolf-Rayet stars have visual absolute magnitudes brighter than -6. The distance to the cluster is found by fitting the observed colour-magnitude diagram for the cluster stars to the so-called zero-age main sequence.

ANDRILLAT (1955) studied the spectra of eight Wolf-Rayet stars in Cygnus and she suggested that the average absolute magnitude for WC stars is -3.1 while for WN stars it is -2.7. In a following paper (ANDRILLAT, 1958) Mme Andrillat reviewed the situation and attempted to estimate the distances of some Wolf-Rayet stars from the amount of reddening caused by interstellar extinction. She concluded that the visual absolute magnitudes of Wolf-Rayet stars fall in the range -1.3 to -5.1 mag.

It is well known that the Wolf-Rayet spectrum in double-lined spectroscopic binaries of a Wolf-Rayet star and an O or early B star appears to be somewhat "drowned". It certainly is not dominant. Thus one may conclude that the average visual absolute magnitude of a Wolf-Rayet star is fainter than -4. ONDERLIČKA (1958) has surveyed the information about the absolute magnitudes of the Wolf-Rayet stars and he finds that the visual absolute magnitudes of the WC and the WN stars lie between -1.5 and -5.5.

FEINSTEIN (1964) has observed 10 Wolf-Rayet stars in the Carina region on the UBV system of photometry. He gives the visual magnitude V, and the colours $B-V$ and $U-B$ for each star. He also assembles from various sources similar data for another 18 stars. When the photometric data are plotted in a two-colour diagram, the WN stars are found to lie along a different reddening path than do the WC stars. There is a considerable scatter among the points. Some of it is due to the differing line strengths and some due to the presence of the spectrum of an O or B type companion, but the two reddening paths appear to be distinct. The WN path intersects the normal sequence near spectral types B0.5 and B1 whereas the WC reddening path intersects the normal main sequence near B3 to B5. Feinstein deduces the following intrinsic colours (including emission lines) and visual absolute magnitudes for Wolf-Rayet stars:

Type	$(B-V)_0$	$(U-B)_0$	M_V
WN	-0.28	-1.00	-3.8
WC	-0.17	-0.60	-1.5

Westerlund and Smith (1964) have done photographic photometry in red, visual, and blue colours for 58 Wolf-Rayet stars in the Large Magellanic Cloud. The photographic work is calibrated by photoelectrically measured sequences. They find that the average value of $B–V$ for WN stars is about -0.30 while for the WC stars it is near 0.00. Westerlund and Smith point out that the WC stars fall in a different part of the two-colour diagram than do the WN stars, but their data do not establish two separate reddening paths such as found by Feinstein. Upon adopting a modulus of 18.7 magnitudes for the Large Magellanic Cloud, Westerlund and Smith find that the average visual absolute magnitude for WN stars in associations is -5.2, whereas for the WN stars which appear to be field stars it is -4.3. The few WC stars which are observed have an average absolute magnitude of -5.3. The seven single-spectrum WN stars in the 30 Doradus complex have an average absolute magnitude near -6.4.

The studies of galactic Wolf-Rayet stars suggest visual absolute magnitudes of the order of -3 ± 1 magnitude. The work of Westerlund and Smith suggests that the average visual absolute magnitude of Wolf-Rayet stars in near to -5. A few Wolf-Rayet stars appear to be as bright as -6. The photometry on which these estimates are based includes the strong emission lines which fall within the filter pass bands. The intrinsic $B–V$ colours of WC stars appear to be redder than those for WN stars. This reddening is certainly in part due to the very strong emission lines which fall in the V pass band in the case of WC stars.

2.2 *Masses*

Seven double-lined spectroscopic binaries consisting of an O or B type star and a Wolf-Rayet star have been studied for radial velocity. They are listed in Table 28 together with the mass ratio deduced from the relative amplitudes. Only the designation OB is entered for the companion in the two cases for which the spectral type of the companion star has not been published. The average mass ratio is 0.34, which means that on the average the Wolf-Rayet star has only one third the mass of its O or B type companion. This ratio appears to be well determined. Not one of the Wolf-Rayet components appears to be as massive as or more massive than its companion.

The average companion is a late O type star or an early B type star. According to

TABLE 28
Seven Wolf-Rayet spectroscopic binaries

HD number	Spectral type	$\mathfrak{M}$ (WR)/$\mathfrak{M}$ (OB)	Orbit
152270	WC7 + O8	0.28	Struve (1944b)
186943	WN5 + OB	0.28	Hiltner (1945a)
211853	WN6 + B0	0.39	Hiltner (1945a)
228766	WN7 + OB	0.21	Hiltner (1951)
168206	WC7 + O	0.33	Hiltner (1945b)
190918	WN5 + B0	0.27	Wilson (1949)
193576	WN6 + B1	0.39	Wilson (1942)
		0.56	Keeping (1947)
		0.39	Münch (1950)

PEARCE (1957) we may therefore expect the absorption-line companion to have a mass between 12 and 24 solar masses. We conclude that the average mass for a Wolf-Rayet star lies between 4 and 8 solar masses. Stars of these masses are expected to have spectral types between about B8 and B3 when they are on the main sequence and to have visual absolute magnitudes lying between about 0.0 and -2. Thus if the bolometric corrections for Wolf-Rayet stars do not differ by a large amount (more than one magnitude) from the bolometric corrections for B stars, the Wolf-Rayet stars are overluminous for their masses.

2.3 *Radii*

The best available information comes from the analysis by KRON and GORDON (1950) of the light curves of the eclipsing variable V444 Cygni (HD 193576). According to the light curves (HILTNER, 1949a; KRON and GORDON, 1950) the Wolf-Rayet star acts as though it is bigger when it is occulting the O star than when it is eclipsed. The Wolf-Rayet star appears to have a large envelope that is more or less opaque, the opacity having little if any dependence on wave length, but which contributes very little light to the system. A possible model of the system consists of an O star with a radius about $10R_\odot$, a luminous disc with a radius of $7R_\odot$ and an electron scattering envelope with a radius of $17R_\odot$.

One cannot use the observed luminosities of Wolf-Rayet stars, uncertain as they are, together with an effective temperature in order to find a radius because there is no reliable way of estimating a brightness temperature which will give a measure of the emergent flux in the visual pass band. Furthermore, much of the measured flux is given by monochromatic radiations, not by a continuous spectrum which may be approximated by a black-body energy distribution.

2.4 *Temperatures*

Since the spectra of Wolf-Rayet stars are dominated by emission lines from the third and fourth spectra of the light elements one may expect that the electron temperature in Wolf-Rayet atmospheres lies in the neighbourhood of 30000 °K to 50000 °K. In order to find a characteristic temperature for the atmosphere one should have a theory of line formation which relates the observed line strengths to temperature. However, an entirely satisfactory theory of this sort does not exist.

Various definitions of temperature may be used. For instance, an electron temperature is that value of the temperature which when inserted in the Maxwell velocity probability function gives the observed velocity distribution of the electrons. One can define a brightness temperature for a selected spectral region which gives a black-body emission equivalent to the observed (or indirectly deduced) continuous flux distribution over the selected wave-length range. Furthermore, one can define the effective temperature as that temperature which in thermodynamic equilibrium will give a total radiation field equal to the total radiation field of the star.

Our knowledge of Wolf-Rayet spectra and of the physical processes occurring in Wolf-Rayet atmospheres is rather rudimentary. Nevertheless estimates of temperature have been made in several ways in order to characterise roughly the energetics of the processes occurring in Wolf-Rayet atmospheres. The results will now be summarised.

Some of the first systematic attempts to deduce temperatures for the Wolf-Rayet stars were made by BEALS (1934) who applied Zanstra's theory for the planetary nebulae. This theory is valid only in optically thin nebulae. Since Wolf-Rayet atmospheres are not optically thin, the resulting temperatures are maximum values. The results have been summarised by BEALS (1940) who gives temperatures ranging from 59000 °K to 110000 °K for seven stars. In this type of analysis when lines of ions of high ionisation potential are used, higher temperatures are found than when lines from ions of low ionisation potential are used. This temperature difference has led to the idea that the level of ionisation and excitation decreases outward quite significantly in Wolf-Rayet atmospheres.

ALLER (1943) has measured the intensities of emission lines in some Wolf-Rayet spectra and from this data has attempted to estimate what the excitation temperature would be were the gas and radiation in local thermodynamic equilibrium. He has noted that in interpreting Wolf-Rayet spectra it is necessary to consider the possibility of ionisation and excitation by monochromatic processes from low lying levels in addition to ionisation and excitation from the ground level of the ion. Aller found excitation temperatures ranging from about 20000 °K to 130000 °K depending upon the ionic spectrum which was studied.

ZANSTRA and WEENEN (1950) and WEENEN (1950) have also considered how the apparent relative intensities of the emission lines in Wolf-Rayet stars might be interpreted in terms of excitation and ionisation temperatures and the abundance of the elements. They note, as does Aller, that the physical conditions in a Wolf-Rayet atmosphere are far removed from those in planetary nebulae. They attempt to estimate abundances by assuming that the radiation field in Wolf-Rayet atmospheres at wave lengths shorter than the Lyman limit of hydrogen can be represented by an appropriate temperature and a black-body energy distribution. Their adopted temperatures are similar to those suggested by Aller, that is, in the neighbourhood of 80000 °K. The relative abundances of the light elements come out to be rather different from the normal abundances suggested by the analysis of B type stars. For instance, they suggest that in WC stars the ratio by number of C to He is 50 × normal. It is very doubtful if these estimates should be given much weight because closer examination of Wolf-Rayet spectra strongly suggests that the observed line strengths result from a number of processes occurring in the extended atmospheres of the Wolf-Rayet stars and that one is not justified in using such a simple physical picture of line formation as Zanstra and Weenen do.

MIYAMOTO (1952) has also considered how the process of line formation in Wolf-Rayet atmospheres may be represented theoretically and he has attempted to deduce the electron temperature in Wolf-Rayet stars. He points out that conditions in Wolf-Rayet atmospheres probably resemble more closely those in a Be atmosphere than those in a planetary nebulae. Consequently the temperatures which he finds (they range between 30000 °K and 41000 °K) are lower than those suggested by Beals.

BAPPU (1958) has estimated excitation temperatures for HD 192103, WC7, and for HD 184738, WC8, from the relative intensities of three C IV lines using theory suitable for hydrogenic spectra. He finds a mean temperature of 48000 °K for HD 192103 and 27400 °K for HD 184738.

One of the most recent attempts to estimate the electron temperature in Wolf-Rayet envelopes from the apparent strengths of the emission lines is due to RUBLEV (1964). Rublev attempts to generalise the method introduced by ALLER (1943), taking into account interstellar absorption. He studied the Pickering series of He II making use of equivalent widths estimated from the spectral tracings published by UNDERHILL (1959a) of HD 192103 and of HD 192163. He finds that the electron temperature in the atmosphere of HD 192103 is about 25000 °K while in the atmosphere of HD 192163 it is 23000 °K or 28000 °K depending upon what is assumed for the colour temperature of the continuous spectrum. Rublev concludes that at such low temperatures collisional excitation will not be important. BAPPU (1958), however, has concluded that collisional processes are important in Wolf-Rayet atmospheres. He reaches this conclusion from a consideration of the relative strengths of He II λ4200 and He II λ4541 in the spectra of HD 192103 and HD 184738.

The above-mentioned studies have attempted to estimate characteristic temperatures for Wolf-Rayet stars from the apparent intensities of some of the emission lines. The application of these ideas has been and still is greatly hampered by the lack of good observational data for a representative selection of stars and for a representative selection of spectral lines.

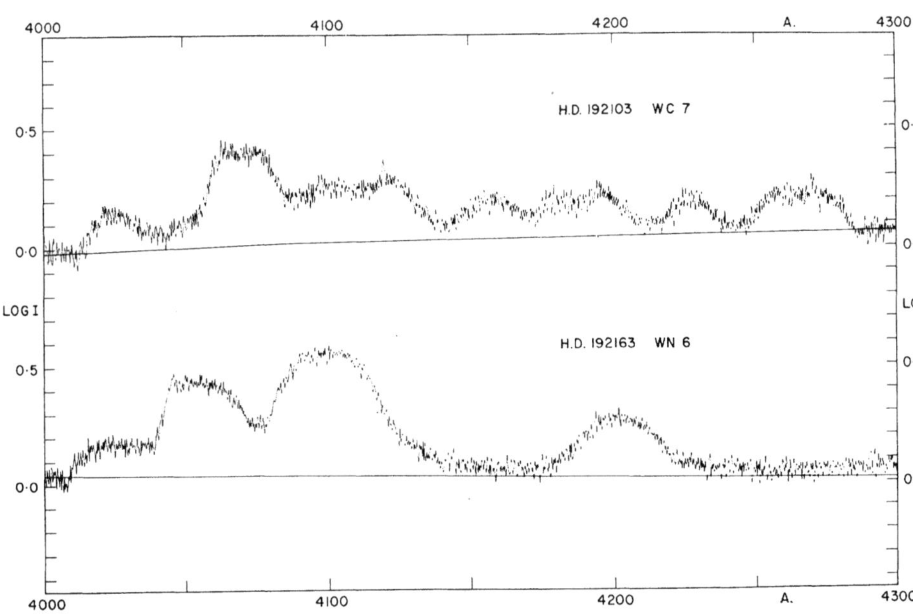

Fig. 41. Log I tracings of the spectrum of HD 192103 and of HD 192163 in the region λ 4000 to λ 4300.

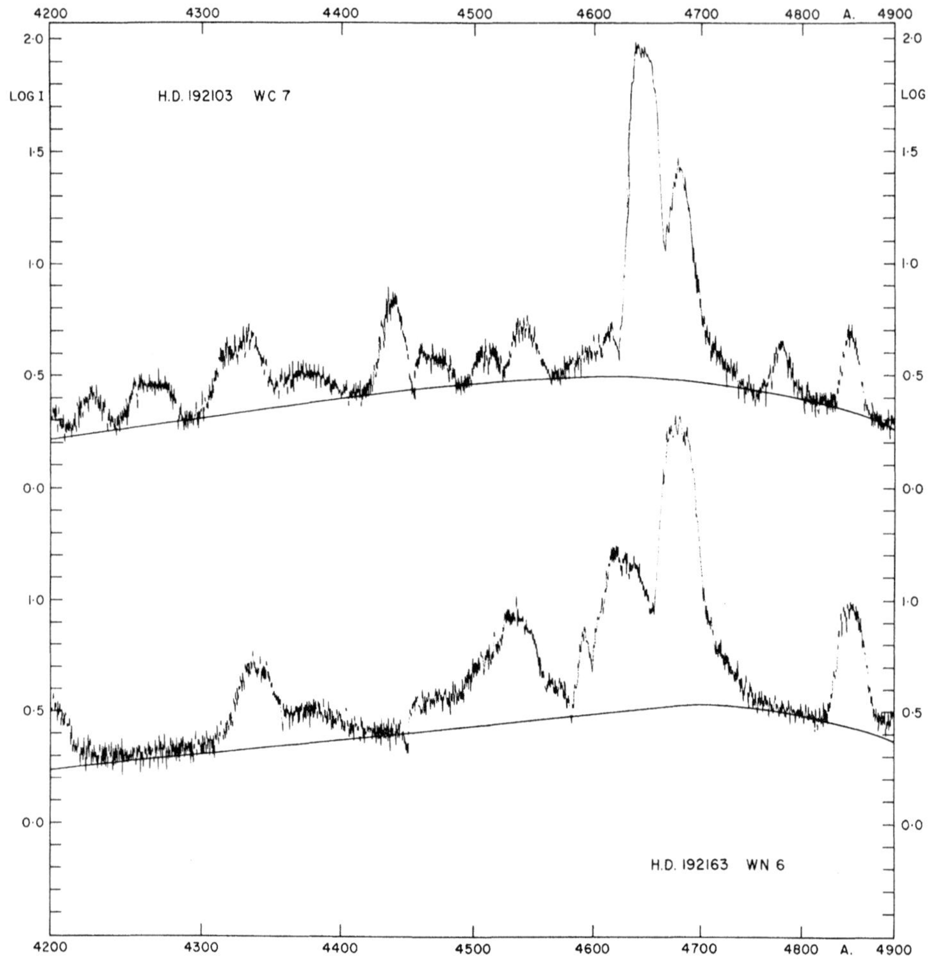

Fig. 42. Log I tracings of the spectrum of HD 192103 and of HD 192163 in the region λ 4200 to λ 4900.

Another way to find a characteristic temperature for a Wolf-Rayet star is to estimate the colour temperature by fitting the observed continuous spectral intensity distribution to a black-body curve over a wave-length range of a few hundred angstroms. GERASIMOVIČ (1929) was one of the first to do this and some interpretations of his work suggest a mean temperature of 17000 °K for eight Wolf-Rayet stars. The effects of interstellar reddening were not allowed for in his work, thus his temperatures are certainly too low. In principle this method of estimating temperatures for the Wolf-Rayet stars should work well. In practice it turns out that it is difficult to be sure where the continuous spectrum of Wolf-Rayet stars lies because of the rather great number of weak and medium strong emission lines in Wolf-Rayet spectra, see for instance Figures 41 and 42 which give part of the log I tracings of Wolf-Rayet spectra published by UNDERHILL (1959). It is clear from these tracings that it is not easy to determine accurately the shape of the continuous spectrum even on slit spectra having a dispersion of 16 Å/mm or 30 Å/mm. On spectra of lower dispersion and lower spectral purity it will be even more difficult to estimate where the continuous spectrum lies. At temperatures greater than 20000 °K a very slight inflection in the intensity distribution between 4000 Å and 5000 Å will lead to rather widely differing estimates of colour temperature. Wolf-Rayet spectra appear to be equally full of emission features of all intensities at longer and at shorter wave lengths than shown here.

VORONTSOV-VELYAMINOV (1945, 1958) has attempted to determine colour temperatures for Wolf-Rayet stars by comparing the apparent energy distributions between 3800 Å and 4800 Å of some Wolf-Rayet stars with the energy distribution of A0 stars. In general he finds colour temperatures less than 16000°. Although it is somewhat uncertain how much these temperatures should be increased in order to allow for interstellar reddening, VORONTSOV-VELYAMINOV (1958) concludes that one cannot avoid the conclusion that the colour temperatures of Wolf-Rayet stars fall much below the electron temperatures – some 60000° to 90000° – estimated by means of the Zanstra theory valid for planetary nebulae. ANDRILLAT (1955) has attempted somewhat similar measurements on low dispersion spectrograms of Wolf-Rayet stars extending over the spectral region 3900 Å to 8000 Å and she has attempted a correction for interstellar reddening. Her temperatures fall between 8000 °K and 19500 °K; the average value is about 10000 °K.

About the only conclusion one can draw from studies of this type is that the electron temperature in Wolf-Rayet atmospheres lies in the neighbourhood of 30000 °K or higher, whereas the observed energy distribution in the continuous spectrum, although poorly known, suggests temperatures not higher than 20000 °K. The low colour temperatures found for Wolf-Rayet spectra are probably fictitious for, as was first noticed by Wyse (see remark by ALLER, 1943), Wolf-Rayet stars are quite bright at wave lengths shorter than 3600 Å. It is much easier to obtain a well exposed spectrogram of a Wolf-Rayet star at $\lambda < 3600$ Å than it is to obtain one of a late B type star. The colour temperatures of late B type stars are in the neighbourhood of 12000 °K.

2.5 *Variability in light and spectrum*

In this section the information available about the variation in light and spectrum of apparently single Wolf-Rayet stars will be summarised. The eclipsing binary stars which have been studied in most detail are known to show irregular fluctuations in light and spectrum, see Section 4. It is difficult to say what part of the observed variations is due to gas streams in the binary systems and what part to intrinsic variation of the stars themselves. We shall presume that the variations noted for Wolf-Rayet stars not known surely to be components of double or multiple systems are typical of the intrinsic variations which occur in Wolf-Rayet stars.

No Wolf-Rayet stars are known to show as large intrinsic variations in light as for instance have led to the definition of classes of variable stars such as the T Tauri stars or the Cepheid variables. The small apparent range of the light fluctuations for the Wolf-Rayet stars may be merely a result of the wave-length range in which we observe. For instance, it is known that the long period variables show much greater amplitudes in blue light than they do in the infrared.

Ross (1961) observed nine Wolf-Rayet stars photoelectrically in three colours (yellow, blue, and ultraviolet) for several months. He selected stars not known to be eclipsing variables that were accessible from Berkeley, California. It turned out that the stars HD 50896, HD 191765, HD 192103, HD 193077, and perhaps HD 192163 showed irregular variations with an amplitude less than 0.1 magnitude. The variations appeared in all three colours. The star HD 50896 showed the most definite variation in light for it decreased by 0.08 magnitude in five months. The stars HD 45166, HD 165763, HD 192641, and HD 193793 were not observed to vary in light during the period of observation.

The spectral types of these stars are given in Table 27. The star HD 50896 may well be a double-lined spectroscopic binary, but there is not sufficient evidence available at this time to make a definite statement on this point. Both O. C. Wilson (1948a) and Smith (1955) have reported spectral variations for this star which are reminiscent of what is seen for the binaries HD 193576 (V444 Cygni) and HD 214419 (CQ Cephei), and O. C. Wilson has reported some variations in radial velocity. It seems not improbable that HD 50896 is a binary system enveloped in streams of gas, but more information is needed.

Bappu (1951) has reported central reversals at He II λ4686 and at He II λ5411 for HD 191765 (BD $+35°4001$) as well as variations in the shapes of the emission features attributed to N V λ4603, N III λ4640, He II λ4541, and He I λ4471, but he gives no quantitative results. No such variations or central reversals are visible on the spectrograms available to the author. Bappu (1951) also reports similar variations, though no central reversals, for HD 192163 (BD $+37°3821$). Again such variations are not visible on the moderate dispersion spectrograms available to the author. No spectral variations have been reported for HD 165763, HD 192103, HD 192641, HD 193077, or HD 193793 although both HD 192641 and HD 193793 show composite spectra. Some radial velocity variations have been reported for HD 193793 (McDonald, 1947) but no period has been established.

The star HD 45166 has shown major spectral changes (Anger, 1933, Neubauer and Aller, 1948) and there is some doubt if it should really be classified as a Wolf-Rayet star for the emission lines are sharper than is usual for a Wolf-Rayet star and more like the emission lines seen in an Of star. In 1922 the spectrum showed broad emission lines of He II λ4686 (strong), N IV λ4058 (weak), and N III λ4634-40 (trace); in 1933 the N IV and He II lines were about as before but N III λ4634–40 had become nearly as strong as He II λ4686. In 1948 Neubauer and Aller described the spectrum as intermediate between Of and WN. Such gross spectral changes are unknown among other Wolf-Rayet stars. According to O. C. Wilson (1948b) the star HD 197460 also appears to be intermediate in spectral type between WN and Of; it is probably a single-lined spectroscopic binary, for the radial velocity from He II λ4686 shows a range from +60 to −120 km/sec. The standard WN8 star HD 177230 also appears to be a single-lined spectroscopic binary (WILSON, 1948b).

STRUVE (1944a) has studied the radial velocities of 20 early type stars in and near the cluster NGC 6231 and he gives a description of the spectrum of HD 151932, WN7. He notes that although this star is not a spectroscopic binary some interesting changes in the spectrum appear on his spectrograms. The most conspicuous changes are in the strength of the shortward displaced absorptions at N V λ4603 and He II λ4541. Struve gives visual estimates of the varying strength of these absorptions. SMITH (1955) has given descriptions of changes seen in the spectra of some southern Wolf-Rayet stars including γ_2 Velorum whose spectrum according to PERRINE (1920) varies.

It appears to be definitely established that the strength of the absorption features vary in some Wolf-Rayet stars, and that the light also varies irregularly by a small amount, but more quantitative information is needed before anything other than a qualitative interpretation of these data can be given.

3. The interpretation of Wolf-Rayet spectra

Short sections of spectrograms of early type stars near Hα, Hβ, and Hγ are shown in Figure 43. The spectral differences displayed here give a brief summary of the characteristic differences between the different types of star.

Strip A gives the spectrum of ι Herculis, a normal B3V star. Note the Stark broadened wings of the hydrogen lines and the relatively sharp He I lines at λ4388 and λ6678. Strip B gives the spectrum of 48 Persei, a well known Be star of type B3. This star is rotating with a projected velocity of about 230 km/sec. Consequently the

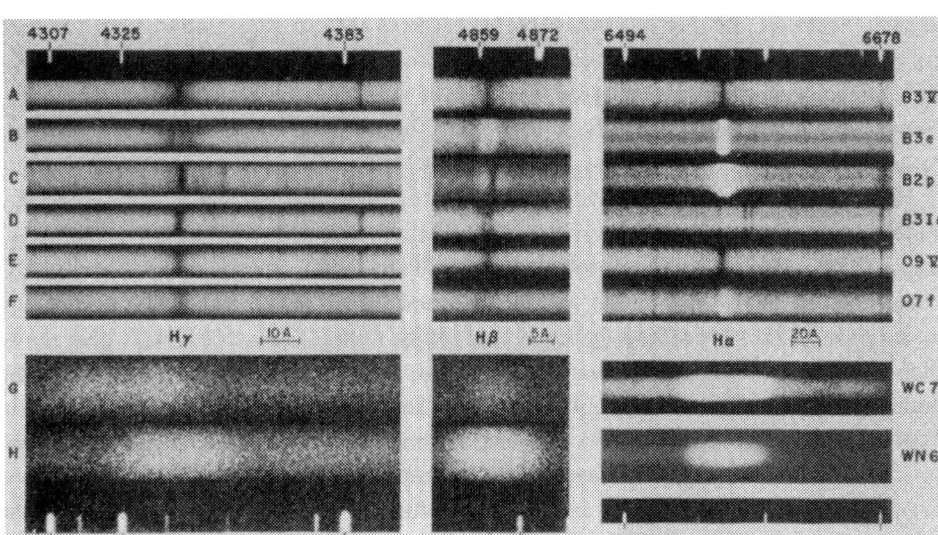

Fig. 43. Spectra of early type stars in the region of Hα, Hβ, and Hγ.

profiles of the He I and H absorption lines are not so sharp as in the case of ι Herculis. Double emission lines separated by an absorption core are seen at Hβ and Hγ. The resolution of the spectrograph is not sufficient to show the expected central reversal at Hα. Strip C gives the spectrum of the shell star ζ Tauri, B2p. The emission lines and the shell absorption lines vary in this star. In the present spectrograms the strong double emission at Hα, Hβ, and Hγ are clearly visible as well as strong absorption cores due to the shell. Since this star is rotating at almost 400 km/sec, the profiles of the stellar He I lines are shallow and broad; the sharp core of He I λ6678 is formed in the shell. Such a core does not occur for He I λ4388. A fairly sharp line due to Fe II λ4385 is visible near the position of He I λ4388. This line is formed in the shell.

Strip D gives the spectrum of the B3Ia supergiant 55 Cygni. Spectral features typical of supergiants are the steep sides of the absorption lines at Hβ and Hγ and the P Cygni-like emission at Hα (see Chapter XIV). Also note the strong C II absorption lines near Hα at λ6578 and λ6582. Strip E gives the spectrum of 10 Lacertae, O9V. The hydrogen lines have wide, Stark broadened wings and rather prominent cores which appear to be due to absorption in an extended atmosphere where randomly directed velocities of the order of 15 km/sec occur. The He I lines at λ4388 and λ6678 are rather strong and moderately sharp. Strip E gives the spectrum of 9 Sagittae, O7f. This star is rotating, thus the profiles of the absorption lines at Hβ, Hγ, and He I λ4388 and λ6678 are rather diffuse. There is a moderately strong, somewhat diffuse emission feature at Hα.

Strips G and H give spectra of the Wolf-Rayet stars HD 192103, WC7, and HD 192163, WN6, respectively. It is clear that Wolf-Rayet spectra differ greatly from ordinary stellar spectra such as are displayed in strips A to F. The Wolf-Rayet spectra consist of rather wide emission bands; only a very few absorption features can be detected. The emission feature near Hγ in the WC star extends to short wave lengths because a strong line due to C III λ4326 blends with the emission due to Hγ and He II. Another blend extending to longer wave lengths occurs at Hα. It is with C IV λ6592. A better idea of the complexity of Wolf-Rayet spectra is given by the log *I* tracings shown in Figures 41 and 42. It is clear that Wolf-Rayet spectra are full of broad, rather rounded emission lines and that few absorption features can be recognised.

Comprehensive lists of observed lines with identifications have been given by BEALS (1930); PAYNE (1933); SWINGS (1942); SMITH (1955); EDLÉN (1956), and UNDERHILL (1959a, 1962c) for the spectral region λ3050–λ6800. Some observations in the near infrared have been made by SWINGS and JOSE (1950), by SMITH (1955), by ANDRILLAT (1955, 1964) and by CODE and BLESS (1964). These observers give many identifications. These studies indicate that apart from lines of H, He I, and He II, the emission lines of C II to C IV and O II to O VI dominate in the WC stars, whereas in WN stars the lines of N II to N V dominate while lines from the carbon and oxygen ions are weak or absent. However, it is not true that *no* lines from the nitrogen ions are found in WC stars (see, for instance UNDERHILL, 1962c) nor that *no* lines from the carbon ions are found in WN stars. Both SWINGS (1942) and ALLER (1943) showed independently that the blended strong C IV lines at λ5801 and λ5812 are present in

WN stars. UNDERHILL (1959a) has shown that lines from the O III spectrum are definitely present in HD 192163, WN6, and that O IV and O V are probably present. It seems very probable that the observed intensity patterns of the emission lines in Wolf-Rayet spectra are largely caused by departures from a black-body intensity distribution of the far ultraviolet radiation from Wolf-Rayet stars. A rough sketch of what the far UV ($\lambda < 912$ Å) may be like has been given by UNDERHILL (1958b) who has pointed out how this may produce spectra like what are observed. It would appear that WN spectra correspond to higher temperatures than WC spectra do.

3.1 *The interpretation of the emission line spectrum*

EDLÉN (1956) has noticed that in WC stars many more lines in the hydrogen-like series of the C IV spectrum can be observed than in laboratory spectra. This is because the density in Wolf-Rayet atmospheres is sufficiently low that the upper levels of the C^{+++} ions are not destroyed by Stark broadening. Thus we conclude that the electron density is moderately low in Wolf-Rayet atmospheres. However, it is not *very* low, for forbidden lines are not seen in Wolf-Rayet spectra. M. JOHNSON (1954) has suggested that the width of the emission lines in Wolf-Rayet spectra is caused by Stark effect. This suggestion is untenable because then one would not find the long series of C IV lines which are observed in WC spectra.

Most of the lines in Wolf-Rayet spectra appear to be from a recombination spectrum. The lines are rounded and have half-widths suggesting that the atoms have an average velocity of the order of 500 to 1000 km/sec. The width of the emission lines varies quite considerably from star to star and in the WC sequence this width is used as an additional classification criterion. The intensities of the stronger lines, such as He II λ4686 and λ3203 may be affected by self-absorption; see UNDERHILL (1959a) for some quantitative measurements on this point. MÜNCH (1950) has pointed out that the broadening effects resulting from electron scattering could produce line profiles of the general shape that is observed, and THOMAS (1949) has drawn attention to the broadening effects of turbulent motion and to the contribution that such motion can make to atmospheric support. Any quantitative theory of the shapes of the lines in Wolf-Rayet spectra should take the suggestions of Thomas and of Münch into consideration.

The first suggestion (BEALS, 1929) about the formation of the broad emission lines of Wolf-Rayet spectra was that the lines originated in a rapidly expanding envelope. BEALS (1931) worked out the expected profiles in a simple case and demonstrated that the computed results resembled a few of the observed emission features in Wolf-Rayet spectra. CHANDRASEKHAR (1934) put the theory of emission lines formed in an expanding envelope on a sound mathematical basis and examined two particular cases to see how the resulting profiles depended on the adopted physical situation. It became clear that any line formed in an expanding envelope should have a shortward displaced absorption component (formed in the part of the shell projected on the stellar disc) and that if the velocity of expansion was large the emission line should be flat topped. The longward edge of the profile would be modified in a characteristic manner if the shell was small enough with respect to the stellar photosphere that the occultation by the photosphere of the atoms moving away from the observer was significant. SOBOLEV (1947) also considered the formation of emission lines in a moving envelope and attempted to take some account of the dilution effects which might occur. It is obvious from elementary considerations that the geometrical shape of the moving shell and the relative velocities of the atoms which are assumed to occur will have an important effect on determining exactly what shape a line profile will have. BAPPU and MENZEL (1954) have explored some expects of this problem and RUBLEV (1961, 1962, and 1963) has investigated other rather schematic models.

These authors can all find models which will explain some features of Wolf-Rayet spectra, but the concept of a rapidly expanding spherical shell will not explain all the spectral features which are observed. O. C. WILSON (1942) has pointed out some particular shortcomings of the expanding shell hypothesis (with velocities of expansion about 1000 km/sec) for interpreting the observed phenomena in spectroscopic binaries containing Wolf-Rayet stars. One is forced to conclude that most of the emission lines are formed in a rather small atmosphere where the line broadening is caused by large random motions of the gas and perhaps by electron scattering. However, one cannot deny that a few shortward displaced absorption components similar to those predicted by the expanding envelope theory are observed, and that one or two emission lines have the flat-topped profile associated with a rapidly expanding envelope. The particular spectroscopic character of the lines which show definite expanding-envelope characteristics permits one to obtain a solution to the problem of defining a simple model for a Wolf-Rayet atmosphere. These lines are discussed in the next section.

The best model for a Wolf-Rayet atmosphere which fits all the observed spectroscopic details, at least qualitatively, is that a Wolf-Rayet star has a moderately extended, moderately dense atmosphere in chaotic motion above a compact photosphere. This atmosphere is probably equivalent to the luminous disc of Kron and Gordon. We must assume that the star is surrounded by a low density, rapidly expanding atmosphere which is significantly opaque (and thus observable) chiefly in lines which are strengthened under conditions of high temperature and moderate dilution. This rapidly expanding atmosphere does not make a significant contribution to most profiles.

3.2 *The spectrum from the expanding shell*

We have seen in Chapter VII that in a low density gas at high electron temperature where the radiation field is dilute, i.e. not equivalent to the radiation field corresponding to thermodynamic equilibrium at the electron temperature, lines from the triplet series of He I will be specially strengthened in comparison to lines from the singlet series. At high temperatures one expects lines from the 2^3S and 2^3P levels of He I to be particularly strong in absorption. A similar distribution of intensities is expected in the equivalent lines from the C III and N IV spectra. In particular one expects λ3187, λ3888, λ4471, and λ5876 of He I, λ4650 of C III, and λ3478–82 of N IV to be unusually strong in absorption.

It is just the above lines which are observed to have shortward-displaced absorption components in Wolf-Rayet spectra. Strong shortward displaced absorption components are not observed for all intrisically strong lines. The conclusion is that Wolf-Rayet stars do indeed have a rapidly expanding envelope but that it contains only sufficient material to become easily visible in lines which are strengthened at the expense of the other lines by the non-thermal-equilibrium conditions which exist in the shell. Lines which are specifically excited in emission by monochromatic fluorescent processes may be also expected.

Spectra of HD 193793, WC6 +O6, and of HD 192103, WC7, showing the shortward displaced absorption component of He I λ3888 are shown in Figure 44. In the

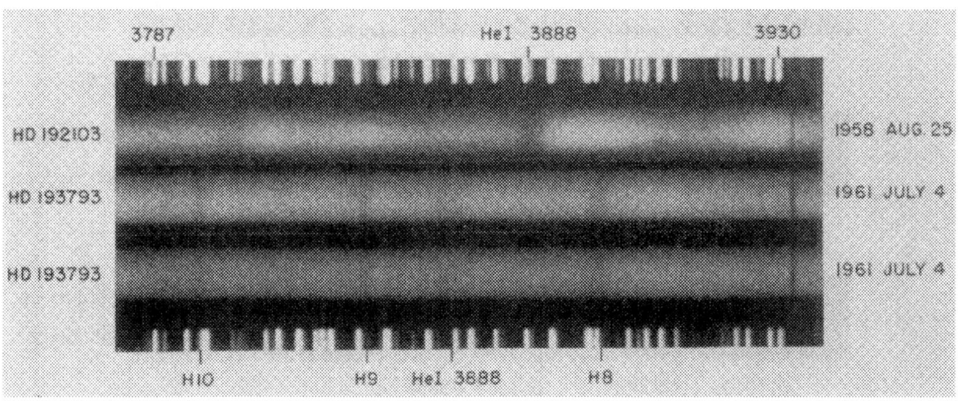

Fig. 44. Spectra of HD 193793, WC6 + O6, and of HD 192103, WC7, showing the shortward displaced He I λ3888 absorption component.

case of HD 192103 the displacements is about −1200 km/sec and the displaced absorption line is moderately sharp. In the case of HD 193793 the displacement is about −2650 km/sec and the absorption feature is rather diffuse. It is difficult to see on the reproduction because the O type spectrum of the companion star dominates in this spectral region. Note the absorption lines due to H8, H9, and H10. A sharp interstellar Ca II K line is visible in each spectrograms at λ3933. Similar displaced absorption lines can be detected at the proper displacements from the other He I and C III lines listed above. In WN stars the C III line is not definitely visible but a displaced absorption component is seen shortward of N IV λ3478–82.

The C III λ4650 absorption can be seen in Figure 42 as a steep decline and dip on the shortward side of the emission feature at λ4650 in the spectrum of HD 192103. It is impossible to say if a weak C III absorption is present in the spectrum of HD 192163, WN6, or not.

Shortward displaced absorption dips at N V λ4603 and λ4619 occur in WN stars, see the tracing for HD 192163 in Figure 42. They illustrate another type of shortward displaced absorption component which is found in Wolf-Rayet spectra.

The similar absorption components for the C IV lines λ5801 and λ5812, which correspond to the same transition as the N V lines, are shown in Figure 45 for the stars HD 192103, HD 192641, and HD 193793. These absorptions appear as steps on the shortward edge of the blended C IV emission component rather than as separate dips. They cause a longward displacement of the apparent wave length of the blended C IV line in WC stars, but not in WN stars. The positions of the absorption components are marked in Figure 45. In each case a densely exposed and a lightly exposed spectro-

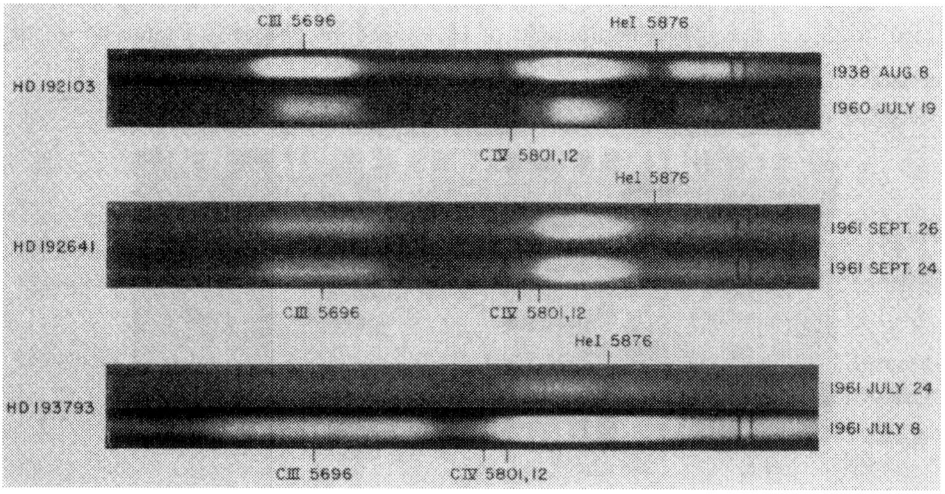

Fig. 45. Spectra of HD 192103, HD 192641, and HD 193793 in the spectral region λ 5600 to λ 5900

gram is shown in order to make the absorption components more readily visible. The shortward displaced He I λ5876 line is also indicated. The sharp absorption lines are due to interstellar Na I. The flat-topped emission feature due to C III λ5696 may also be seen.

The lines C IV λ5801, λ5812, N V λ4603, λ4619, and O VI λ3812, λ3834 come from the transition 3^2S–3^2P^0. The level 3^2S lies in each case not far above the 2^2P^0 levels which are coupled directly to the ground level. In each spectrum the 2^2P^0 levels lie relatively close to the ground level, considering the electron temperatures we are concerned with in Wolf-Rayet atmospheres. Thus, in effect, the 3^2S levels in C IV, N V, and O VI have rather lower excitation potentials than the lower levels of other transitions which may be observed in the spectra of C IV, N V, and O VI. It appears that these levels are sufficiently populated in the expanding shell that shortward-displaced absorption components in the 3^2S–3^2P^0 transitions of C IV and N V may be observed. No case is yet known in which the O VI lines have been observed in absorption, but it must be remembered that few Wolf-Rayet spectra have been studied in detail at sufficiently high resolution to detect such absorption lines were they present.

The important point about the shortward-displaced absorptions of C IV, N V, and possibly O VI is that these transitions come from *normal* levels. Thus the relative populations of the levels allow one to estimate relative temperatures and to arrange the Wolf-Rayet stars in a temperature sequence. The excitation potentials and ionisation potentials increase from C IV to N V to O VI. The conclusion is that the general level of excitation is lower in the expanding envelopes of WC stars than it is in the expanding envelopes of WN stars. This conclusion is in agreement with the conclusion of UNDERHILL (1958b) reached from a study of the possible shape of the extreme ultraviolet spectra of Wolf-Rayet stars. If Wolf-Rayet stars were found in which the O VI lines at λ3812 and λ3834 appeared in absorption, they would have the highest temperatures.

Since the lower excitation potential of the C IV lines λ5801 and λ5812 is 37 volts whereas that of He II λ4686 is 48 volts and that of the N V lines is 56 volts, one would expect weak absorption for He II in a shell which gave C IV λ5801 and λ5812 in absorption. Strong shortward-displaced He II absorption should occur in a WN star where the shortward displaced N V absorption lines are seen. It has been suggested that this type of absorption, which can cause a not very obvious mutilation of an emission line profile, is the cause of the well known positive displacement in spectroscopic binaries of velocities measured using He II λ4686. All but two of the double-lined spectroscopic binaries which have been studied are of type WN and the γ velocities estimated from the positions of He II λ4686 are usually positively displaced with respect to the γ velocity found from the normal absorption stellar spectrum of the companion.

The emission line C III λ5696 in WC spectra is the only truly flat-topped emission line in Wolf-Rayet spectra. It probably originates mainly in the expanding shell as a result of a fluorescent process involving He II λ303. This fluorescent process is explained in Chapter XV where the processes causing particular lines to appear in emission in Of spectra are described. According to the hypothesis for the production of C III λ5696 selectively in emission (UNDERHILL, 1957) a population of C^+ions is required as well as radiation in the line He II λ303.78. Thus one would expect C III λ5696 to appear preferentially in expanding shells of not too high temperature. This line occurs only in WC spectra and not in WN spectra. This observed fact is consistent with the idea that the level of excitation is higher in WN atmospheres and shells than in WC atmospheres and shells. In the most densely exposed spectrogram of HD 193793 shown in Figure 45, some structure, which is probably due to gas streams in the binary system, is visible in the middle of the very broad C III λ5696 emission feature. Beals was lead to his suggestion of an expanding atmosphere for Wolf-Rayet stars chiefly as a result of his study of the profile of C III λ5696. O. C. WILSON (1942) considered chiefly the behaviour of lines which present ideas suggest are largely formed in the inner atmosphere where the motions are not directed in one direction and he could find little support for the concept of an expanding atmosphere. One gains the impression that some of the He I lines that have strong shortward displaced absorption components have also flat-topped emission components but the evidence is not so clear as for C III λ5696.

TABLE 29
The Expansion velocities in Wolf-Rayet stars and novae

Wolf-Rayet star spectrum	Expansion velocity	Nova	Expansion velocity
HD 184738 WC8	580 km/sec	Herculis 1934	310 km/sec
HD 192103 WC7	1200	Cygni 1948	548
HD 192641 WC7	1300	Lacertae 1950	872
HD 192163 WN6	1400	Aquilae 1945	1135
HD 191765 WN5	1600	Aquilae 1918	1380
HD 193793 WC6	2700	Sagitarii 1936	2120

The velocities of expansion found for Wolf-Rayet stars are large. Some results, chiefly from the work of the author, are given in Table 29 where they are compared with the velocities of expansion shown by various novae. It is clear that the velocities of expansion shown by Wolf-Rayet stars are larger than what are sometimes observed for novae. No evidence exists for any of the Wolf-Rayet stars listed that there has been any significant change in the velocity of expansion during the years within which observations have been made. Most of the variations in line strength mentioned in Section 2.5 may be explained by postulating that the *amount of material* in the expanding shell changes from time to time rather than that the velocity of expansion changes by a significant amount.

It is clear that Wolf-Rayet stars must be loosing mass, for the observed velocities of expansion exceed the velocity of escape. We have

$$v_{\text{esc}} = (2GM/R)^{\frac{1}{2}} \tag{XIII-1}$$

or, when the mass and radius are measured in solar units,

$$v_{\text{esc}} = 618\,(M/R)^{\frac{1}{2}}\,\text{km/sec}. \tag{XIII-2}$$

For an average Wolf-Rayet star the mass may be about six solar masses and the radius of the inner atmosphere may be about six solar radii. Thus any material moving with a velocity in excess of about 600 km/sec will escape. The observed velocities of expansion are usually greater than 600 km/sec.

In order to estimate how much mass a Wolf-Rayet star is loosing, one must estimate how much material is crossing the outer boundary of the expanding shell per unit time. For an order of magnitude calculation let us assume that the radius of the shell is $16R_{\odot}$, the velocity of expansion is 1000 km/sec and that the density in the shell is 10^{-12} g/cm^3. The latter value is one tenth the density in the outermost part of a model atmosphere constructed to represent a main sequence O9 star. The selected typical values of radius, velocity and density lead to a loss of 4×10^{27} g per year or one solar mass in 5×10^5 years. Clearly Wolf-Rayet stars are not loosing mass at a rate that is fast in comparison with the time expected for them to pass through their chief internal energy generation evolution. At the above rate it would take a Wolf-Rayet star about 2×10^6 years to loose enough mass so that it could become a white dwarf. If the density in the expanding shell is an order of magnitude lower than what has been estimated, the rate of mass loss will be an order of magnitude smaller.

4. Wolf-Rayet spectroscopic binaries

About half of the known Wolf-Rayet stars are members of binary systems. ROBERTS (1962) has prepared a list of the known binaries. The system HD 192641, WC7 + Be should be added to this list. Seven double-lined stars for which orbits exist are listed in Table 28. It is known that many of these stars show intrinsic variations in light of small range and that spectral variations may be seen when spectrograms of moderate or high dispersion are inspected. However, little quantitative data is available except for HD 193576 (V444 Cygni) and for HD 214419 (CQ Cephei) both of which are eclipsing variables.

The star HD 193576 was the first Wolf-Rayet spectroscopic binary to be discovered (WILSON, 1939). After the announcement by O. C. Wilson of the radial velocity variation, S. Gaposchkin examined Harvard patrol plates and discovered that the system was eclipsing. In the next ten years photometric and spectrographic studies were made of this system by a number of astronomers. The photometric results have been summarised and discussed by KRON and GORDON (1950) and the spectrographic studies have been summarised and coordinated by MÜNCH (1950). These authors give full references to earlier work. We shall point out here some of the highlights of the resulting model for a Wolf-Rayet star.

Light curves are available at $\lambda 3500$, $\lambda 4500$, and $\lambda 7200$ (HILTNER, 1949a, KRON and GORDON, 1950) and they indicate that the star is surrounded by an extended electron-scattering envelope, for the depth of minimum does not vary with wave length and the minimum is wider when the OB component is eclipsed (primary minimum) than when the Wolf-Rayet star is eclipsed. The spectral type of the OB companion is not well known. Münch gives reasons for concluding that it is later than O6; following ROBERTS (1962) the type B1 is listed in Table 28. Probably the true spectral type is intermediate between these extremes. Since spectral types in the range B1 to O6 are largely assigned according to the relative intensities of absorption lines occurring in extended atmospheres, it is not surprising that it is not possible to assign a certain spectral type to the companion star in a system such as V444 Cygni where there is evidence of much gas around both components. In such a case the empirical classification criteria are strongly dependent on chance situations such as how much gas happens to be in the system close to the stars at any moment.

The relative dimensions of the orbits and of the stars can best be estimated from the analysis by Kron and Gordon. The results differ depending upon what values are adopted for the radial-velocity orbit. The following dimensions have been found using the orbits by WILSON (1942) and by KEEPING (1947).

Quantity	O. C. Wilson orbit	Keeping orbit
radius of O star orbit	10.5 $R_{\odot}$	14.7 $R_{\odot}$
radius of WR star orbit	26.2	26.0
separation of the stars	36.7	40.7
radius of O star	9.3	10.3
radius of WR core	2.1	2.3
radius of WR disc	7.0	7.8
radius of WR envelope	16.	18.

Münch has derived a radial-velocity orbit which is essentially the same as that by O. C. Wilson. In a spectrum so full of blended absorption and emission features as the spectrum of V444 Cygni, it is very difficult to determine the effective wave lengths to use so that the measured displacements of the absorption and emission features represent the orbital motions of the two stars. Wilson and Münch measured their spectrograms in the normal manner; Keeping attempted to resolve some of the problems of blending by measuring on microphotometer tracings. It is clear that whichever set of results is adopted, we have an internally consistent set of parameters with which to describe the relative sizes of the stars and of their orbits. The inclination of the orbit is 78°.4. Consequently the results of O. C. Wilson and of Münch indicate that the mass of the Wolf-Rayet star is 10.7 solar masses and that of the O star is 25.6 solar masses.

Münch has discussed the problems raised by the spectral variations which are observed throughout the orbital period and from cycle to cycle. He has pointed out that V444 Cygni is a very complicated system consisting of two massive stars embedded in an extensive envelope in which large outward directed motions of the order of 1000 km/sec occur. In particular, striking changes occur in He II λ4686, in lines of N III, of N IV, of N V, and of He I, especially λ3888. These changes can best be interpreted as the result of gas streams in the system. SAHADE (1958) has drawn attention to narrow, relatively sharp emission features which can be seen superposed on the broad He II λ4686 emission line, and he has suggested that there is gas in the system near the Lagrangian point L_1. This gas is considered to give rise to the observed sharp emission features. The additional fact that sharp displaced absorption components occur, particularly near primary minimum, indicates the presence of gas streams in the system. Furthermore, the emission lines which are strengthened by monochromatic fluorescent processes in Of atmospheres (see Chapter XV) are enhanced at some phases in the orbital period. Clearly the spectrum of the Wolf-Rayet star is not a pure recombination spectrum and it cannot be interpreted by the simple hypothesis of only an expanding envelope. Münch suggests that many of the spectral details can be understood by postulating an atmosphere in which the level of excitation and ionisation increases outward, and where broadening of the emission lines by electron scattering is important. There may well be deceleration of the particles as they move outward from the stellar surface.

The star HD 214419 (CQ Cephei) is a second very famous Wolf-Rayet eclipsing binary. Its spectral type is WN5+(?) B1. The period is 1.641 days or 39.4 hours. The light curve is continuously variable, rather as for β Lyrae, and there are two nearly equal minima. The most comprehensive photometric study is that by HILTNER (1950) who has observed light curves at λ3550, λ4686, and λ5300. Hiltner has concluded

1. that the primary eclipse corresponds to the eclipse of the Wolf-Rayet component,
2. that the ultraviolet and violet light curves are deeper at both minima than the λ5300 curve,
3. that the primary eclipse is the narrower eclipse (at this time the Wolf-Rayet star is eclipsed),
4. that both minima show marked asymmetries which may be associated with the spectral changes which are known to occur but which are not yet well documented,
5. that the He II λ4686 emission is stronger by about 0.04 magnitude when the Wolf-Rayet star is eclipsed than when it is in front of its companion,
6. that the intensity of He II λ4686 is intrinsically variable,
7. that no changes in polarisation greater than 0.005 occur.

Clearly the system CQ Cephei is another binary system which is enveloped in gas. The gas contributes not only absorption features to the combined spectrum but also emission lines. HILTNER (1944b) has given a preliminary spectrographic orbit, but the system is not favourable for interpretation since the spectrum varies from cycle to cycle. Owing to the faintness of the star it is difficult to obtain high dispersion spectrograms with sufficient time resolution to follow the spectral changes around a cycle. Also it is necessary to observe the star from several places on the earth if a more or less continuous series of observations is to be obtained throughout one cycle.

BAPPU and SINVHAL (1955) have measured equivalent widths of He II λ6560, He I λ5857, He II λ5411, and N IV λ4058 at different phases of CQ Cephei and they find that all of these lines increase in intensity at primary and secondary minima just as He II λ4686 does. On some nights they found that all lines were increased in intensity over the general average. The profile of He II λ4686 shows a sharp central peak and wide wings at primary minimum when the Wolf-Rayet star is eclipsed, but it is more rounded when the Wolf-Rayet star is in front at secondary minimum. Bappu thinks that the observations of CQ Cephei could be explained by postulating that the two stars have a common envelope.

The double-lined spectroscopic binary γ_2 Velorum, HD 68273, is apparently bright and thus is a candidate for serious study. Some brief notes about its spectrum have appeared (PERRINE, 1920; SMITH, 1955; SAHADE, 1958; ALLER, FAULKNER, and ROOD, 1962; ALLER and FAULKNER, 1963) and CODE and BLESS (1964) have given a detailed description of the spectrum, in particular in the infrared. The spectrum of γ_2 Velorum is variable and there is considerable evidence that gas streams are important in this binary system; however, no comprehensive radial-velocity study of this star has yet been published.

5. Some unsolved problems raised by the Wolf-Rayet stars

One of the most interesting questions that has not yet been answered about Wolf-Rayet stars is to determine at what stage of evolution these stars are. A related problem is to develop an adequate theory of line formation so that the observed spectra can be interpreted quantitatively and without contradiction in terms of temperature, pressure, density, and state of motion of the gases forming the stellar atmosphere. Intuition suggests that if one could determine the physical state of the atmosphere, which in principle can be found from an analysis of the observed relative intensities and line shapes in the spectrum, one could make a secure statement concerning the stage of development of the stars which are placed in the classes WC and WN.

The Wolf-Rayet stars are a rather inhomogeneous group of objects which have the following general properties:

1. They are associated with O and B stars and about half are known to be members of binary systems. Some are associated with nebulosity.

2. The masses of Wolf-Rayet stars are not excessively large. The average Wolf-Rayet star which is a member of a double-lined binary system has a mass between four and eight solar masses.

3. The average Wolf-Rayet star has a visual absolute magnitude in the neighbourhood of -3 to -4, but there appears to be a handful of Wolf-Rayet stars which are as bright as -6. These occur in the 30 Doradus complex and in the cluster NGC 6231. Any reduction in the adopted distance of the 30 Doradus complex and of NGC 6231 would reduce the absolute magnitudes of the Wolf-Rayet stars which appear to be members of these stellar groupings. The absolute magnitude -6 is not typical of most Wolf-Rayet stars. Even at -3, the Wolf-Rayet stars are overluminous for their masses in comparison to the luminosities expected for stars of the same mass which lie on the main sequence.

4. Small intrinsic, irregular variations in light are observed for many Wolf-Rayet stars and the spectra seem to vary irregularly in many cases. The chief spectral variations are caused by changes in the apparent strengths of the few absorption lines which are observed rather than by changes in the general emission spectrum.

5. Wolf-Rayet stars are losing mass at a rate in the neighbourhood of one solar mass in 5×10^5 years.

The spectroscopic data have not yet been analysed in a way to give a description of the physical conditions in the stellar atmosphere that is free from contradictions. However, the following statements appear to be plausible:

1. The electron temperature in Wolf-Rayet atmospheres lies in the range 30000° to 50000°. Furthermore, the electron temperature is lower in WC atmospheres than in WN atmospheres.

2. Most of the emission lines are formed by recombination in a relatively small, moderately dense atmosphere where the line broadening is caused by the large random motions of the ions and perhaps by electron scattering. The average velocities of the ions are in the range 500 to 1000 km/sec.

3. A less dense, rapidly expanding envelope exists outside the inner atmosphere. Here the electron temperature is high, but the radiation field is not in thermal equilibrium. Consequently dilution effects are observed and particular lines may appear in emission as a result of monochromatic fluorescent processes. The density in this expanding atmosphere is not very low since this envelope produces absorption in strong lines from relatively low-lying normal levels as well as in lines from levels which are overpopulated with reference to thermodynamic equilibrium due to the presence of a metastable level in the level scheme of the atom or ion and to the non-equilibrium conditions. The expanding atmosphere is not visible in all spectral lines.

It seems clear that any theory of the formation of the emission and absorption lines in Wolf-Rayet spectra should take into account collision processes as well as purely radiative interactions between atoms, ions, and radiation. The intrinsic variations in spectrum and light which are observed indicate that Wolf-Rayet atmospheres are mechanically unstable and that perhaps there is heating by mechanical energy. Code and Bless have noted that protons moving with speeds near 1000 km/sec have an energy of 5 kev while helium ions (alpha particles) moving at these speeds have energies of 21 kev. At such large velocities atomic collisions may be as important in establishing the stage of ionisation and excitation in the Wolf-Rayet atmosphere as purely radiative processes. Code and Bless have made an order of magnitude calculation and they are able to show that in an average WC7 atmosphere 60 per cent of the ionisation may be caused by protons, the rest chiefly by alpha particles, rather than entirely by radiative processes as is assumed in the papers summarised in Section 2.4.

It seems clear that Wolf-Rayet spectra may be expected at all stages of stellar development where a mechanically unstable atmosphere is generated and the radiation field corresponds to a high temperature. One question that has so far not been answered is the origin of the large motions which occur in the atmospheres of Wolf-Rayet stars.

In spite of the large motions which occur, Wolf-Rayet stars appear to be in a state of quasi-equilibrium. The spectra do not change significantly in 30 years or longer. LIMBER (1964) has suggested that the required velocity field is produced by forced rotational instability. That is he assumes that Wolf-Rayet stars are rotating at the limit of rotational instability and that as a result of continuing gravitational contraction in a *post-main-sequence* stage they continually eject matter tangentially from the equator with speeds equal to the critical velocity there.

A difficulty with any theory based on symmetrical phenomena such as rotation and contraction is to account for the *fluctuations* in light and spectrum which occur for many Wolf-Rayet stars. STOTHERS (1965) has also suggested that Wolf-Rayet stars are in a post-main-sequence stage of development. He would like to identify them with an instability strip in the Hertzsprung-Russell diagram in which, at lower masses, the β Cephei stars lie. Stothers assumes (contrary to all observational knowledge) that Wolf-Rayet stars are very massive stars. His results imply that the observed atmospheric motions are the result of an instability against radial pulsation.

Ideas of the sort suggested by Limber and by Stothers conflict seriously with the observed fact that all direct evidence about the masses of Wolf-Rayet stars indicates that the Wolf-Rayet stars are considerably less massive than their OB companions. The companion stars are believed to be on or near the main sequence and normal theory would suggest that the less massive Wolf-Rayet stars are less evolved.

A much more credible idea concerning the position of the Wolf-Rayet stars in the Hertzsprung-Russell diagram is that they are very young stars just approaching the main sequence via the Hayashi contraction sequence. This initial contraction is believed to generate much mechanical motion in the atmosphere. It is generally thought that the T Tauri stars (Herbig, 1962b, Kuhi, 1964) are examples of low mass stars at this stage of development. There are large motions in the atmospheres of T Tauri stars. According to Iben (1965) a star of 15 solar masses takes 6.170×10^4 years to reach the main sequence; a star of 5 solar masses takes 5.759×10^5 years, while a star of 1 solar mass takes more than 5×10^7 years.

If the point of view is adopted that Wolf-Rayet stars are young, rather massive stars (say, about six solar masses) approaching the main sequence for the first time, no difficulty is met with the observed statistics about OB–WR binaries. The observed more massive component is also the most evolved; it has reached the main sequence. Furthermore one can speculate, following theories of the heating of the solar chromosphere and corona, that some of the high excitation seen in Wolf-Rayet spectra is due to heating from acoustic energy generated by the field of mechanical energy in the extended envelope.

The above point of view leads also to an explanation of the fact that no Wolf-Rayet stars are observed in the Orion and Taurus clouds although many T Tauri stars are found there. This state of affairs could arise if some event stopped the formation of massive stars in the Orion and Taurus clouds sufficiently long ago that all the Wolf-Rayet stars have evolved onto the main sequence. They can then no longer be recognised by their spectra but appear as B type stars. The implication follows that star formation may still be in progress in regions such as Cygnus, Cepheus, and Carina where Wolf-Rayet stars now are found.

The spectral phenomena which are selected to characterise Wolf-Rayet stars result in a selection of stars with large atmospheric motions and a high level of excitation in their atmosphere. It is possible that a Wolf-Rayet spectrum could result at each evolutionary stage in the life of a star where large atmospheric motions and an extended atmosphere at high electron temperature are generated either by convection, shock waves, or rotational instability. If this is true, the Wolf-Rayet stars do not form a pure population from the point of view of evolutionary development. There may be young Wolf-Rayet stars and old Wolf-Rayet stars. It certainly seems to be unwise to conclude that all Wolf-Rayet stars are post-main-sequence objects.

CHAPTER XIV

SUPERGIANTS AND P CYGNI

1. The early type supergiants

The early type supergiants are relatively easy to recognise because their absorption lines have a steep-sided blunt shape. The wings of the hydrogen lines are narrow and frequently an emission feature is seen at Hα in the spectra of the Ia supergiants. The Hα feature usually consists of a fairly sharp emission component accompanied by a shortward displaced absorption component. Such a composite line profile is called a P Cygni line profile, see Section 4. Emission lines are occasionally visible at Hβ and at He I λ5876 and λ6678. In the late O type supergiants weak, rather sharp emission lines appear at C III λ5696 and at N III λ4634–41. The He II line λ4686 is partially filled in by emission in these stars. In the O9 and O9.5 supergiants the sharp emission lines are superimposed on rather broad faint emission features. R. WILSON (1956) has noticed that this is also true for the Hα and Hβ emissions in the B1Ia supergiant HD 190603.

The MK standard O and B supergiants (MORGAN and JOHNSON, 1953) are listed in Table 30. The V magnitude is given where available; in the few cases where the V magnitude could not be found, the visual apparent magnitude m_V from the *Henry Draper Catalogue* is listed in italics. Two of the MK supergiants, namely HD 38775 κ Orionis B0.5Ia and HD 9311 B5Ib, probably are not supergiants. High dispersion spectrograms of κ Orionis show that this star has broad lines typical of moderate

TABLE 30

The O and B type standard supergiants

HD Number	Name	α (1900)	δ (1900)	V	Spectral type	Source for magnitude	Remarks
210809		22ʰ07ᵐ9	+ 51°56′	7.56	O9I	2	
30614	α Cam	4 44.1	+ 66 10	4.38	O9.5Ia	3	
195592		20 27.2	+ 43 59	7.08	O9.5Ia	4	
37742	ζ Ori	5 35.7	− 2 00	1.72	O9.5Ib	3	
209975	19 Cep	22 02.1	+ 61 48	5.10	O9.5Ib	4	
37128	ε Ori	5 31.1	− 1 16	1.71	B0Ia	1	
167264	15 Sgr	18 09.3	− 20 46	5.38	B0Ia	5	
204172	69 Cyg	21 21.7	+ 36 14	*5.84*	B0Ib	6	
38775	κ Ori	5 43.0	− 9 42	2.04	B0.5Ia	1	8
194839		20 22.8	+ 41 03	7.49	B0.5Ia	5	
192422		20 09.7	+ 38 28	7.09	B0.5Ib	5	
2905	κ Cas	0 27.3	+ 62 23	4.15	B1Ia	1	

TABLE 30 (*concluded*)

HD Number	Name	α (1900)	δ (1900)	*V* Mag.	Spectral type	Source for magnitude	Remarks
216411		22 47.6	+ 58 28	7.20	B1Ia	5	
24398	ζ Per	3 47.8	+ 31 35	2.83	B1Ib	1	
91316	ρ Leo	10 27.5	+ 9 49	3.85	B1Ib	2	
190919		20 02.2	+ 35 24	*7.30*	B1Ib	6	
190603		20 00.7	+ 31 56	5.65	B1.5Ia	5	
194279		20 19.7	+ 40 26	7.01	B1.5Ia	5	
193183		20 13.8	+ 37 55	*7.12*	B1.5Ib	6	
14143		2 12.2	+ 56 43	6.66	B2Ia	2	
41117	χ^2 Ori	5 58.0	+ 20 08	4.63	B2Ia	1	
13841		2 09.8	+ 56 34	7.40	B2Ib	2	
13866		2 10.0	+ 56 15	7.52	B2Ib	7	
206165	9 Cep	21 35.2	+ 61 38	4.72	B2Ib	1	
42087	3 Gem	6 03.7	+ 23 08	5.76	B2.5Ib	5	
14134		2 12.1	+ 56 40	6.55	B3Ia	2	
53138	o^2 CMa	6 58.8	− 23 41	3.01	B3Ia	3	
198478	55 Cyg	20 45.5	+ 45 45	4.82	B3Ia	1	
43384	9 Gem	6 10.8	+ 23 46	6.28	B3Ia	1	
13267		2 04.6	+ 57 11	6.39	B5Ia	2	
58350	η CMa	7 20.1	− 29 06	2.41	B5Ia	3	
167838		18 11.9	− 15 28	6.73	B5Ia	5	
36371	χ Aur	5 26.2	+ 32 07	*4.88*	B5Iab	6	
9311		1 26.6	+ 60 10	*7.26*	B5Ib	6	9
164353	67 Oph	17 55.6	+ 2 56	4.00	B5Ib	3	
15497		2 24.6	+ 57 15	7.03	B6Ia	2	
183143		19 23.0	+ 18 05	6.87	B7Ia	5	
15452		2 $15^{\mathrm{m}}.9$	+ 56 56	7.00	B8Ia	2	
34085	β Ori	5 09.7	− 8 19	0.08	B8Ia	1	
199478		20 52.4	+ 47 02	5.66	B8Ia	5	
12301	53 Cas	1 55.6	+ 63 54	5.58	B8Ib	5	
208501	13 Cep	21 51.5	+ 56 08	5.79	B8Ib	5	
17088		2 39.5	+ 57 19	7.50	B9Ia	5	
21291		3 21.0	+ 59 36	4.23	B9Ia	1	
202850	σ Cyg	21 13.5	+ 38 59	4.28	B9Iab	6	
21389		3 21.9	+ 58 32	4.53	A0Ia	5	
223960		23 48.9	+ 60 18	6.91	A0Ia	5	
46300	13 Mon	6 27.5	+ 7 24	*4.50*	A0Ib	6	
87737	η Leo	10 01.9	+ 17 15	*3.58*	A0Ib	6	

Sources for Magnitude and Remarks

1. Morgan, W. W. and Johnson, H. L.: 1953, *Astrophys. J.* **117**, 313.
2. Johnson, H. L.: 1955, *Ann. Astrophys.* **18**, 292.
3. Eggen, O. J.: 1961, *Roy. Obs. Bull.* No. **41**.
4. Hiltner, W. A. and Johnson, H. L.: 1956, *Astrophys. J.* **124**, 367.
5. Hiltner, W. A.: 1956, *Astrophys. J. Suppl.* **2**, 389.
6. *Henry Draper Catalogue.*
7. Johnson, H. L. and Morgan, W. W.: 1955, *Astrophys. J.* **122**, 429.
8. This star is probably not a supergiant; see the notes to Table 10.
9. This star is probably not a supergiant; see M_V from Hγ, Table 10.

rotation. It is possible that the star had a weak shell when the classification spectra were taken. The spectrum of HD 9311 also is not typical of high luminosity. This star may belong to luminosity Class III, but further study is necessary to settle this point. The visual absolute magnitudes of the supergiants are discussed in Chapter III.

2. The variability of supergiants

The radial velocity of early type supergiants appears to vary in an irregular manner with a range of about 30 km/sec. A detailed study of the variations in the case of χ^2 Orionis, 55 Cygni, and 67 Ophiuchi has been made by UNDERHILL (1960) who has shown that the velocity changes are not periodic in character and has suggested that the variations are due to motions of the atmosphere. ABT (1957) has shown that the A and F type supergiants also vary in radial velocity in a semi-regular manner. The best documented case is α Cygni, A2Ia. PADDOCK (1935) observed α Cygni at high

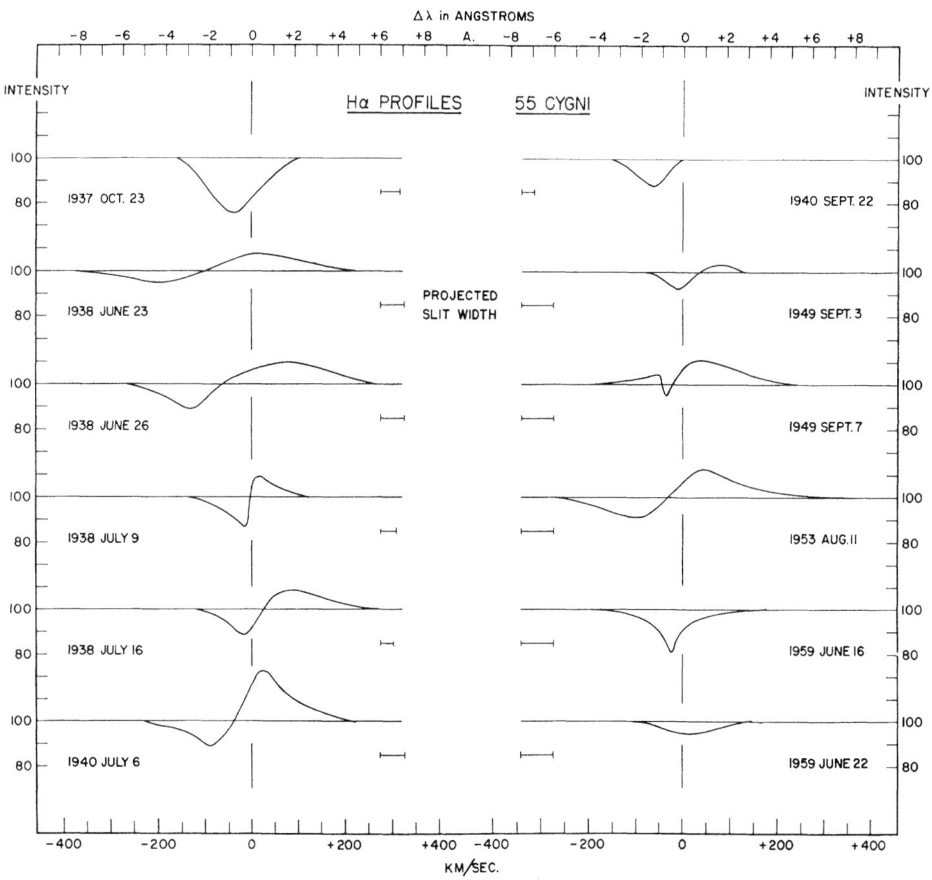

Fig. 46. The variable profile of Hα in the spectrum of 55 Cygni, B3Ia.

dispersion on every opportunity for three years and found that the radial velocity varied from +6 km/sec to −9 km/sec. However, the variation is not periodic. It seems more likely that the observed variations in radial velocity are due to motions of the stellar atmosphere than to orbital motion. It would be interesting to know how the line profiles and equivalent widths vary as the radial velocity varies.

The profile of the Hα line varies in an irregular manner in the Ia supergiants of type B. An example of what may occur is shown in Figure 46 which displays some changes observed for 55 Cygni. The spectral purity is different on the various spectrograms. It is clear that the strength of the emission component and its shape changes from time to time. The Hα profiles of the Ib supergiants do not show emission features and they are more stable in shape. In Figure 47 the profile of Hα is shown for ρ Leonis, B1Ib, 67 Ophiuchi, B5Ib, and for ι Herculis, B3IV. The line shape is determined by Stark broadening in the case of the main-sequence star. The profiles of the Ib supergiants may be interpreted in terms of rotation of the star or in terms of macroturbulence, that is one says that the line form is given by the statistical distribution of large-scale motions in the stellar atmosphere.

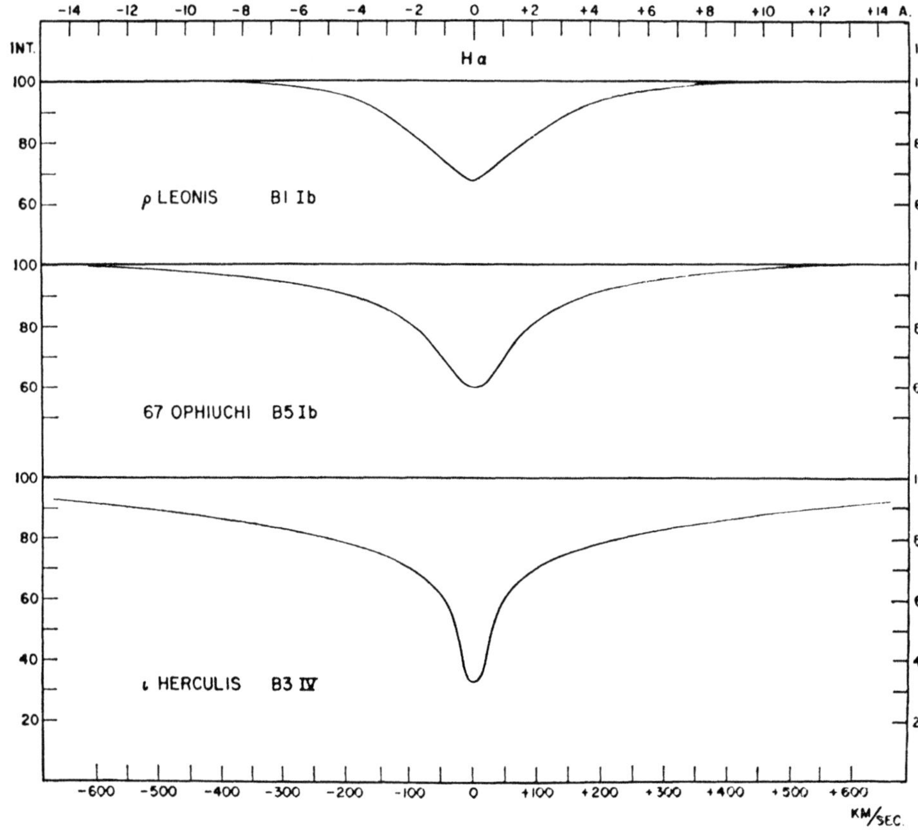

Fig. 47. The profile of Hα in ρ Leonis, B1Ib, 67 Ophiuchi, B5Ib, and ι Herculis, B3IV.

In the case of the Ia supergiants the absorption dip at Hα is always displaced shortward and the emission peak, if present, occurs on the longward side of the line. Such composite profiles may be interpreted directly in terms of an expanding atmosphere. The expected symmetrical normal absorption line due to Hα formed in the normal reversing layer is not visible. It must be assumed that it has been filled in by emission. An interpretation of the Hα profile of the Ia supergiant α Cygni has been put forward by GROTH (1960). He assumes that part of the shortward displaced absorption dip is the remaining part of the photospheric absorption feature. He reconstructs the photospheric feature (using the predictions from a model for the atmosphere of α Cygni), and from the differences with the observed profile he obtains the profile of the emission and absorption component from the shell. Since the emission component is longward displaced, Groth implies that the outer atmosphere of α Cygni may be falling in towards the stellar surface. The longward displacement may also be caused by mutilation of an originally symmetrical emission component by the strong shortward displaced absorption core.

It is very likely that the OB supergiants fluctuate slightly in light, but no series of observations is available for the apparently bright supergiants which are studied spectroscopically.

3. The interpretation of the spectra of supergiants

Standard analyses have been performed for the supergiants 55 Cygni, B3Ia, and ζ Persei, B1Ib, see Chapter XI. It is somewhat doubtful how much trust should be put in the numbers which come from these analyses because the appearance of the spectrum, especially in the case of the Ia supergiants, suggests that a plane parallel layer of gas in hydrostatic equilibrium does not make a good model for a supergiant atmosphere. The distinctive features of the spectrum, that is the line shapes and the relatively great intensities of lines which are strong at low densities, indicate that the atmospheres are greatly extended and in chaotic motion.

The equation of hydrostatic equilibrium may be written as

$$\frac{\mathrm{d}P_{\mathrm{g}}}{\mathrm{d}\tau} = \frac{g}{\kappa} - \frac{\mathrm{d}P_{\mathrm{r}}}{\mathrm{d}\tau} - \frac{\mathrm{d}P_{\mathrm{turb}}}{\mathrm{d}\tau}. \tag{XIV-1}$$

Here P_g is the gas pressure, P_r is the radiation pressure and P_{turb} is a pressure term which represents the action of the atmospheric motions in supporting the stellar atmosphere. The optical depth, τ, is measured using a characteristic absorption coefficient κ, and the acceleration of gravity is g. If a supergiant atmosphere is to exist for any length of time, equation XIV-1 should be satisfied.

The MK classification system recognises supergiants only to type O9. From equation XIV-1 one may show that supergiants corresponding to significantly higher temperatures would not be expected. The condition that a stellar atmosphere exist is that the term $\mathrm{d}P_g/\mathrm{d}\tau$ should be greater than zero. Let us set $\mathrm{d}P_g/\mathrm{d}\tau$ equal to zero and furthermore let us ignore the last term on the right in our first approximation and see what value of g is required in order to have a stellar atmosphere in mechanical equilibrium at high temperature. If this value of g is in the neighbourhood of 10^4 cm/sec^2 a main-sequence star is expected. If it is in the neighbourhood of 10^2 the star will be a supergiant. These are empirical statements drawn from a consideration of the value of g at different places in the Hertzsprung-Russell diagram and from fitting observed hydrogen-line profiles with those computed from model atmospheres of OB stars.

Let us consider a star which has a temperature $T_0 = 30000°$ in its outer layers. Such a temperature might correspond to about type O8. Then if one assumes a grey-body temperature law it turns out that $dP_r/d\tau$ is of the order of 3500. Furthermore one will expect κ to be less than or equal to 1.0. In order to have mechanical stability g must be greater than 3500. The conclusion is obvious. Stars having temperatures as large as 30 000° in their atmospheres cannot exist unless they have values of g typical of main-sequence stars. The presence of the $dP_{turb}/d\tau$ term makes this argument even stronger.

Supergiants of spectral types earlier than O9 cannot exist, for their atmospheres will dissipate because the term $dP_g/d\tau$ will be negative when g is less than about 5×10^3. Model atmosphere computations indicate that when g is greater than 10^3 the profiles of the Balmer lines of hydrogen will be Stark broadened and like what has been empirically selected as representing main-sequence stars. Theory and the MK classification system are more or less in accord up to spectral type O9. The difficulty arises when one attempts to reconcile with the above estimates the suggestions of Kopylov (1958b) and Hack (1963) (see Chapter III) that supergiants of spectral types earlier than O9 exist and that they can be discovered by spectral classification.

The use of the terms supergiant and main-sequence star implies that after having sorted the stars according to effective temperature (spectral type) one can then separate the stars with an expanded photosphere (large radius) from those with a normal photosphere (normal radius). This separation is done by examining the relative strengths and shapes of selected spectral lines. It is based upon the fact that the pressure will be lower in the extended atmosphere than in the normal atmosphere, and thus certain changes in the ionisation equilibria and in the effect of Stark broadening on the hydrogen and helium lines will be evident. In the B spectral types the profiles of the empirically selected luminosity lines are more and more determined by the state of motion in the extended atmosphere than by pressure effects as one goes to earlier types. Thus to a considerable extent the physical parameter which is empirically chosen to enable us to select stars of different luminosity is the amount of motion in the atmosphere. Up to about type O9, the Stark broadening of the Balmer lines of hydrogen remains a very useful and essential means of separating stars of different luminosities. However, at spectral types earlier than O9, the observed strong lines of H, He I, and He II are formed chiefly high above the main photosphere and here the dominant broadening factor is chaotic motion, not Stark effect. Thus, since large chaotic motions can occur in an extended shell around a main-sequence star as well as in the extended atmosphere of a supergiant, the apparent strengths of the generally selected "luminosity criteria" do not provide a true discrimination in luminosity at the earliest spectral types. They merely enable one to isolate stars with extended envelopes. What is lacking is a spectroscopic criterion which can be used to infer the radius of the photosphere of the star rather than the radius of the extended atmosphere which is observed in line frequencies. The additional criteria selected by HACK (1963) are no better in this respect than the well established criteria. Unfortunately, as was pointed out in Chapter III, the available *independent* information about the distances of the O stars is not precise enough to enable us to settle the question of the luminosities of the O stars without recourse to the theory of line formation in stellar atmospheres. This theory suggests that stellar atmospheres in which g is small and the radiation pressure gradient and turbulence pressure gradient is high cannot exist.

4. P Cygni

4.1 *The spectrum of P Cygni*

This star (HD 193237) is one of the first stars discovered in which the spectrum is dominated by fairly sharp emission lines which are accompanied by strong, shortward displaced absorption lines. Except for the emission lines, the spectrum resembles that of a B1 supergiant. This star is a type star; relatively sharp emission lines accompanied by shortward displaced absorption components are called "P Cygni lines".

The star P Cygni is different from the normal supergiants in that it has changed brightness by several magnitudes during historical time.

A summary of the detailed history of the known light variations of P Cygni has been given by BEALS (1950). The high-lights are as follows. On August 8, 1600 P Cygni was noted to be at the third magnitude where it stayed until 1606. Before that time it was not known as a naked-eye star. By 1620 it was at about the 6th magnitude. Between 1626 and 1654 P Cygni was not visible to the naked eye. In 1655 it brightened again to the 5th magnitude and then to magnitude 3.5 where it remained until 1659. It declined in brightness until in 1662 it was no longer visible. In 1665 the star was near the 6th magnitude and it fluctuated in light sometimes being fainter, sometimes brighter. In 1715 P Cygni reached magnitude 5.2 and since this date no major changes in brightness have occurred.

An intensity tracing of the spectrum of P Cygni in the spectral region 3925 Å to 4000 Å is shown in Figure 48. The vertical scale is linear; the bottom line of the drawing corresponds to "clear plate". This tracing was made from a spectrogram at 2 Å/mm obtained by HERBIG (1962c) with the coudé spectrograph of the 120-inch telescope of the Lick Observatory. Several typical features are visible in the section of tracing which is displayed here. The emission and absorption profile of N II λ3995 is a typical "P Cygni line". The peak of the emission corresponds in wave length nearly to the laboratory wave length while the absorption core is displaced 90 km/sec shortward. The weaker N II line at λ3956 shows a similar structure. There is, however, only a very weak emission component for the He I line at λ3926 (2^1P–8^1D) which is quite strong in absorption. In this case, the absorption component is also displaced about 90 km/sec shortward.

The complex structure near Hε is composed of a shortward displaced absorption component due to O II λ3973, an emission peak and shortward displaced absorption component due to Hε and a similar structure from He I λ3965 (2^1S–4^1P^0). There is also a very sharp absorption component (probably double) due to interstellar Ca II λ3968. An interstellar absorption line due to CH^+ is visible at λ3958 and a complex line due to Ca II can be seen at λ3933. There appear to be at least two Ca II K components formed in the expanding atmosphere. These lie shortward of the interstellar feature. There is no emission from Ca II λ3933. The absorption components due to Ca II λ3968 are difficult to detect owing to blending with the emission from He I λ3965 and the Hε absorption. They may correspond to the inflections seen on the longward side of the He I emission. It is obvious that the spectrum of P Cygni is very complicated and that high dispersion spectrograms are required to resolve the different components. Weak emission wings appear on either side of Hε. They may correspond to the faint broad underlying emission features found by R. WILSON for the supergiant HD 190603.

BEALS (1950) has given a detailed description of the spectrum of P Cygni and has collected a list of lines and identifications. Many lines are composed of emission and absorption components; for some lines only emission is seen; for others only absorption is seen. Beals has noted that the ratio of the emission intensity to the absorption intensity is greatest for spectra of low ionisation potential. The Balmer lines of hydrogen are the strongest emission lines. The spectra of He I and of N II are moderately strong in emission as may be seen from Figure 48. Spectra which have a

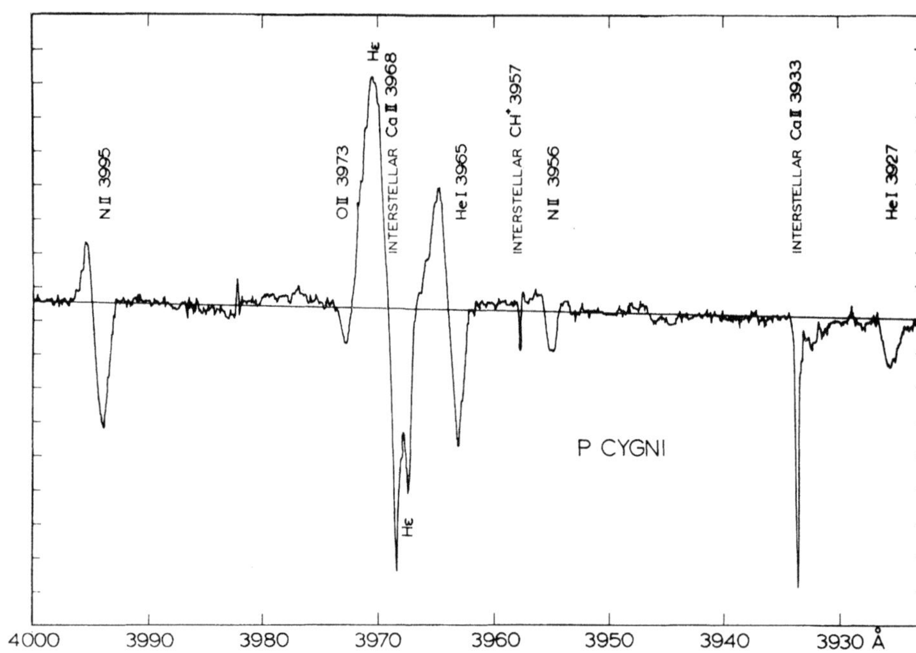

Fig. 48. An intensity tracing of the spectrum of P Cygni in the region 3925 Å to 4000 Å.

high ionisation potential such as N III and Si IV appear only in absorption. The N III multiplet λ4634–41 which appears in emission of Of stars as a result of a cycle involving the emission of He II λ303 is very weak in emission. Another typical Of emission line, C III λ5696, is also weak in emission though strong in absorption. Beals does not report He II λ4686 as being present in absorption or in emission. It seems that few He^+ ions are present in the extended atmosphere around P Cygni and that the flux in He II λ303 is not large.

4.2 *Peculiarities in the spectrum of P Cygni*

The relative intensities of the emission and absorption lines formed in the shell of P Cygni present several puzzles. Beals has noted that the Na I lines at λ5890 and λ5896 appear weakly in emission and that the emission is crossed by strong interstellar absorption lines. There are also shortward displaced absorption components formed in the extended stellar atmosphere. There is, however, no emission seen at the resonance lines of Ca II. Beals suspected that an emission line might occur at Ca II λ3933 and that it was absorbed by the interstellar lines. However, the intensity tracing shown in Figure 48 demonstrates that this cannot be true, because the interstellar line is sharper than the emission features from the shell. If there was a Ca II emission line as strong as those observed for Na I, there would be a narrow emission peak on the longward side of the combined stellar and interstellar absorption feature. Nothing of this sort is observed.

The absence of Ca II emission and the presence of Na I emission is puzzling. In the solar spectrum and in the spectra of G and K type giants the Ca II lines often appear in emission, but the Na I lines are not seen in emission, which is opposite to what occurs in the atmosphere of P Cygni. In the extended atmosphere of P Cygni there is a mixture of atoms, ions, and radiation and the radiation field must certainly contain very strong emission lines of the Lyman series of hydrogen and the resonance 1^1S–n^1P^0 series of He I as well as emission continua from hydrogen and neutral helium. The irregular intensity distribution of the radiation field falling upon the atoms and ions in the shell very probably causes some departures of the ionisation and excitation balance from what would be expected in a normal stellar atmosphere in thermal equilibrium.

The Na I emission is probably the result of ionisation of the Na atoms by radiation which has energy greater than 5.138 volts, that is having wave lengths shorter than about 2413 Å. The Na^+ ions recombine with electrons and then cascade to the ground energy state emitting the Na D lines. The ionisation potential of Ca^+ is 11.87 volts and ionisation from the ground term may be caused by radiation having a wave length shorter than about 1044 Å. Thus we would expect much ionisation of Ca^+ to be caused by Lyman β (λ1025) as well as by radiation of shorter wave lengths. If the resulting Ca^{++} ions then recombine with electrons we would expect them to emit lines of the Ca II spectrum as they cascade to the ground level. The lowest three terms of Ca II are 4^2S, 3^2D, and 4^2P^0. When the excited ion reaches the level 4^2P^0 it may cascade to the ground level, 4^2S, emitting the well known H and K lines, or it may cascade to the 3^2D level emitting the so-called infrared triplet of Ca II at λ8498, λ8542, and λ8662. The 3^2D levels are metastable. Atoms can only move from these levels as a result of collisions or by radiative transitions upwards. It is believed that in the case of the sun collisional transitions from 3^2D have much to do with determining the strength of the emission lines seen at H and K. ATHAY and ZIRKER (1962) have discussed also the effect of ionisation from the 3^2D and 4^2P^0 levels by Lyman α. In the case of P Cygni the possibility of ionisation from the 3^2D levels by absorption of a quantum of Lyman α radiation should be taken into account. The energy of Lyman α is 82258.9 cm^{-1}, just slightly more than the ionisation energy of 82097.8 cm^{-1} from the $3^2D_{3/2}$ level or 82037.1 cm^{-1} from the $3^2D_{5/2}$ level. Ionisations from the 3^2D levels by absorption of Lyman α quanta will, however, only make it more probable that quanta in the H and K lines of Ca II appear in emission in the spectrum of P Cygni. Such quanta are not observed nor is there emission in the lines of the infrared triplet (MERRILL, 1934). Therefore, the recombination process $Ca^{++}+e$ cannot be very efficient in the atmosphere of P Cygni.

One way of preventing Ca^{++} ions from recombining at a rapid rate with electrons would be to raise the Ca^{++} ions to excited states. Another way is to carry the ionisation one stage further. Since the ionisation potential of Ca^{++} is 51.21 volts, which is an energy rather higher than the average energy seen in the spectrum of P Cygni, it does not seem likely that the third ionisation of calcium from the ground state of the Ca^{++} ion is the chief process preventing an efficient recombination of Ca^{++} with electrons. Now the lowest excited states of Ca^{++} lie some 25 to 57 volts above the ground level. This energy is only slightly greater than the primary ionisation energy of neutral helium. It seems likely that there may be a significant radiation field at the above energies arising from the He I 1^1S continuum. Consequently, it seems plausible to attribute the absence of emission at Ca II λ3933 and λ3968 to an inefficient recombination process $Ca^{++}+e$, the process being made inefficient because many of the Ca^{++} ions are in excited states. The action of such an "inefficiency factor" is not expected to be important in the case of the Na^+ ions since the first excited states of the Na^+ ions lie nearly 33 volts above the ground level, and the spectrum of P Cygni is not expected to produce much radiation of this energy. Thus the Na^+ ions will likely remain in their ground state and a normal recombination spectrum is expected for Na I.

The emission and absorption intensities in the Si II spectrum also are not as may be expected from comparison with laboratory spectra and with normal stellar spectra. Usually lines from multiplet number 1 (see the *Revised Multiplet Table*, MOORE, 1945) at λ3853, λ3856, and λ3862, from multiplet 2 at λ6347 and λ6371, and from multiplet 3 at λ4128 and λ4130 are quite strong in absorption, while lines from multiplet 4 at λ5957 and λ5978 and from multiplet 5 at λ5041 and λ5056 are weaker. All of these lines have comparable strengths in a hollow cathode discharge. In P Cygni, however, rather strong emission features accompanied by weak shortward displaced absorption components occur at the lines of multiplets 2, 4 and 5 while both emission and absorption lines are weak or absent at the wave lengths of multiplets 1 and 3.

STRUVE and ROACH (1939) have drawn attention to the fact that the Si III lines of the multiplet 4^3P^0–4^3D at λ3791, λ3796, and λ3806 and the 4^3D–5^3F^0 line at λ3487 appear only in emission whereas the well known 3^3S–4^3P^0 multiplet at λ4552, λ4568, and λ4574 appears strongly in absorption with only weak emission wings on the longward side. This behaviour is probably a dilution effect like those which are known to occur for He I. The lines from the 4^3S level of Si III may be strengthened in absorption in an extended atmosphere where the radiation field is not in thermal equilibrium and the electron temperature is high. The 4^3S level of Si III lies above the 3^3P^0 levels which are metastable.

The lines from the metastable 2^3S levels and from the 2^1S level of He I are greatly strengthened in absorption in the spectrum of P Cygni. This is expected for an atmosphere where the radiation field is dilute. From the dilution effects in the He I spectrum STRUVE and ROACH (1939) estimate that in the case of P Cygni the radius of the expanding shell is about 2.5 times the radius of the stellar photosphere.

More recently GHOBROS (1962) has made a detailed quantitative study of the strength of the hydrogen and helium lines in the spectrum of P Cygni taking dilution effects into account. He finds that the relative strengths of the helium emission lines can be explained by adopting a radiation temperature of 25000°, an electron temperature of 10000°, and a dilution factor near 0.03 for the emitting shell. Thus this estimate of the radius of the emitting shell is about the same as that of Struve and Roach. Ghobros also attempts to interpret the observed strengths of the He I absorption lines in terms of simple curve-of-growth theory for the formation of absorption lines in a stationary extended atmosphere. However, since the Doppler motions in the expanding atmosphere of P Cygni present us with a different physical situation than that envisaged by the theory, little weight can be put on Ghobros' results in the case of absorption lines.

The relative intensities of the emission component of the Balmer and Paschen lines of hydrogen in the spectrum of P Cygni have been discussed by MERRILL and WILSON (1934) and by BURBIDGE and BURBIDGE (1955). These authors are able to show that the observed emission intensities are compatible with an electron temperature in the extended atmosphere near 15000° and a radius for the emitting shell of the order of 2 to 3 times the stellar photosphere.

Attention has been directed in the past to the unexpectedly great strength in absorption and emission of the line at 4396 Å. Struve and Roach, however, have pointed out that this line is probably a blend of O II λ4395.95 and Fe III λ4395.76. In this case the apparent strength is not anomalous, because the lines of Fe III are strong in the spectrum of P Cygni.

4.3 *The radial-velocity displacements in the spectrum of P Cygni*

The radial-velocity displacements of the emission lines remain almost constant at about −13 km/sec (BEALS, 1950; ADAMS and MERRILL, 1957). They do not appear to have changed appreciably since 1900. This velocity probably represents the motion of the star. There is some scatter in the results, but no more than can be attributed to the difficulties of measuring the blended absorption and emission features which are seen. The displacements and intensities of the absorption lines do appear to change from time to time and sometimes two absorption components are distinctly seen, sometimes not. The doubling of the absorption lines is most clearly seen in the lines of hydrogen, in those He I lines which are strengthened by dilution effects, and in the strongest Fe III lines which arise from metastable levels.

HERBIG (1962c) has noted that on high-dispersion (2 Å/mm) spectrograms of P Cygni obtained with the coudé spectrograph of the 120-inch reflector on August 8 and 9, 1960, rather sharp absorption lines of Ni II appear between λ3370 and λ3770. These lines have no emission wings and they have a displacement of -213 km/sec which is in the neighbourhood of the largest shortward displacements found for the absorption components of P Cygni. The chief P Cygni absorption component lies at about -130 km/sec for most lines. Herbig also has noted that some of the strongest Ni II lines have a weaker component at about -230 km/sec. Clearly the expanding shell of P Cygni is composed of atoms moving at different velocities which sometimes produce multiple absorption components.

It is interesting that BEALS (1950) has found that the strongest Na I absorption components lie at about -205 km/sec, while the weaker components lie at -130 km/sec. The absorption lines from N II and O II appear to be less negatively displaced, lying, respectively, near -100 km/sec and -80 km/sec, while the lines from Si IV, N III, and C III lie in the neighbourhood of -30 km/sec. The simplest inference is that the velocity of expansion is greatest in the cool, outermost part of the shell. In making this interpretation we assume that the level of excitation and ionisation in the expanding atmosphere decreases outwards and that the gas is accelerated outwards. The correlation which is observed is merely that the absorption lines showing the greatest velocities of expansion may be observed at the lowest level of ionisation and excitation in the gas. Probably the true explanation of the observations will take into account also the number of atoms moving at each velocity.

Any theory of the formation of spectral lines in a moving atmosphere such as that of P Cygni or the atmospheres of Wolf-Rayet stars should take account of correlations of the above sort. MÜNCH (1950) has suggested that a similar correlation between velocity of expansion and level of excitation exists in the atmosphere of the Wolf-Rayet component of V444 Cygni. Like the Wolf-Rayet stars, P Cygni is a star which appears to be in quasi-equilibrium even though it is ejecting material with large velocities. We do not know what is the number of particles ejected each second at each velocity, nor whether this density changes significantly with time. We do know, however, that the whole process is not something which occurs in a matter of hours or days as in the case of a nova eruption, but it is something that has reached a stable stage for a period of time at least as long as 50 years.

4.4 *The visual absolute magnitude of P Cygni*

The visual absolute magnitude of P Cygni is not very certain. From the intensities of the interstellar lines and from considering that P Cygni may be associated with the cluster NGC 6871, Beals came to the conclusion that P Cygni is at a distance of

1200 parsecs. Accordingly its visual absolute magnitude is −6.3 if it is assumed that the visual extinction is 0.8 mag per kiloparsec. Because of the emission lines in its spectrum, one cannot use the *B–V*-colour and an adopted intrinsic colour to obtain an accurate estimate of the reddening and thus of the extinction. ARKHIPOVA (1963) has measured the spectrophotometric gradients of 20 early type stars with emission-line spectra using objective prism spectra and she has estimated colour temperatures. She makes a correction for interstellar reddening. Arkhipova concludes that P Cygni and HD 242257, HD 45910, HD 51480, HD 169515, HD 207757, and MWC 374 suffer some intrinsic reddening as well as reddening by interstellar extinction. This intrinsic reddening presumably is caused by the extended shell. The intrinsic reddening is small in amount, but it appears to be real. All of the above stars have abnormal spectra, thus it is perhaps too much to expect that they should have the same intrinsic colours as normal stars of the assigned spectral types.

CHAPTER XV

Be STARS, SHELL STARS, AND Of STARS

1. Stellar rotation and the B stars

The fact that some stars rotate rapidly was discovered by SCHLESINGER (1909, 1913) who found a small distortion of the radial-velocity curve during the ingress to and egress from eclipse of the principle components of δ Librae and of λ Tauri. This he interpreted as a rotational effect. This discovery was followed up and consolidated by studies of other binary stars and it was soon evident from the observations that rapid rotation of the star produced diffuse absorption lines like those observed in many single B stars. A summary of the observational steps in the detection of stellar rotation has been given by HUANG and STRUVE (1960). These authors have also given a summary of the theoretical procedures which have been developed to calculate the observed profile of an absorption line formed in the atmosphere of a rapidly rotating star. Reference should be made to that article for the details of the analysis. It is sufficient to note here that an estimate of the speed of rotation of a star may be obtained by measuring the total width of a strong line, one not normally broadened by Stark effect, or by comparing the observed shallow, dish-shaped profile with a grid of profiles computed from a simple representation of a stellar atmosphere rotating at various speeds. The parameter which is derived is the projected speed of rotation of the star, $v_R \sin i$.

Here v_R is the equatorial velocity of the star and i is the inclination of the axis of rotation of the star to the line of sight. When the projected velocity of rotation is small, that is less than about 25 km/sec, it is difficult to prove that the derived value of $v_R \sin i$ does indeed represent rotation of the star. The observed line shape may be due to physical causes such as Stark effect or collision broadening, or to geometrical causes such as motions in the atmosphere resulting from pulsation, expansion, turbulence, etc., rather than to rotation.

Since SHAJN and STRUVE (1929) made the first systematic studies of the rotational velocities of O, B, and A stars, considerable effort has gone into surveying the rotational velocities of stars of different spectral types. TREANOR (1960) has listed the systematic investigations done up to 1959, BOYARCHUK and KOPYLOV (1964) have prepared a catalogue of all rotational velocities published before 1962. There are 2558 stars of types O5 to G5 in their catalogue.

The quoted velocities of rotation are not all on the same scale and it is known that systematic differences exist between the results of various workers. These systematic differences arise because different methods are used to estimate the projected speeds of rotation.

The results of these surveys may be summarised briefly as follows. Rapid rotation of single stars occurs chiefly in spectral types O, B, A, and F. The Be stars rotate most rapidly, velocities as high as 500 to 600 km/sec having been observed by SLETTEBAK (1949) for a few stars. The highest rotational velocities observed among absorption-line B and A stars are of the order of 400 to 500 km/sec; the highest velocities among the F stars are around 200 to 300 km/sec. Among the stars of types F5 to G5 only the giant stars show appreciable rotational velocities, the maximum value being about 100 km/sec.

The axes of rotation of the stars appear to be randomly oriented in space. The fact that Be stars seem to have larger than average velocities of rotation is probably in part a selection effect. The spectroscopic criteria by which we identify a star as Be are most readily visible in those stars where the axis of rotation is perpendicular to the line of sight. When a star is "pole-on", the hydrogen-emission lines are narrow and not so easily detected on low-dispersion spectra as when the star is "equator-on". Also it seems likely that rapid rotation helps a star to form the extended envelope in which the emission lines arise.

GUTHRIE (1963) has investigated the effect of rotation on the Balmer line intensities of B type stars and has shown that much of the intrinsic spread in the curves relating criteria based on the strength of the Balmer lines to visual absolute magnitude is due to the apparent changes in line strength and shape resulting from rotation. MCNAMARA (1963) has also come to the same conclusion from a study of stars in the Orion region.

In a second paper GUTHRIE (1965) has investigated in some detail how the spectra of "pole-on" B stars differ from the spectra of B stars inclined at large angles to the line of sight. Rapid rotation is expected to distort the shape of a star. This distortion in form will have the result that the temperature and the effective value of gravity are higher at the pole of the star than at the equator. In a rapidly rotating star the expected changes in physical conditions may change the spectrum formed in the polar areas significantly from that formed in the equatorial regions. There may even be an extended envelope in the equatorial region. This envelope usually simulates the spectrum of a star of later spectral type but higher luminosity. When such an envelope contains much matter, an additional set of absorption lines is visible. Then we say we have a shell star.

Although most of the statistical studies of the rotational velocities of B stars have been concerned with field stars, a few investigations have attempted to find out if the distribution of rotational velocities is different in stars of special groups. One such investigation is the study of stars in the Orion region by MCNAMARA and LARSSON (1962) and by MCNAMARA (1963). They estimated the projected rotational velocities of 50 stars of types B0 to B3 and of 33 stars of types B5 to B9. Both groups of stars are found to have a lower rotational velocity, on the average, than the field stars. A summary of the material of McNamara and Larsson for the Orion stars and of SLETTEBAK (1954) and of SLETTEBAK and HOWARD (1955) for field B stars is given in Table 31. The average rotational velocity, $\langle v \rangle$, for each group has been determined

TABLE 31
Average rotational velocities of B type stars

Spectral type	Orion stars			General field stars		
	$\langle v \sin i \rangle$	$\langle v \rangle$	n	$\langle v \sin i \rangle$	$\langle v \rangle$	n
	km/sec	km/sec		km/sec	km/sec	
B0–B1	96	122	20	209	266	11
B2	106	135	20	130	165	45
B3	126	160	10	147	187	44
B5–B7	171	218	23	203	257	35
B8–B9	143	182	10	161	205	29

assuming that the axes of rotation are distributed at random. Smaller values of $\langle v \rangle$ are found for spectral types B2 and B3 among the general field stars than are found for types B0 and B1 and for types B5 to B7. This apparent reduction in the average speed of rotation at types B2 and B3 is probably a selection effect owing to the exclusion of Be stars from the analysis (VAN DEN HEUVEL, 1965). It is not proper to exclude the Be stars from the analysis of rotational velocities because they appear to be the same as normal stars in their general properties, and indeed stars are often observed to change from Be to B, the hydrogen emission disappearing. Since the frequency of spectroscopic binaries among the Orion stars is higher than normal, it is tempting to think that the lower values of $\langle v \rangle$ among the Orion stars may be the result of the formation of binary systems in place of rapid rotation of single stars.

2. The Be stars

Rapid rotation appears to produce favourable conditions for the ejection of material into a circumstellar envelope. The parts of this envelope which are projected against the stellar disc will produce absorption lines while the light emitted from the parts which are opaque only in line frequencies will appear as emission lines in the stellar spectrum. The strongest emission lines are those due to hydrogen. Sometimes Fe II lines are also seen. Forbidden lines are not usually present. Any B type star showing hydrogen-emission lines in its spectrum is called a Be star. If, in addition, there are absorption lines from other elements visible superposed on the normal B type spectrum the star is called a shell star. A shell spectrum simulates a later type than the type of the photospheric layers.

The Be stars have exercised a fascination for spectrographic observers ever since they were discovered and the astronomical literature is full of papers detailing the secular changes in spectrum which occur for many Be stars. MERRILL and BURWELL (1933, 1943, 1949, 1950) have prepared a catalogue of Be and Ae objects, and they give bibliographies for many stars as well as notes about the spectral changes which have been seen. Because most spectrographic telescopes are concentrated in the northern hemisphere, the Be stars north of $-23°$ have been observed in most detail.

However, surveys such as that by JASCHEK, JASCHEK and KUCEWICZ (1964) of southern Be stars brighter than magnitude 6.5 are now being carried out. The catalogues and bibliographies of Merrill and Burwell are invaluable for giving information about observations made before 1950. More recent information about Be stars can be found by searching the literature.

The occurrence of hydrogen emission lines in B type spectra is quite common, about 20 per cent of the stars in the several comprehensive catalogues of B stars being listed as Be. From the information compiled by Merrill and Burwell one can readily see that the emission lines change in intensity in most stars, the changes occurring usually at intervals of 5 to 10 years. A detailed quantitative documentation of the changes in seven Be stars between 1953 and 1963 has been given by LACOARRET (1965). Occasional Be stars show changes in their spectrum within a matter of days, but usually significant changes occur at a more leisurely pace. A few Be stars have been noted to change slightly in magnitude as their spectrum changes, but the changes are small. Clearly the extended atmospheres which give rise to the emission lines are unstable and they form and dissipate in an irregular manner without causing a serious disturbance to the star. Since it is improbable that the rate of rotation of the star is varying in an irregular manner, the ejection of the envelope must be triggered by a physical cause that can vary without upsetting seriously the internal structure of the star. One thinks of a variable radiation pressure due, perhaps, to local heating just below the photosphere.

The hydrogen emission lines are usually double with a central absorption core. The simple picture of a rapidly rotating B star seen "equator-on" which has an extended, moderately dense atmosphere confined chiefly to the equatorial plane will serve to explain qualitatively most observations. The rapidly rotating star produces wide, dish-shaped hydrogen absorption lines. Upon these spectral features are superimposed sharper emission components from the lobes of the extended atmosphere, the side lobe which is turning away from the observer giving an emission component displaced longwards from the centre, the other lobe giving an emission component displaced shortwards. The part of the equatorial gas layer projected against the stellar disc will produce an undisplaced absorption core.

The observation that the quantity $\Delta\lambda/\lambda$ is constant for the emission lines of most Be stars, where $\Delta\lambda$ is the total width of the emission feature, implies that the displacement of the emission peaks is due to Doppler effect. This observation is a strong reason for believing in the simple geometric model sketched above for a Be star.

The relative intensities and shapes of the emission and absorption features in Be spectra may remain the same for years. On the other hand, fluctuations in the total intensity of the emission feature and of its shape may occur. The most conspicuous type of change which occurs is a V/R change. In this case the relative intensity of the shortward, V, and longward, R, components changes. Such changes are not strictly periodic in character. They are accompanied by a shift in velocity of the absorption core. The core is displaced shortward when the longward emission component is strong and longward when the shortward emission component is strong. It appears that when V/R changes are observed, the gas in the extended atmosphere has a variable radial component of motion as well as the component of motion due to the rotation of the star. Studies of the spectral changes in Be stars and in shell stars by Struve and by others strongly suggest that the particles in the extended stellar atmosphere rotate with conservation of angular momentum. By measuring the velocity separation of the two emission components one may estimate the speed of rotation of the gas layer in which the emission originates. Comparison of this speed with the speed of rotation of the star (which is estimated from the shape of the absorption lines) gives an estimate of the radius of the emitting layer. In this way one finds that the average extended atmosphere has a radius near 3 R_*.

The Be stars are generally distributed among the B stars and do not appear to differ fundamentally from normal B stars. They occur in clusters, for instance in the Pleiades, as components of spectroscopic binaries (ζ Tauri, φ Persei), of eclipsing binaries (V367 Cygni), as components of systems with composite spectra (HD 50820 which is K+Be and HD 192641 which is WC7+Be). It appears that the instability which produces the Be phenomenon is a property of the star itself and that it is not fundamentally affected by the position of the star in space.

MENDOZA (1958) has observed MK spectral types and UBV colours and magnitudes for 175 Be stars north of $-20°$ and has shown that the visual absolute magnitude of B1e and B2e stars is about -3.6. He finds that the Be stars are not concentrated in associations but are scattered rather uniformly along the Milky Way. He suggests that this distribution and the fact that the Be stars appear to be between about one-half and one magnitude above the main sequence indicates that these stars are older than the stars with purely absorption-line spectra which define the main sequence.

The hydrogen-emission spectrum of Be stars is a recombination spectrum formed in an extended envelope in which differential motions exist. The problem of obtaining quantitative information about the physical state of the emitting gas from a study of the line intensities and shapes is challenging and a number of astronomers have contributed to this study. PAGEL (1960) has summarised much of the theoretical treatment. As survey of the field has also been given by KOURGANOFF (1958). If the full problem is to be solved, it is necessary to consider the interaction of a dilute radiation field with an ionised gas in which there may be differential radial motions as well as rotation. Some theoretical treatments have emphasised the spectroscopic phenomena occurring in a dilute radiation field and have laid less emphasis on the motion of the gas. For instance, STRUVE and WURM (1938) have studied the formation of absorption lines in a stationary gas with a dilute radiation field. More recently KOGURE (1959a, 1959b, 1961) and POTTASCH (1961) have studied the formation of the H emission spectrum in stationary Be atmospheres. Others, such as BEALS (1929) and ROTTENBERG (1952) have emphasised the effects of motion and have paid less attention to the details of the radiative transfer problem. SOBOLEV (1947) has attempted the difficult task of setting up a theory which takes account both of motion in the atmosphere and of dilution effects. DOAZON (1965) has applied Sobolev's theory to interpret the variable emission line spectrum of HD 50138, B5e. This star seems to have an unusually extended atmosphere, for the [O I] lines at λ6300 and λ6364 are observed. Also this star is known to show radial-velocity variations which appear to indicate radial motion in the shell. It does not appear to be a "normal" Be star.

3. Shell stars

In a review paper STRUVE (1942) summarised the major results of his series of observations of the spectra of stars with extended atmospheres and he discussed the meaning of the observations in terms of an extended atmosphere around the star. Although more quantitative information is now available about the changes which occur in Be and shell spectra than was available in 1942, the interpretation of shell spectra has not changed greatly since that time.

Struve noted that in shells, monochromatic radiation processes and dilution effects are important. The strengthening in absorption when the radiation field is dilute of lines arising from metastable levels, or from levels lying immediately above metastable levels, particularly in the He I spectrum, has been discussed in Chapter VII. Lines arising from metastable levels are typically seen in shell spectra. The He I lines usually are strong only if the spectral type of the underlying star is B2 or earlier. Absorption lines from normal levels are usually rather weak in shells. STRUVE and WURM (1938) have directed attention to the weakness of Mg II λ4481 ($3^2\mathrm{D}$–$4^2\mathrm{F}^0$) in absorption in most shell stars in comparison to the strength of lines of Fe II, Ni II, Cr II etc. The latter lines arise from metastable levels which have excitation potentials less than 5 volts. The excitation potential of the lower levels of λ4481 is 8.86 volts and it takes two radiative transitions, namely 3^2S–3^2P^0 followed by 3^2P^0–3^2D, to place atoms in the 3^2D levels by radiative excitation, for the 3^2D levels of Mg II cannot combine directly with the ground level, 3^2S. In a diluted radiation field in a gas where the excitation temperature is low, the population of the 3^2D levels will not be large. Struve and Wurm also directed attention to the weakness of Si II λ4128 and λ4130 (3^2D–4^2F^0) in shell spectra and suggested that this weakening was also a dilution effect similar to that in the Mg II spectrum. This case, however, is not quite the same as that of Mg II because the 3^2D levels of Si II can combine directly with the ground levels of Si II, for the ground levels are 3^2P^0. The excitation potential of the 3^2D levels of Si II is high, 9.84 volts. Thus if the excitation temperature is low in the shell and the radiation field is dilute, one may expect Si II to be weak in absorption. However, other Si II multiplets from levels of lower excitation potential also appear in shell spectra, namely Si II λ3853, λ3856, and λ3862 (3^2D–4^2P^0) and Si II λ6347 and λ6371 (4^2S–4^2P^0). Sometimes these lines are quite strong in absorption (see Section 3.1) even though their lower excitation potentials are 6.86 and 8.12 volts respectively. Struve and Wurm were unable to observe these lines. The true explanation of the observed intensities of the Si II multiplets in shell spectra appears to involve more than dilution effects. The dilution factor in normal shells lies in the range 0.01 to 0.1.

Another point to which Struve drew attention was the tendency of all shells to vary in an irregular manner. Occasionally the variation is nearly periodic for a certain time. Shells form and dissipate without having a marked effect on the underlying star. Sometimes shells appear to be stationary, but when the observations have been carried out over a long period changes are usually seen to occur. For instance, Struve thought in 1942 that the shell of ζ Tauri was stationary, but since 1958 the shell of this star has been very active with velocities of expansion greater than 50 km/sec occurring and a variety of complex spectral changes being seen. This shell is no longer "stationary". One of the still unsolved problems of Be and shell stars is the mechanism which triggers the formation and dissipation of a shell. The origin of the observed variable radial motions is not certainly known.

Sometimes, when a shell is rather active, a considerable range of radial velocities may be detected at one time in the shell. Then some of the absorption lines may have asymmetric profiles. Most frequently the asymmetric profiles are steep on the longward side with a wing tailing off to shorter wave lengths. Once absorption lines with the above asymmetric shape are seen, the radial velocity shown by the Balmer lines becomes more negative with higher series members. This systematic increase of velocity of approach as one goes up the Balmer series towards the Balmer limit was discovered by MERRILL and SANFORD (1944) in the star 48 Librae during one of its active periods. Merrill calls it the Balmer progression. The simplest interpretation of the observations is that material is ejected with a large velocity and that the material is decelerated as it moves outward from the star. Since the higher members of the Balmer series are intrinsically weak lines, one looks through to the deepest layers of the shell in these wave lengths. Here the outward velocity is highest. The leading members of the series, Hα, Hβ, and Hγ are so strong in absorption that one can see only through the outermost layers of the shell in these wave lengths. Here the outward velocity is reduced. Occasionally the absorption lines in shell stars have a reversed asymmetry, the shortward side being steep and a wing tailing off to the longward side. Following this the Balmer progression is reversed and Hα shows the least velocity of approach to the star; the higher members of the Balmer series show greater positive velocities Since the density and electron pressure in a shell is low, one can usually follow the Balmer series to at least $n=30$ in a shell spectrum. Occasionally the individual Balmer lines in a shell spectrum can be distinguished to $n>40$. Although the emission feature at Hα in shell stars has roughly the same shape as in Be stars (it usually consists of two peaks separated by an absorption core), it is very much wider. The wings of Hα in shell stars can usually be followed to more than ± 500 km/sec (about ± 11 Å) from the line centre; in normal Be stars the emission wings extend only to about ± 200 km/sec or ± 4 Å. It seems probable that the width of the Hα emission feature in shell stars is due to the velocity distribution of the H atoms in the shell. Some astronomers however, have suggested that the hydrogen atoms in the shell have rather small components of velocity in the line of sight and that the wide Hα profile is the result of radiation damping. Mme RINGUELET-KASWALDER (1963) has studied the relative widths of the emission features at Hα and Hβ in a few Be stars and in some shell stars. She confirms that the width of the Hα emission is correlated with the rotational velocity of the star in the case of the rapidly rotating Be stars, but that in the shell stars

TABLE 32
Some bright shell stars

HD number	Name	MWC number	α (1900)	δ (1900)	Spectral type	V (m_V)
5394	γ Cas	9	0^h50^m7	+ 60°11′	B0ne	2.64
10516	φ Per	16	1 37.4	+ 50 11	B0ne	4.03
22192	ψ Per	69	3 29.4	+ 47 51	B5ne	4.21
23862	Pleione	75	3 43.3	+ 23 51	B8ne	5.08
37202	ζ Tau	115	5 31.7	+ 21 05	B2ne	2.99
41511	17 Lep	519	6 00.5	− 16 29	Aeq	(5.0)
45725	β Mon A	143	6 24.0	− 6 58	B3ne	(4.7)
45910	AX Mon	145	6 25.2	+ 5 57	B2e	6.76
50138		158	6 46.7	− 6 51	B8e	(6.6)
58978		179	7 22.8	− 22 53	B2ne	(5.5)
147983	48 Lib	239	15 52.6	− 13 59	B3pe	(4.7)
179343		978	19 07.1	+ 2 27	Ae	(6.8)
183656		988	19 25.5	+ 3 14	B6pe	(6.3)
193182		632	20 13.8	+ 39 16	A0e	(6.6)
195325	1 Del	1019	20 25.5	+ 10 34	A0e	(5.9)
205637	ε Cap	373	21 31.5	− 19 54	B3e	4.60
217050		394	22 52.7	+ 48 09	B3ne	5.48
217675	o And	–	22 57.3	+ 41 47	B6.5	(3.6)
218293		397	23 02.6	+ 49 40	Ae	(6.8)

such a relation is not always evident, particularly if the shell spectrum is variable. Then the Hα profile is exceptionally wide.

A short list of bright shell stars is given in Table 32. The MWC number is the serial number of the star in the catalogues of MERRILL and BURWELL (1933, 1943, 1949, 1950). The V magnitude given is from the measurements by MENDOZA (1958). These stars demonstrate most of the types of variation of spectrum, radial velocity, and light that are known to occur for shell stars. Usually the magnitude of a shell star does not change by a significant amount while the spectrum is changing, but few reliable photometric studies have been done of shell stars. The shell star which has suffered the most conspicuous change of light since modern stellar spectroscopy began is γ Cassiopeiea. In about 1936 the star brightened by nearly one magnitude and then faded by about 1.5 magnitudes. A brief summary of the spectral and light changes which occurred is given by Struve and by McLaughlin in the notes by MERRILL and BURWELL (1943). In 1964 the spectrum of this star was the same as that of any rapidly rotating B0 star and the shell appears to have completely disappeared.

Each shell star is different from every other shell star. Much has been written about the spectral changes and radial-velocity changes which have been observed in some cases. Rather than attempt a general survey of many unlike objects, we will now give a survey of the principle spectroscopic changes which have been observed for ζ Tauri and for 48 Librae. These stars are typical of all shell stars. Since the writer has access to spectrograms of each star she can speak with some authority about the spectroscopic changes which occur for these stars.

3.1 *The spectrum of ζ Tauri*

This star is a spectroscopic binary with a period of 132.91 days, but only one set of lines, those of the rapidly rotating B2 star and its shell have ever been seen. The most recent discussion of the radial velocity variations of ζ Tauri is that by UNDERHILL (1952b). Here a summary is given of all previous work and a set of orbital elements is derived which represents the motion around the unseen companion.

Radial-velocity observations of ζ Tauri have been made at irregular intervals since 1902. From 1902 until 1952 no large systematic deviations of the observed radial velocities from binary motion occurred although the velocities from the hydrogen lines have fluctuated by amounts of the order of 5 to 10 km/sec from the orbital motion. These fluctuations probably indicate small irregular components of motion in the radial direction in the shell. None of these fluctuations are of sufficient magnitude to suggest that the shell was expanding rapidly or collapsing during the periods of observation. Since 1958, however, the shell has become more active. The observations of HERMAN and DUVAL (1963) in France and of GÖGKÖZ, HACK and KENDIR (1963) in Italy have shown that inflow and outflow velocities as large as 100 km/sec have occurred between 1961 and 1963. In October and November 1964 the stronger shell lines were displaced approximately 40 km/sec shortward from the velocities expected owing to orbital motion. Also at this time the line profiles were asymmetrical. During this active period the profiles of the absorption lines formed in the shell changed shape in a few weeks time. Clearly the shell of ζ Tauri is no longer stationary, as it seemed to be until 1952 at least, but has entered a stage of activity. No change in the apparent magnitude of ζ Tauri has been reported.

The shell spectrum of ζ Tauri was strong in 1914 (BALDWIN, 1941) and in 1950. The spectral changes occurring in the years 1950 to 1952 have been studied by the author (UNDERHILL, 1952a) who suggested that they were due to an increase in pressure in the shell. Since 1950 regular spectrographic observations of ζ Tauri have been obtained by Margherita HACK (1957) and her colleagues (GÖGKÖZ, HACK, and KENDIR, 1963; HACK, AYDIN and ISLIK, 1964) in Italy, and by Mme R. Herman (HERMAN and DUVAL, 1963) in France. These observations show that the shell lines weakened from 1950 to 1952 and then strengthened again until 1958–59. In 1961 a complex structure of absorption and emission components was observed at Hβ. Since 1961 the shell spectrum of ζ Tauri has been quite variable and some strikingly asymmetric line profiles have appeared as well as some unusually strong absorption lines.

In Figure 49 the spectrum of ζ Tauri as it was on October 19, 1964 is compared with that of the B8Ia supergiant β Orionis over the spectral range λ 5850 to λ6600. The underlying star of ζ Tauri has a spectral type not later than B2 and it is rotating at about 350 km/sec. (It is extraordinarily difficult to specify the precise spectral type of a shell star owing to the complex spectrum.) All lines due to the normal photospheric layers of ζ Tauri should be wide and shallow. None can be clearly seen in Figure 49 except, perhaps, faint wings at He I λ5876. The outstanding points of this spectrogram are the following:

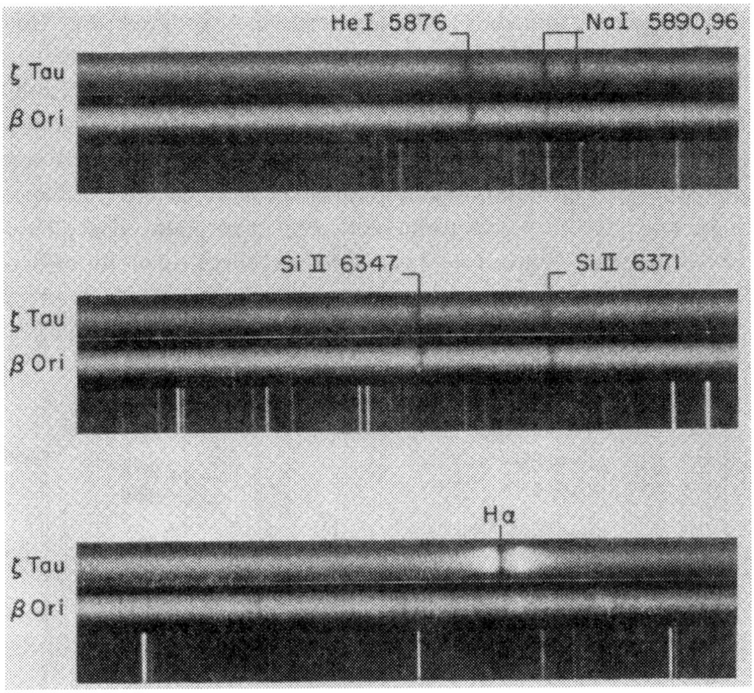

Fig. 49. The spectrum of ζ Tauri on October 19, 1964, compared with that of β Orionis, B8Ia, in the region λ 5850 to λ 6600.

1. At He I λ5876 there is a moderately sharp, moderately strong core due to the shell. This line is very like that in β Orionis in strength and shape and it is not clearly asymmetrical. The shell lines due to He I have intensified steadily since 1950 as HACK (1957) and GÖGKÖZ, HACK, and KENDIR (1963) have demonstrated. Apparently the level of excitation in the shell as well as the amount of material there has been increasing in recent years.

2. The Na D lines are strong in absorption and their profiles are asymmetrical, being steep on the longward side and winged shortward. Such strong lines from Na I have never before been reported for ζ Tauri. They indicate that the excitation and ionisation temperature is low in part of the extended shell even though lines of He I and Si II are formed in other parts of the shell. There seems to be a considerable range of velocities present in the shell at this time. On October 19, 1964, the velocity of expansion indicated by the Na I lines was −14 km/sec, though the other lines (H, He I, Si II, and Fe II) indicated an expansion velocity of about −40 km/sec. The Na D lines are very weak in the spectrum of β Orionis, just as is expected for a B8 star.

3. The multiplet 4^2S–4^2P^0 of Si II at λ6347 and λ6371 is strong in absorption and the profiles are asymmetric. The equivalent widths of these lines are about the same as in β Orionis or in α Cygni, A2Ia. Spectrograms covering the blue-violet spectral

region are now being studied. Preliminary estimates indicate the Si II lines at λ3853, λ3856, λ3862 and at λ4128, λ4130 are about as strong as in the spectrum of β Orionis or of α Cygni. Thus it appears that the strengths of the He I and Si II lines formed in the shell of ζ Tauri in October, November 1964 are similar to those in late B or early A type supergiants. They are not significantly weaker. It appears that the level of excitation and ionisation in a large part of the shell of ζ Tauri in October 1964 was quite like that in the atmosphere of a B8 supergiant. The dilution effects mentioned by Struve and Wurm for Mg II λ4481 and for Si II are not evident.

4. The emission line at Hα is very wide and strong in the spectrum of ζ Tauri and there is a strong, asymmetric central absorption. The Hα profile of ζ Tauri is shown in

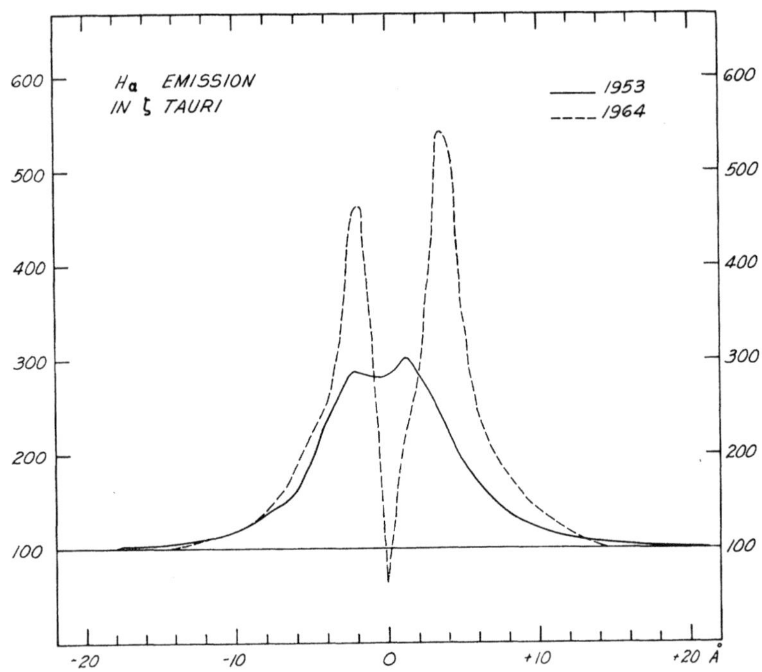

Fig. 50. The profile of Hα in ζ Tauri in 1953 and in 1964.

Figure 50 for the years 1953 and 1964. The emission is much stronger in 1964 than it was in 1963, but the total width of the emission feature has not changed. (The spectrograph used in 1953 had insufficient resolution to show the sharp absorption core which was most likely present since sharp absorption cores are present at this date at the other H lines.) The apparent asymmetry of the Hα absorption core in 1964 is reversed from what is visible at the Na I and Si II lines. The observed profile can be considered to be the result of combining a shortward displaced, asymmetric (shortward wing) absorption line with a symmetrical emission profile which moves with the orbital velocity of the star. It is possible to make a model which reconciles the observed profiles of Hα and of the other lines when one remembers that the major part of the shell appears to be expanding with a velocity of about 40 km/sec and that at Hα 45 km/sec corresponds to a displacement of one angstrom.

The conclusion to be drawn from the spectrogram of October 19, 1964, is that at this time the state of excitation and ionisation in a large part of the shell of ζ Tauri was comparable to that in a B8 supergiant atmosphere. There was, in addition, a considerable range of velocities present and there was an outer, cooler shell where the strong Na D lines are formed.

In general the spectra of Be stars and of shell stars do not change rapidly, months or years being required before a significant change is seen. The spectrum of ζ Tauri, however, was very variable in November and December 1964. The spectra displayed in Figure 51 demonstrate this. Here the absorption profiles of H11 and H12 can be

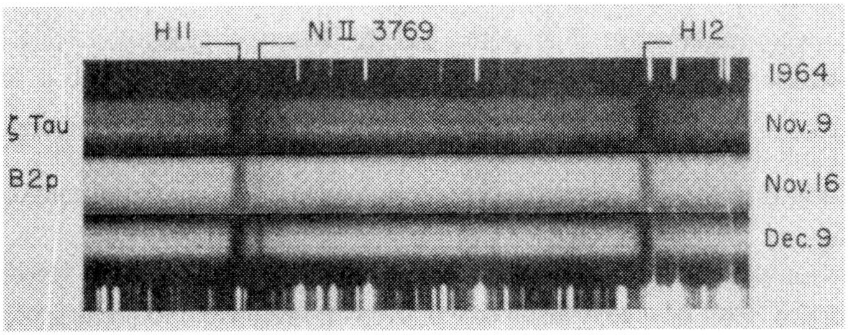

Fig. 51. An example of the short-term changes which occurred in the spectrum of ζ Tauri in November and December 1964.

seen and also the strong Ni II line at λ3769. This line arises from the $a^4P_{2\frac{1}{2}}$ level which is metastable. On November 9 the absorption lines were broad and they had a rather steep longward side. By December the lines had become narrower and perhaps, deeper. They are not so asymmetric as on November 9. (The iron arc comparison spectrum is scattered across the stellar spectrum at wave lengths shorter than H12 on the spectrograms of November 16 and December 9.)

In his review paper of 1942 STRUVE suggested that the chief difference between supergiants and shell stars was that the continuous opacity in the ordinary spectral region is greater in supergiants than in shell stars. In the case of a supergiant, one cannot see through to the "photosphere" below the extended atmosphere. The changes which have been observed in 1964 in the spectrum of ζ Tauri give qualitative support to Struve's ideas. Possibly the spectroscopic phenomena we have been observing in ζ Tauri are a beginning of the transition of ζ Tauri into a supergiant. If the part of the shell which gives the He I and Si II lines could be maintained and made more dense, the most rapidly moving atoms being lost and not renewed, and if the cool outermost shell where the Na D lines are formed were evaporated, we would be left with an object that had a spectrum like a B type supergiant. In order to understand completely the difference between a shell star and a supergiant one must have a detailed theory of how the outer atmosphere is ejected and maintained at the observed density and state of motion. It is most important to know the origin of the radial motions which are observed in stellar atmospheres, for they appear to control the density and level of excitation and ionisation in the extended atmosphere of the star. There is much in common between the spectra of Be stars, shell stars, and Wolf-Rayet stars, for the state of motion in the atmosphere is an important factor in determining what we see in each case. The range of velocities which is encountered appears to be greatest in the hotter objects.

3.2 *The spectrum of 48 Librae*

The shell star 48 Librae has been observed intensively for over thirty years. During this time the spectrum and radial velocity have undergone some remarkable changes. From the time the first observations were made for the *Henry Draper Catalogue* and slit spectra were obtained in 1904 at the Yerkes Observatory until at least 1931, the spectrum appears to have been that of a rapidly rotating B star with a weak shell which gave sharp hydrogen absorption cores. In 1935 it was discovered that the velocities from the hydrogen lines and from the metallic lines were strongly shortward displaced and that the shell lines had become stronger. The shell has remained active ever since.

The major papers about the spectrum of 48 Librae are those by STRUVE (1943) and HILTNER (1944a) who gave some radial-velocity results and who directed attention particularly to the strong shell spectrum and to the asymmetric absorption-line profiles which were visible in 1943, the longward side of the strong lines from metastable levels and the hydrogen lines being steep, the shortward side consisting of a wing which decreased in intensity more slowly; by MERRILL and SANFORD (1944) who made a detailed radial-velocity study of the star from 1904 to 1944 as well as discussing the line shapes; by MERRILL (1953) who continued the detailed radial-velocity study from Mount Wilson spectrograms to 1952; and by UNDERHILL (1953b) who presented radial-velocity observations from 1950 to 1953 and attempted also a quantitative analysis of the strength of the lines formed in the shell. The star 48 Librae was kept under observation at the Dominion Astrophysical Observatory until 1962 and most of the spectrograms have been measured for radial velocity. The equivalent widths of about 100 shell lines have been measured on the spectrograms for the years 1950 to 1956 in order to evaluate the changes in line strength which have occurred.

The radial-velocity results from the Mount Wilson and the Victoria spectrograms obtained between 1938 and 1962 are displayed in Figure 52. The velocity of Hδ, as read by Merrill from plots of the Balmer progression and the mean of the velocities of Hγ, Hδ, and Hε measured on Victoria spectrograms are shown as filled circles; the Fe II velocities are shown as open circles; and the velocities from the Ti II lines are shown as crosses. These velocities are sometimes rather widely different from each other. Intuitive reasoning such as that followed by STRUVE (1943) suggests that the Hδ velocities represent the motion of high levels of the atmosphere, the Fe II velocities

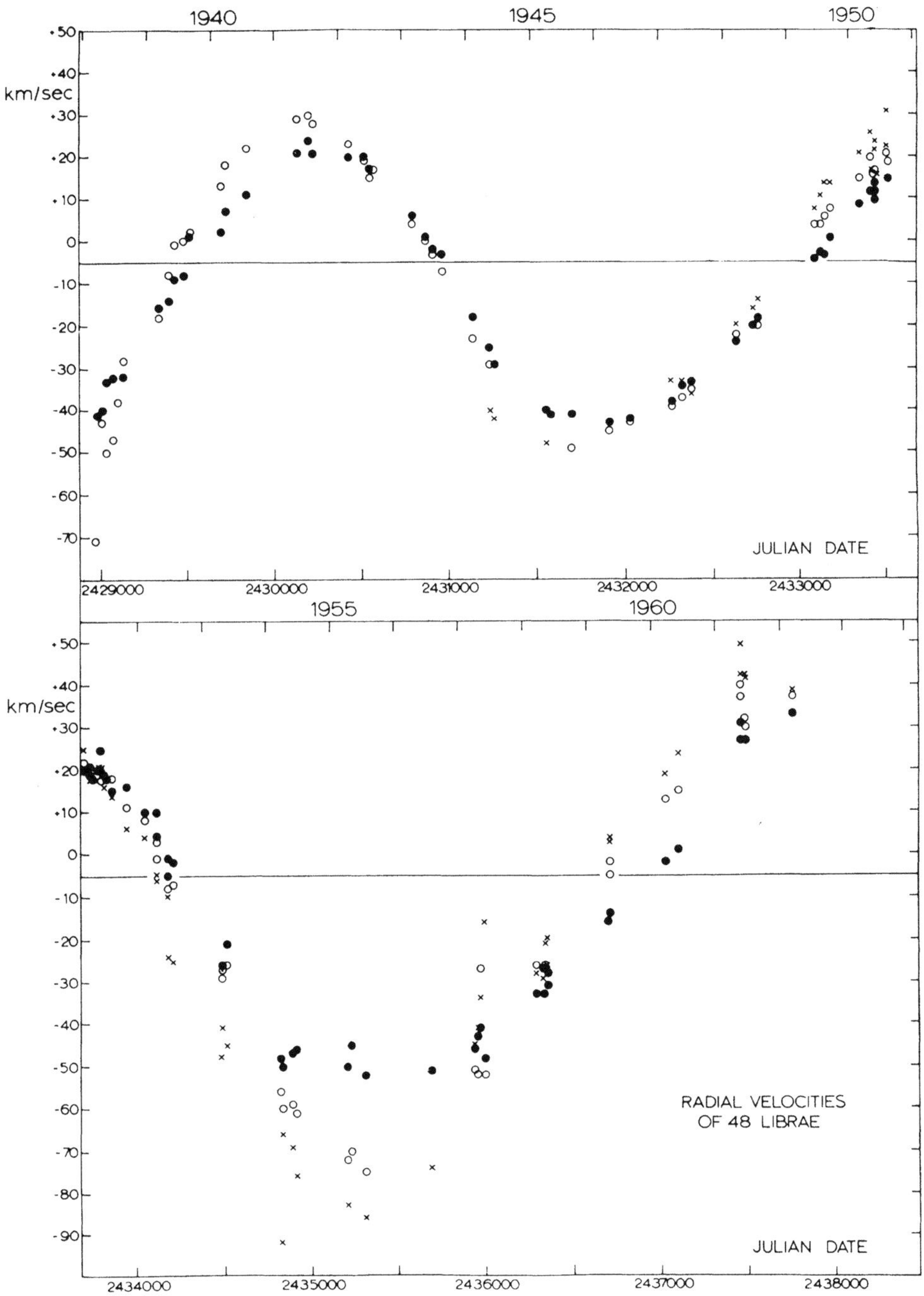

Fig. 52. The radial-velocity changes of 48 Librae between 1938 and 1962. The velocities from Hδ are indicated by dots, those from the Fe II lines by open circles, and those from the Ti II lines by crosses.

represent layers of intermediate height, and the Ti II velocities represent rather deep layers.

Before 1931 the radial velocity of 48 Librae was more or less constant in the neighbourhood of -5 km/sec which is the velocity expected for a member of the Scorpio-Centaurus stream. Since 1935 the radial velocity has oscillated with a range of about 100 km/sec and a period of about 10 years. The radial velocities plotted in Figure 52 show the range in velocities that occurs at any one time. This range is largest when the curve is at maximum or minimum. The change in velocity shown by Hδ is the smallest; that by the Ti II lines is the largest. When there is a large spread in velocities, the Balmer progression is steep. At minima the Balmer lines with $n > 30$ have velocities similar to the velocities of the Ti II lines, lines with $n = 25$ have velocities similar to the velocities of the Fe II lines, while the velocities of the first members of the Balmer series are less negative. At maxima the Balmer progression is reversed, the Balmer lines with the largest values of n having the greatest positive values of radial velocity.

Struve (1943) has pointed out, as is confirmed by later spectrograms, that the profiles of the Ti II lines are more symmetric and diffuse than those of Fe II or H. He suggests that the Ti II lines are formed deeper in the shell, thus closer to the stellar photosphere, than are the Fe II and other strongly asymmetric lines.

The line profiles are not always asymmetrical. The correlation of the line asymmetries with phase in the radial-velocity variations may lead to a better understanding of the velocity fields in the shell. Merrill and Sanford (1944) have noted that when the velocity was about zero km/sec in 1940 the lines were winged longward, whereas in 1943, when the velocity was again about zero km/sec, the lines were winged shortwards. Merrill (1953) states that the lines were winged longward just before the velocity maximum of 1951 and that just after maximum they were winged shortward. The spectrograms obtained at Victoria indicate that the same pattern of asymmetries occurred in 1952 (zero velocity going negative) and in 1958 (zero velocity going positive). The lines are symmetrical for most of the time corresponding to the minimum of the curve. Very high dispersion is required to show clearly the asymmetries in line shape which occur. What happens is that the line shapes of the moderately strong lines foreshadow the range of velocities which will be shown by the lines from different ions about one quarter period later. The observed asymmetries and the velocity displacements are not those predicted theoretically for a moving plane parallel atmosphere. In that case an outward moving atmosphere produces a line sharp on the shortward side with a wing to longer wave lengths.

The relative variation in equivalent width of lines of Fe II, Ti II, and Si II is shown in Figure 53 for the years 1950 to 1956. The lines were strongest in 1951 when the shell showed a maximum positive velocity (presumably the gases of the shell were falling towards the stellar photosphere) and the lines have weakened steadily as the shell stopped contracting and began to expand. The changes seen in the shell spectrum suggest that the excitation temperature in the shell remained between 8600° and 9300° during the time of the radial-velocity oscillations. Thus it seems probable that the observed changes in equivalent width are caused more by changes in the number of atoms projected against the stellar disc than by changes in the level of excitation.

The Na D lines are strong in absorption in the shell spectrum of 48 Librae, just as for ζ Tauri, but no thorough study has been made of them. MERRILL and SANFORD (1944) have obtained a few observations at high dispersion. They were able to show

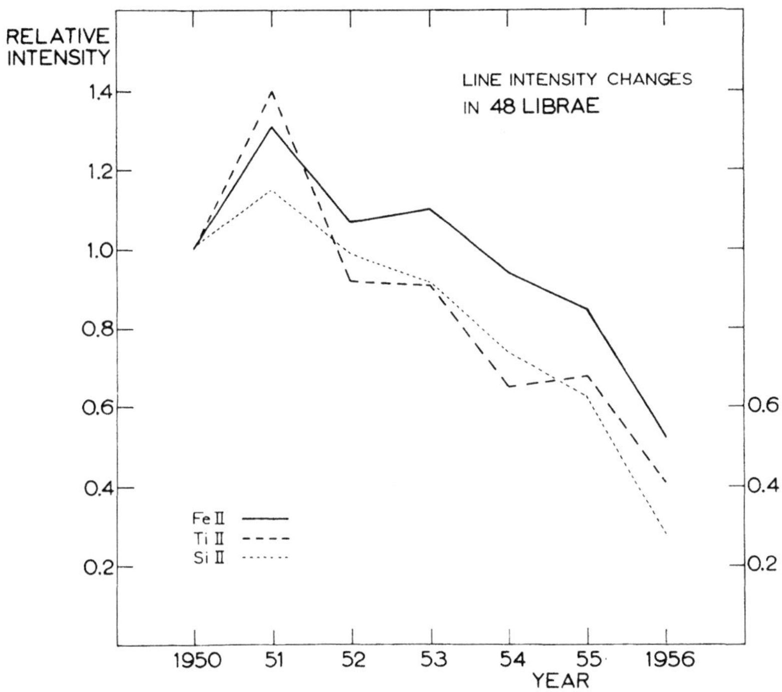

Fig. 53. The relative variation in equivalent widths of lines of Fe II, Ti II, and Si II in the spectrum of 48 Librae between 1950 and 1956.

that there is an interstellar component which gives a radial-velocity of −14 km/sec and another component which yields a velocity agreeing with that from the other metallic lines. They suggest a circumstellar envelope for the place of origin of the Na D lines. A similar, but weaker shell absorption component is seen at Ca II K as well as an interstellar line. Strong Na D lines can also be seen in the spectra of the shell stars 1 Delphini and HD 193182.

The general shape and intensity of the Hα emission line has not changed between 1950 and 1956. What has happened is that the absorption core has moved from a displacement of +25 km/sec to −85 km/sec. This has caused a significant V/R change. The profiles of Hα, Hβ, Hγ, and Hδ in 1950 are shown in Figure 54. The profiles for the Be star 48 Persei are also shown for comparison. Although the peak intensities of the emission features in 48 Persei are about the same as those in 48 Librae, the width of

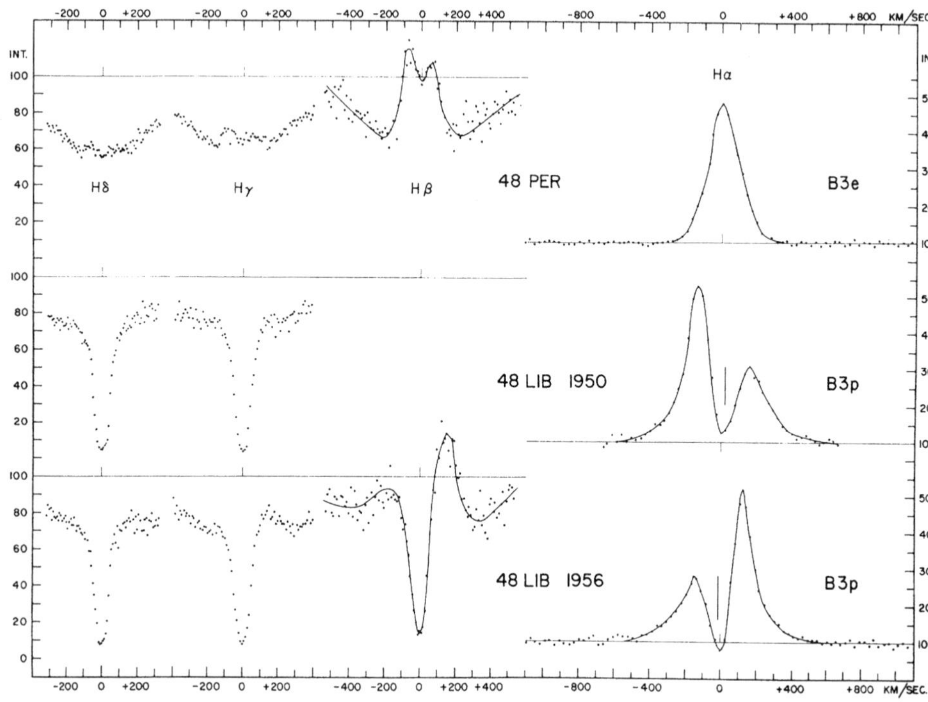

Fig. 54. Profiles of Hα, Hβ, Hγ and Hδ in 48 Librae and in the Be star 48 Persei. The vertical scale for the Hα profiles is more compressed than that for the other profiles.

the emission features is quite different. There are no strong central absorption cores in the spectrum of 48 Persei.

4. The Of stars

The latter f was introduced by J. A. PEARCE (1930) to indicate that the lines N III λ4634, λ4640, λ4641 and He II λ4686 appear in emission in some O stars. The hydrogen lines visible on blue-sensitive plates are not in emission in most Of stars. However, there is usually a weak emission line at Hα and at C III λ5696. The lines N V λ4603 and λ4619, N IV λ4057, Si IV λ4088 and λ4116, and He I λ5876 have been observed in emission in a few Of stars of very early spectral type. There are also rather broad, faint unidentified emission lines at 4485.7 Å and 4503.7 Å. These emission lines are difficult to detect, but the observations by WOLFF (1963) place their reality beyond doubt. No lines from the ions which are expected to be present in O type atmospheres can be found at these wave lengths in tables of atomic spectra, nor do these lines appear in Wolf-Rayet spectra. Furthermore, no absorption lines have been found in stellar spectra at these wave lengths. Thus one must conclude that λ4486 and λ4504 arise from transitions which have a low probability of occurrence in sources as dense as laboratory plasmas or ordinary stellar atmospheres. The observations indicate that

λ4486 and λ4504 strengthen with increasing temperature from B0Ia to O7 and that they appear only when the star has an extended atmosphere. The few available observations suggest that the intensity of λ4486 and λ4504 decreases in stars of types O6 and O5. The spectral type region in which the unidentified emission lines λ4486 and λ4504 are strongest is about the same as that in which the emission line at λ5696 is strongest. The latter line is attributed to C III. Thus as source of these unidentified lines one would think of ions of an abundant element having ionisation potentials in the neighbourhood of those of the carbon ions. Possibly the lines come from transitions in a set of levels formed by two electrons being excited. Such energy levels

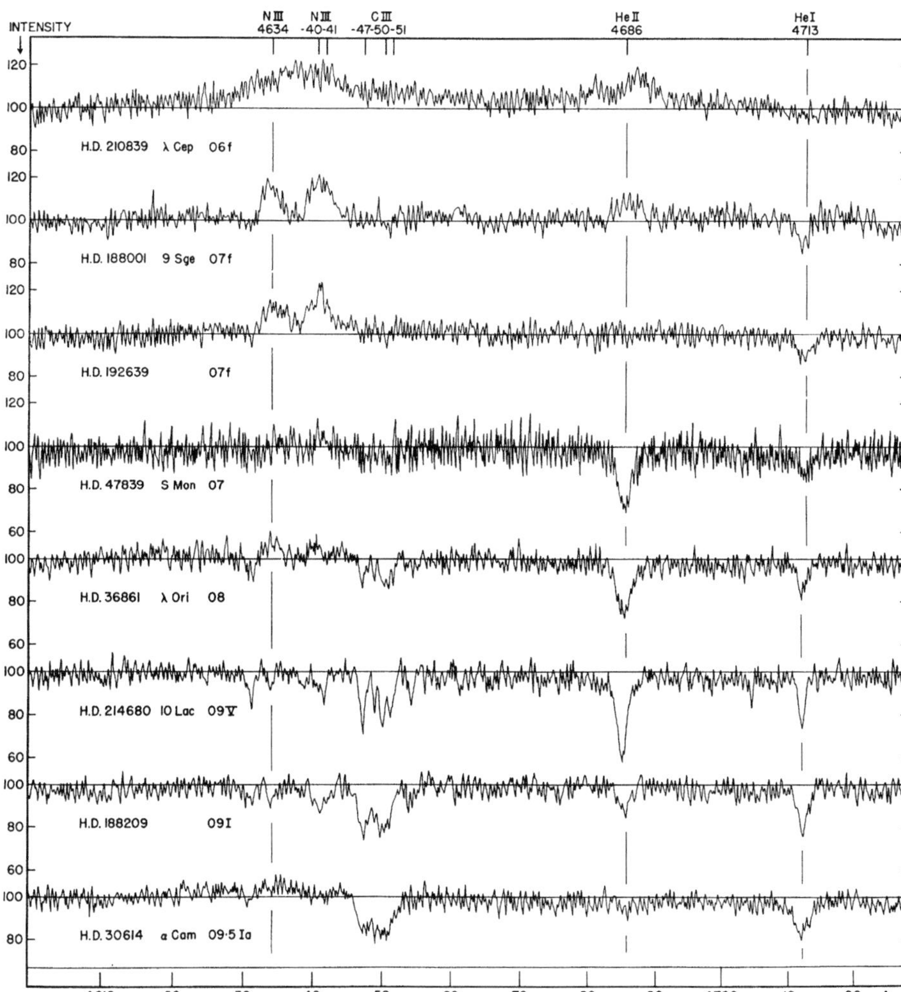

Fig. 55. Intensity tracings of the spectra of eight O type stars showing the strength of the N III λ 4634, 40, 41 and He II λ 4686 emission features.

are not well known from laboratory studies, for often they lie above the primary ionisation limit of the carrier ion. The resulting lines are too broad and weak to be observed readily in the laboratory because of the great probability of auto-ionization. Doubly excited levels are known for C^+ and for C^{++} but lines at 4485.7 Å and at 4503.7 Å have not been observed for C II by GLAD (1952) nor for C III by BOCKASTEN (1955).

Rectified intensity tracings of the spectra of some O stars in the region 4600 Å to 4730 Å are shown in Figure 55. The emission at N III λ4634,40,41 is quite strong in the three Of stars and the emission at He II λ4686 fills in the expected absorption line and rises above the level of the continuum in two cases. Very much weaker emission occurs at λ4634,40,41 in the O stars S Monocerotis, O7, and λ Orionis O8; He II λ4686 appears in absorption in these stars. Only after a detailed study of the He II lines from several series could one tell if there is some emission in He II λ4686 in S Monocerotis and λ Orionis. No emission is evident in the spectrum of 10 Lacertae and the group of O II and C III lines near λ4650 are strong in absorption. The O II and C III absorption lines are strong and wide in the spectra of the supergiants HD188209 and α Camelopardalis, but the weakness of the absorption dip at λ4686 and the course of the spectrum near λ4640 indicates that emission in the Of lines occurs in supergiants. The material of Figure 55 demonstrates that all strengths of the typical Of emission features occur in O stars. The emission lines in λ Cephei are comparable in width to those in some Wolf-Rayet stars.

About 13 percent of the O stars in the catalogues by MORGAN, CODE, and WHITFORD (1955) and by HILTNER (1956) are listed as Of. In these catalogues the f designation is given only to stars of type O8 and earlier. However, detailed quantititave measurements on moderate and high dispersion spectra (WILSON, 1957, UNDERHILL, 1958b) indicate that some emission at the N III λ4640 multiplet occurs in practically all O type stars. The pure absorption-line O stars and the Of stars do not form separate groups; there is a continuous gradation in the strength of the N III λ4640 emission.

STRUVE and SWINGS (1940), OKE (1954), MANNINO and HUMBLET (1955), WILSON (1957), SLETTEBAK (1956) and KUMAJGORODSKAJA (1960, 1962) have studied the line strengths in a number of Of stars, and UNDERHILL (1958a, 1959b) has given a detailed study of the spectrum of 9 Sagittae, O7f. From these studies it appears that the emitting shells around the Of stars are variable and moderately extended, rather like the shells of Be stars. In fact the Of stars seem to be related to the absorption-line O stars rather as the Be stars are related to the absorption-line B stars. Very probably the cause of the shell is the same in both cases.

It is not clear that the Of stars are significantly more luminous than the pure absorption O stars of the same spectral subdivision. ROMAN (1951) from a study of some stars in Cygnus suspected that the Of stars might be more luminous than the O stars, but UNDERHILL (1955) was not able to substantiate this claim. The extended atmosphere in which the emission features originate is not necessarily a criterion of high luminosity and the f characteristics alone will not distinguish between a shell star and a supergiant.

The emission at the hydrogen lines is probably a pure recombination spectrum. At the He II lines, it may also be a recombination spectrum except for the rather strong emission at He II λ4686 which is considerably stronger than the emission at any other observable He II line. BOWEN (1935) has suggested a cycle involving Lyman-α of hydrogen, which would account for the strong emission at He II λ4686. The other emission lines appearing in Of spectra probably are excited by selective processes. SWINGS (1948) has summarised the relevant information regarding many of these lines. Here we shall mention only the exciting mechanisms for the N III λ4640 multiplet and for the C III λ5696 line. These lines are the selectively excited lines most frequently observed in Of spectra. The N III lines appear to be excited by the cycle proposed by BOWEN (1935) to account for the occurrence of these lines in the spectra of gaseous nebulae. This cycle involves a wave-length coincidence between He II λ303 and an O III line and another coincidence between O III λ374 and a N III line. The result is that the N III lines, λ4634, λ4640 and λ4641 will be excited in emission whenever a strong flux of He II λ303 passes through a high temperature gas of low density. The C III line λ5696 is probably excited by preferential ionisation from the 3^2D levels of C II by He II λ303 quanta (UNDERHILL, 1957). This process, which is the same as ejection from the K and L shells known from γ-ray work, selectively creates C^{++} ions in the 3^1D level which is the upper level of λ5696. The 3^1D level cannot be reached from the ground level of the C^{++} ion by the absorption of one quantum of radiation because such a transition is forbidden by parity selection rules.

CHAPTER XVI

THE BETA CANIS MAJORIS STARS

1. Introduction

The β Canis Majoris stars are sharp-lined stars in the spectral-type range B0.5 to B2 which show radial-velocity and light changes in periods of the order of three to seven hours. The light amplitude is usually only a few hundredths of a magnitude; the radial-velocity amplitudes range from 5 km/sec to 150 km/sec. The 18 stars at present known to be β Canis Majoris variables are listed in Table 33. The period which VAN HOOF (1962a) considers to be the fundamental period of pulsation is listed in column seven. This period is not necessarily identical with the dominant period of the radial-velocity or light variations.

Although maxima in the radial-velocity curves and in the light curves recur at periodic intervals, the amplitude and shape of the velocity curves of most of the stars change slowly. In the case of β Canis Majoris, for instance, two dominant periods, namely 6^h00^m and 6^h02^m, appear to be present. The composite radial-velocity and light curves of β Canis Majoris vary with a beat period of about 49 days. Van Hoof has

TABLE 33
The Beta Canis Majoris stars

HD number	Name	α (1900)	δ (1900)	m_V	Spectral type	Fundamental period
886	γ Peg	$00^h08^m.1$	+ 14°38′	2.8	B2IV	3^h38^m
16582	δ Cet	02 34.4	− 00 06	4.0	B2IV	3 52
19374	53 Ari	03 01.8	+ 17 30	6.1	B2IV	3 40
21803	KP Per	03 25.8	+ 44 32	6.3	B2IV	4 48
29248	ν Eri	04 31.3	− 03 33	4.1	B2III	4 10
44743	β CMa	06 18.3	− 17 54	2.0	B1II	6 02
46328	ξ^1 CMa	06 27.7	− 23 21	4.4	B1IV	5 02
50707	15 CMa	06 49.2	− 20 06	4.8	B2III	4 26
111123	β Cru	12 41.9	− 59 09	1.2	B0.5IV	5 40
126341	τ^1 Lup	14 19.7	− 44 46	4.6	B2IV	4 16
129056	α Lup	14 35.3	− 46 58	2.3	B1III	6 14
147165	σ Sco	16 15.1	− 25 21	2.8	B1III	5 55
157056	θ Oph	17 15.9	− 24 54	3.4	B2IV	3 22
165174	V986 Oph	17 59.6	+ 01 55	6.1	B0.5III	6 56
199140	BW Vul	20 50.1	+ 28 08	6.4	B2III	4 49
205021	β Cep	21 27.4	+ 70 07	3.3	B2III	4 34
214993	12 Lac	22 37.0	+ 39 42	5.2	B2III	4 38
216916	16 Lac	22 51.8	+ 41 04	5.5	B2IV	4 04

analysed the existing information about the light and radial-velocity variations of β Cephei, β Canis Majoris, ξ^1 Canis Majoris, θ Ophiuchi, ν Eridani and β Crucis and found several related periods for each star. Performing an harmonic analysis is one way of representing the observed continuously changing radial-velocity and light curves of the β Canis Majoris stars, but it is doubtful if each period found by van Hoof may be interpreted as a stable pulsational mode.

Little or no change of line shape during a radial-velocity cycle has been observed in about half the stars, namely γ Pegasi, δ Ceti, 53 Arietis, ξ^1 Canis Majoris, 15 Canis Majoris, β Crucis, τ^1 Lupi, α Lupi and β Cephei, whereas in the other cases, KP Persei, ν Eridani, β Canis Majoris, σ Scorpii, θ Ophiuchi, BW Vulpeculae, 12 Lacertae and 16 Lacertae, the line shapes change conspicuously, the lines becoming broad at times and sometimes double. The author has no information about the spectrum of V986 Ophiuchi. The doubling of the lines may be interpreted to mean that at certain phases different parts of the stellar atmosphere are moving with distinctly different velocities. Each star behaves rather differently in this respect and it is impossible to make any valid generalisations. Some of the better observed cases are discussed in Section 3. VAN HOOF (1962b) has suggested that the occurrence of line broadening is associated with free pulsation in the fundamental mode. According to van Hoof, when the fundamental mode is unexcited or only very slightly excited, no changes of line shape are observed during the radial-velocity cycle. The stars which show the greatest line broadening changes tend to show the greatest light changes.

The β Canis Majoris variables are named after the first star of the group which was studied in detail. The first member of the group which was found was β Cephei. FROST (1902) at the Yerkes Observatory discovered that the radial velocity of β Cephei varied in a very short period. Shortly thereafter (FROST, 1906) he was able to show that the period was 4^h34^m. In 1913 GUTHNICK (1913) discovered that the light varied in the same period with an amplitude of 0.05 magnitude. Since the light curve was rather sinusoidal in shape and not like that of an eclipsing variable, and since the short period and small, variable amplitude (18 to 46 km/sec) of the radial-velocity variations made it unlikely that β Cephei was a spectroscopic binary, it was concluded that β Cephei was a pulsating variable. By 1912 four B type stars showing short-period radial-velocity variations have been discovered, namely β Cephei, σ Scorpii, β Canis Majoris, and 12 Lacertae. These stars form the nucleus of the class. HENROTEAU (1918) discovered that the shapes of the lines in the spectrum of β Canis Majoris changed periodically, the period being 6^h02^m which is distinctly different from the period of 6^h00^m which is the dominant period of the radial-velocity changes. Thus by 1918 the major characteristics of this class of variable stars had been discovered. It was not until Struve and his co-workers in the 1950's began an intensive study of the β Canis Majoris stars that the true complexities of behaviour of these stars became evident. A satisfactory theoretical explanation and interpretation of the observed variations in light, radial velocity and spectrum is not yet available. These stars are sometimes called β Cephei stars after the star which was first discovered.

Summary papers have been written about the observed properties of the β Canis

Majoris stars by STRUVE (1955), by LEDOUX and WALRAVEN (1958), and by VAN HOOF (1962a). Rather extensive studies have been made of a few stars, as will be mentioned in Section 3, and some theoretical studies have been made. These are discussed in Section 4. STRUVE (1955) found it useful to divide the stars into two groups according to whether only one period was dominant in the radial velocity or not. Struve calls the period which appears only in the radial velocity P_1; he calls the period which occurs in the radial-velocity variations and with which the line shapes vary P_2. The stars in Struve's Group I show only one period, either P_1 or P_2; those in Group II show both P_1 and P_2. However, recognising only two periods is an inadequate simplification in some cases, as van Hoof has shown. Any interpretation of the β Canis Majoris stars should take account of the fact that these stars may show complex radial-velocity and light variations and that several interfering modes may be present in the atmosphere.

2. Generalisations

2.1 *The period-luminosity relation*

There appears to be a period-luminosity relation among the β Canis Majoris stars, the stars of longest period being intrinsically the brightest. MCNAMARA (1953) first noticed that such a relation existed. BLAAUW and SAVEDOFF (1953), using visual absolute magnitudes derived from membership in associations and from proper motions, found the relation

$$M_V = -10 - 9 \log P \tag{XVI-1}$$

where P is expressed is days. This relation was found from the six stars δ Ceti, β Canis Majoris, σ Scorpii, β Cephei, 12 Lacertae, 16 Lacertae.

PETRIE (1954a) estimated M_V from measurements of the equivalent width of Hγ $W(\mathrm{H}\gamma)$, in the spectra of γ Pegasi, δ Ceti, β Canis Majoris, BW Vulpeculae, β Cephei, 12 Lacertae and 16 Lacertae, and found the relation

$$M_V = 0.4 - (18.1 \pm 2.3)\, P\,, \tag{XVI-2}$$

where the period is expressed in days. The observed visual absolute magnitudes are not accurate enough to permit differentiation between relations (XVI-1) and (XVI-2). However, on theoretical grounds one expects a relationship between M_V and $\log P$. PETRIE (1964) has revised his calibration of $W(\mathrm{H}\gamma)$ in terms of M_V. When the new calibration of $W(\mathrm{H}\gamma)$ is used with Petrie's measured equivalent widths and a least-squares solution is performed, it is found that the period-luminosity relation for the β Canis Majoris stars is

$$M_V = -(8.7 \pm 1.4) - (6.7 \pm 1.9) \log P\,. \tag{XVI-3}$$

Relationship (XVI-2) can be changed to the new system by replacing the constant term by -0.5, for the average ΔM_V per star resulting from the use of the new calibration of $W(\mathrm{H}\gamma)$ is -0.9 mag.

Among the β Canis Majoris stars the longer periods occur with the earlier spectral types in contrast to the situation with the classical cepheids. Equation (XVI-3) corresponds to the statement that the fundamental period of the star varies as $L^{0.37}$, where L is the luminosity of the star.

VAN HOOF's (1962d) suggestion that the fundamental period of β Crucis is 0.2365072 day rather than 0.160474 day, the period tentatively suggested by PAGEL (1956), leads to an estimated visual absolute magnitude for β Crucis of -4.5. This value is more consistent with the appearance of the spectrum, the equivalent width of Hγ, and the membership of β Crucis in the Scorpio-Centaurus cluster than the absolute magnitude -3.4 which is found by adopting the period suggested by Pagel and using relationship (XVI-3). Van Hoof considers Pagel's period to be the first harmonic.

2.2 *Phase relationships and line shifts*

The light curves are nearly sinusoidal in shape and they lag about ninety degrees in phase behind the radial-velocity curves. Thus maximum light occurs near the moment when the radial velocity is crossing the γ-axis on the descending branch of the velocity curve. This is true of stars with single periods and of stars with multiple periods and it suggests that maximum light occurs at minimum radius.

In some of the stars, particularly β Canis Majoris, 12 Lacertae, 16 Lacertae, σ Scorpii and BW Vulpeculae, the radial-velocity curves derived from different elements, for instance H, He I, Si III or O II, are not quite in phase. Usually the H and He I velocities lag behind the Si III and O II velocities. This effect which was discovered by van Hoof (VAN HOOF and STRUVE, 1953) probably indicates that there is a range of velocity through the stellar atmosphere. These stars all show line doubling or at least line widening during the radial-velocity cycle. The van Hoof effect is more pronounced on the descending branch of the velocity curve than on the ascending branch.

Struve, upon several occasions, has drawn attention to possible systematic velocity differences between lines from different ions and levels of ionisation in the spectrum. Such differences are difficult to isolate because one must be sure that the adopted undisplaced wave lengths for the various lines are all rigidly on the same system and that all possible blends have been eliminated or that the adopted wave length has been adjusted correctly to account for the blending. The results of a careful radial-velocity study of high-dispersion spectrograms of nine β Canis Majoris stars have been summarised by STRUVE and ZEBERGS (1960). They find that at any epoch lines of nearly all elements may show departures from the mean velocity which exceed the probable error of measurement. They believe these departures are due to unresolved blends. No evidence is found for systematic motions of one element with respect to another and there is no evidence to suggest that the outer layers of the atmosphere are expanding. The case of σ Scorpii forms an exception, for here the hydrogen lines tend to give positive residuals with respect to the other lines. There is no convincing evidence of differential motion between the layers where strong lines are formed and those where weak lines are formed.

Struve and Zebergs noted that the strong Si III lines almost always showed a systematic negative displacement and they were puzzled about the meaning of this. TORRESON (1960) has given a new list of wave lengths for the Si III spectrum and it may be seen that the new wave lengths are sufficiently different from those adopted by Struve and Zebergs to account for the observed displacements. The Si^{++} ions, therefore, have the same velocities as the other atoms and ions in the atmospheres of the β Canis Majoris stars.

2.3 *Rotation*

The β Canis Majoris stars as a group show very sharp lines and their velocities of rotation are rather smaller than those of most stars in the spectral range B0.5 to B2. MCNAMARA and HANSEN (1961) have measured the projected velocity of rotation of twelve β Canis Majoris stars using high dispersion spectrograms and they have shown the $v \sin i$ does not exceed 40 km/sec for any of the group. They measured the profiles of N II λ3995 and O II λ4415 and fitted them to rotationally broadened profiles using only spectrograms obtained at phases when the lines were sharp. They adopted δ Ceti as a typical non-rotating star and assumed that any differences existing between the line shapes in δ Ceti and the other stars were due only to rotation. Their results are given in Table 34.

McNamara and Hansen note that the average velocity of rotation of the β Canis Majoris stars (assuming a random orientation of their axis of rotation) is 22 km/sec, whereas it is 164 km/sec for non-variable stars of similar spectral type and luminosity

TABLE 34

Characteristics of the light and velocity changes of Beta Canis Majoris stars

Star	Light Variation amplitude*	Light Variation character	Velocity Variation amplitude	Velocity Variation character	Line Profiles $v \sin i$	Line Profiles character
	mag.		km/sec		km/sec	
γ Peg	0.015	const	7	const	0	const
δ Cet	0.025	const	13	const	0	const
53 Ari	none known		5	const		const
KP Per	0.1	var	16–25	var		var
ν Eri	0.02–0.11	var	22–49	var	7	var
β CMa	0.03	const	6–12	var	28	var
ξ^1 CMa	0.01–0.045	var	36	const	0	const
15 CMa	0.01	const	7	const	39	const
β Cru	0.04	const	14	var		const
τ^1 Lup	0.03	var	11	const		const
α Lup	0.03	var	14–20	var		var
σ Sco	0.08	var	80–120	var	30	var
θ Oph	0.06	const	5–20	var	26	var
BW Vul	0.19–0.26	var	$>$ 150	var	26	var
β Cep	0.02–0.05	var	18–46	var	25	const
12 Lac	0.04–0.074	var	15–36	var	29	var
16 Lac	0.06–0.11	var	9–30	var	5	var

* The amplitude is different in different colours. These numbers are typical results.

class. They advance the hypothesis that high rotation inhibits the pulsation of stars, and they point out that the rather sharp-lined stars in the spectral range B0.5 to B2 which are not known to be β Canis Majoris stars may well be rather rapidly rotating stars for which the angle of inclination of the axis of rotation is large, sin $i \to 0$.

Further information concerning the variations in light, radial velocity and line shape which have been observed for the β Canis Majoris stars is given in Table 34. The character of each variation is stated to be constant (const) when there is no report of significant changes during the period of observation. Otherwise it is noted as variable (var). In some cases prolonged observation with equipment capable of resolving very small changes of light, line shape and displacement in short periods of time might indicate that a variation which is now said to be "constant" in behaviour is variable.

2.4 *Emission lines*

Variable emission at Hα has been observed in the case of β Cephei and inferred in the case of BW Vulpeculae (see Section 3), but it has not been reported for the other stars.

The star θ Ophiuchi is quite unusual, for VAN HOOF and BLAAUW (1958) have reported the presence of weak emission lines of Ca II λ3933, He I λ3965 and Mg II λ4481 around the time of maximum velocity. The red region of the spectrum was not investigated. The above lines do not usually appear in emission in B2 stars. Usually the most prominent emission lines in the blue-violet spectral region of B stars are Hβ and Hγ, followed by some lines of Fe II. The emission features which are displaced longwards by some 70 to 100 km/sec from the Ca II and Mg II absorption lines are the strongest emission lines in the spectrum. When the Ca II emission is strongest there is strong asymmetry in the profile of He I λ3965 suggesting the presence of emission here and weak emission appears near N II λ3995. This emission spectrum deserves further study.

3. Observations of individual stars

Here some of the highlights will be given of the information available about the individual stars. Much information is available about a few stars while others require more observation at high dispersion and at short time intervals. Few of the stars appear to repeat their light, radial-velocity and spectral variations exactly. Even these few might show long term variations were they observed in adequate detail. Those stars which undergo the greatest range in light changes and in radial velocity tend to be the most irregular.

γ Pegasi: A full bibliography to 1958 has been given for this star by SCHNELLER (1960). This star is a standard B2 IV star and because of its apparent brightness and the sharpness of its lines, the spectrum of this star has been investigated in detail, see Chapter XI. Nevertheless, the spectral lines are not so sharp as they should be were the atmosphere of this star in hydrostatic equilibrium at the temperatures suggested by normal model atmospheres. In such a case, the total halfwidths of lines of the light elements should be due chiefly to thermal Doppler broadening. Values around 0.2 Å

would be expected. High-dispersion, high-resolution spectrograms, however, indicate a total width at half-intensity near 0.35 Å (UNDERHILL, unpublished). The extra width of the line profiles is probably due to motions of the atmosphere which may be characterised as turbulence. The broadening is a true effect and not due to inadequate spectral purity.

McNAMARA (1955) first demonstrated the radial-velocity variations of γ Pegasi and later observations (McNAMARA, 1956; SANDBERG and McNAMARA, 1960) have shown that the period and radial-velocity amplitude have remained constant.

δ Ceti: A full bibliography for this star has been given by SCHNELLER (1960). The light changes and radial-velocity changes appear to be constant in range and period. This star has rather sharp lines which do not appear to change shape during the radial-velocity cycle. McNAMARA (1955) confirmed earlier suspicions that this star might be a β Canis Majoris star.

53 *Arietis*: The radial velocity of this sharp-lined B2 IV star was measured by MÜNCH and FLATHER (1957) using 10.2 Å/mm coudé spectrograms from the Mount Wilson Observatory and they were able to demonstrate that the period and range of the radial-velocity variations are similar to those of γ Pegasi, a star with the same spectral type. No information seems to be available about possible light variations of this star. LYNDS (1959) observed photoelectrically 36 early type stars which might have been β Canis Majoris stars according to their spectral types, but 53 Arietis was not on his list. This star is one of the "run-away stars" discussed by BLAAUW (1961), see Chapter VI. VAN HOOF and BLAAUW (1964) have confirmed the variations found by Münch and Flather and they suspect that the amplitude and mean velocity vary.

KP Persei: The light variations of this star were discovered by LYNDS (1959) and the radial-velocity variations have been described by STRUVE and ZEBERGS (1959). KLOCK (1965) has given further observations of the light variations. Both the amplitude and the period of the light and of the velocity variations are variable and the spectral lines vary in width. Although this star has nearly the same period as BW Vulpeculae, the light and radial velocity ranges are considerably smaller and the period appears be more variable. Klock has suggested that there is a fluctuation which occurs at intervals of 10 to 15 minutes superposed on the two main periods of 4^h51^m and 4^h45^m. ROSSATI (1963) has also made photometric observations.

ν Eridani: A full bibliography for this star has been given by SCHNELLER (1960). The radial-velocity and light curves are very complex and the line profiles vary greatly during the cycle. VAN HOOF (1961a, 1961b) has attempted an harmonic analysis of the variations and he has found that twelve related periods are required to represent the observations. Clearly the concept of simple pulsation is inadequate to explain the character of this star.

β Canis Majoris: This star does not appear to be known officially as a light variable and it is not listed by SCHNELLER (1960) although FATH (1935) has published a light curve and the radial-velocity variations have been known for many years. VAN HOOF (1963) has published light curves of β Canis Majoris obtained on two nights confirming the character of the light variations found by Fath. The light variations of this star

should be studied more extensively with modern equipment. There is little doubt about the short-period (6^h) light variations found by Fath and VAN HOOF (1963) has shown that variations in the U amplitude show up with the well known beat period of 49 days found in the radial-velocity observations. Unfortunately Fath selected 17 Leporis as his comparison star, and since this star is a shell star it may vary slightly from time to time.

The most recent and most comprehensive discussions of the radial-velocity variations of this star have been made by Struve and his coworkers (STRUVE and SWINGS, 1941; STRUVE, 1950; STRUVE, MCNAMARA, KUNG, VAN HOOF and DEURINCK, 1954). It is quite clear, as Meyer's long series of observations (published only in abstract form, MEYER, 1934, and later completed by STRUVE, 1950) showed, that the amplitude and shape of the radial-velocity curve change and that these changes can be fairly well represented by two sinusoidal curves having periods of 6^h00^m and 6^h02^m. VAN HOOF (1962b), however, has shown that the observations may be represented by at least 12 different oscillations and Struve *et al.* have noted that irregular distortions occur in the velocity curves of β Canis Majoris just as they do for ν Eridani and for 16 Lacertae. MILONE (1965) has given further observations of the irregularities in radial velocity. UNDERHILL (1946) has published a preliminary spectrophotometric study of the spectrum of this star using high-dispersion spectrograms obtained by Struve and Swings with the coudé spectrograph of the McDonald Observatory. In view of the greater understanding now of the variations in the spectra of β Canis Majoris stars, it would be worthwhile to study the light and spectral variations in considerable detail using very high dispersion and short exposure times.

ξ^1 Canis Majoris: A bibliography for this star is given by SCHNELLER (1960). MCNAMARA (1955) drew attention to this star which has very sharp lines and for which there is no change of line shape. VAN HOOF (1962c) has analysed a series of photometric observations taken at the Boyden station in 1961–1962. These observations have revealed secondary oscillations in the light variations which have not been detected in the radial-velocity changes. Van Hoof concludes that the fundamental period is not excited in this star, thus there is no variation of line shapes, and that the main observed period is a coupling between the second and fourth overtone or between the first and third overtones.

15 Canis Majoris: A bibliography for this star is given by SCHNELLER (1960). This star was selected by Henroteau as a possible β Canis Majoris variable. Radial-velocity observations have been made by LYNDS, SAHADE, and STRUVE (1956) over the period November 1955 to January 1956 and WALKER (1956b) has studied the light variations. The velocity curve does not repeat exactly from cycle to cycle and there is some indication that the mean velocity varies. There is no evidence that the spectrum varies. VAN HOOF (1965) has made further photometric observations.

β Crucis: PAGEL (1956), from a study of the radial-velocity variations of this star, suggested that it might be a β Canis Majoris star. VAN HOOF (1962d), from the study of photometric observations obtained at the Boyden station, showed that multiple

periods appear in the light variations and he demonstrated that the period suggested by Pagel was probably the first overtone.

τ^1 *Lupi*: A bibliography is given by SCHNELLER (1960). The spectrum of this star was observed by Pagel at 21 Å/mm and no changes were seen in the line profiles. VAN HOOF, PRETORIUS and PIKOOS (1964) have made photoelectric observations of this star on 29 nights and they find that beats with a period of 8.0 days occur in the light cycle.

α *Lupi*: This star was discovered to be a β Canis Majoris star by PAGEL (1956). RODGERS and BELL (1962) have studied 6.7 Å/mm spectrograms of this star obtained at Mt. Stromlo. The hydrogen lines appear to lag behind the other lines in phase and there is a marginal broadening of the line profiles on the descending branch of the radial-velocity curve. VAN HOOF (1964 and unpublished) has been observing α Lupi for several years and he finds that the amplitude of the short period light variations changes slowly. Small changes in period seem to occur in parallel with the changes in amplitude.

σ *Scorpii*: A bibliography is given by SCHNELLER (1960). This is a star for which conspicuous irregularities in the light and velocity variations are seen. The lines become double on the descending branch of the velocity curve. The bright star appears to be a single-lined spectroscopic binary with a period near 34 days. It also pulsates with a period near 0.25 day and a beat period of 8.028 days appears to be superimposed on the short period. VAN HOOF (unpublished) has rediscussed the radial-velocity material together with photometric observations made in 1962–65. He finds that the orbital period is 33.008 days and that the beat period is 8.252 days. It is possibly significant that the beat period is commensurate with the orbital period. The colour and line-shape changes do not seem to follow the same pattern as in other β Canis Majoris stars.

θ *Ophiuchi*: VAN HOOF and BLAAUW (1958) have studied the spectral variations on 152 moderate dispersion (35 Å/mm) spectrograms obtained on four consecutive nights at the McDonald Observatory while VAN HOOF (1962a) has studied the brightness variations. The radial velocity varies with a period of 3^h22^m but the curve is distorted by secondary fluctuations. Such secondary fluctuations are also present in the light curve and van Hoof has found eight separate periods. The star θ Ophiuchi is the brighter component of a spectroscopic binary, and it is most remarkable in showing emission lines of Ca II, Mg II and He I at some phases, see Section 2.

BW Vulpeculae: A bibliography for this star is given by SCHNELLER (1960). This star is an outstanding member of the β Canis Majoris class of variable stars, for its light range is the largest known and its radial velocity range is also very large. Moreover the amplitude of the radial-velocity variations has been increasing at a rate of about 0.7 km/sec per year since 1928 when Petrie first began observing the star for radial velocity. Petrie's numerous radial-velocity observations have been discussed thoroughly by him (PETRIE, 1954b); ODGERS (1955) has carried on the extensive observational series at the Dominion Astrophysical Observatory and has attempted to develop a theoretical explanation of what is seen.

This star appears to have only one dominant period in contrast to the case of other β Canis Majoris stars of rather large range. The period appeared to be increasing at a rate of 3.7 seconds per century until 1953–54 when it began to decrease. The amplitude of the radial-velocity variations, however, has continued to increase. The radial-velocity variations do not repeat exactly from cycle to cycle. A typical velocity curve is shown in Figure 56. It was obtained by Odgers with the IQ spectrograph on 1954, August 11, using exposures of the order of four minutes in order to obtain good

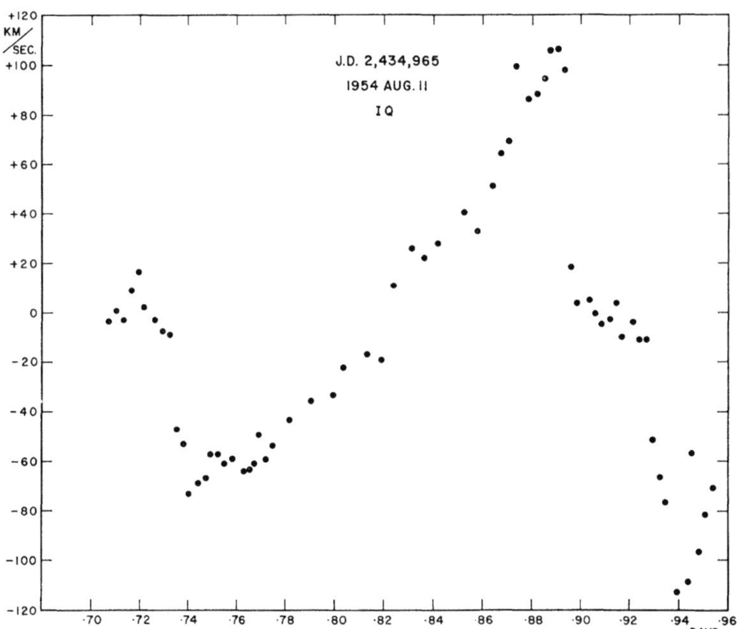

Fig. 56. A radial-velocity curve of BW Vulpeculae obtained on August 11, 1954. The spectrograms were obtained at four minute intervals.

time resolution. The IQ spectrograph has quartz optics and permits observations to 3600 Å; the dispersion is 67 Å/mm at Hγ. Lines of H and of He I were measured for radial velocity.

The major features of the radial-velocity changes of BW Vulpeculae are visible in Figure 56, namely the irregularities during the cycle, the very abrubt drop after maximum and the apparent "stillstand" halfway down the descending branch. These features in lesser degree are characteristic of the radial-velocity curves of most β Canis Majoris variables. McNamara, Struve and Bertiau (1955) using the more powerful spectroscopic equipment of the Mount Wilson Observatory have been able to show that the spectral lines become double after maximum velocity and that several bodies of gas moving at different velocities appear to be present during this part of the cycle.

On low dispersion spectrograms of low spectral purity the doubling appears as a widening of the lines and a stillstand in the velocity curve.

ODGERS (1956) has noted that there appears to be some emission filling in Hα and Hβ at the phase of maximum star, presumably due to emission in the extended atmosphere at this phase. ODGERS and KUSHWAHA (1958) have demonstrated that the theory of shock waves in an isothermal atmosphere will serve to give an explanation of the spectroscopic phenomena which are seen. However, no light is given on the origin of the periodic disturbance which is postulated as the cause of the outgoing wave.

β Cephei: STEBBINS and KRON (1954) have made a six-colour photometric study of β Cephei and have shown that the light curve is of small amplitude and varies with the dominant period of the radial velocity variations. The older radial-velocity variations have been reviewed by B. SMITH (1943) and he has shown that there is some evidence that the mean velocity varies with a period of 50 years. Observations by STRUVE, MCNAMARA, KUNG and BEYMER (1953) have given information about the short-period variations during the years 1950–52. They demonstrate that the amplitude and the mean velocity of β Cephei do vary, but the variations do not appear to be periodic in character. VAN HOOF (1962b) has analysed the radial-velocity variations of β Cephei and has suggested that more than 100 periods are represented. Clearly his results indicate that the observations do not refer to a phenomenon which repeats itself exactly. Many of the amplitudes suggested by van Hoof are less than 2 km/sec; the largest amplitude is 24.5 km/sec and it certainly represents a real phenomenon.

Miss ROMAN (1947) has studied the line profiles on a series of spectrograms having a dispersion of 4.5 Å/mm and she finds that there is no variation of the line width during the radial-velocity cycle. Van Hoof interprets this observation to mean that the fundamental pulsation period is not excited in β Cephei.

SAHADE and STRUVE (1956) have found that the Hα absorption line is flanked by emission borders and that the feature behaves, with respect to phase, as though the emission is relatively at rest and the absorption shifts upon it in accordance with what one would expect from the star's short-period velocity variation. The presence of the Hα emission had been inferred by R. Wilson and Seddon from their studies of B type spectra at the Royal Observatory Edinburgh.

12 Lacertae: A bibliography is given by SCHNELLER (1960). This star is also known as DD Lacertae. It is one of the more irregular members of the class. DE JAGER (1963) has reported the results of a world-wide program to obtain simultaneous light, colour and radial-velocity observations of this star. BARNING (1963) has analysed the light curve and shown that the following four periods appear to be present: 0.193089 day, 0.197358 day, 0.182127 day and 25.85 days. The first two periods are well known from the early photometric and radial-velocity studies. STRUVE and ZEBERGS (1955) have shown that the hydrogen lines lag on the descending branch of the velocity curve and that double lines occur as in BW Vulpeculae and in σ Scorpii when the amplitude is large.

16 Lacertae: A bibliography is given by SCHNELLER (1960). This star is also known as EN Lacertae. It was discovered to be a β Canis Majoris variable by WALKER (1950) and Struve and his colleagues have studied the radial-velocity variations. Both the light curve and the radial-velocity curve vary in amplitude and the mean velocity of the star varies from cycle to cycle. Two short periods, namely 0.17085 and 0.16916 day, appear to be well established. The mean velocity varies with a period of 12.097 days and a semi-amplitude of 24.3 km/sec. This variation is considered to be the result of motion in a binary orbit. STRUVE, MCNAMARA, KRAFT, KUNG and WILLIAMS (1952) have given a set of preliminary elements for this orbit as well as a discussion of the short-period radial-velocity changes. There is evidence of a beat period near 17 days between the two short periods, but the interpretation of the variations is not without some questions. The absorption lines are sharp and they appear to widen slightly on the descending branch of the radial-velocity curve. WALKER (1954) has observed one cycle in three colours, infrared, yellow and ultraviolet, and has found some changes in colour during the cycle. He has tested to see if the light and colour changes and the radial-velocity changes are consistent with the hypothesis of radial pulsations and has come to the conclusion that the observations suggest rather that 16 Lacertae undergoes some sort of non-radial pulsation.

4. Theoretical interpretations of the β Canis Majoris stars

The theoretical interpretation of the β Canis Majoris stars faces several difficulties. Although it has been stated several times in the previous sections of this chapter that these stars pulsate, the concept of purely radial oscillations runs into difficulties in accounting for the inferred period-density relationship (LEDOUX and WALRAVEN, 1958; GURM, 1963). Furthermore it is difficult to infer what drives the postulated radial pulsations. The high-temperature boundary of the instability zone in the HR-diagram (this zone contains the RR Lyrae stars and the Cepheid variables) is believed to be due to a lack of effectiveness at higher temperatures of the hydrogen and helium ionisation zones as energy storage valves. Thus, according to the theory which accounts reasonably well for the variability of the RR Lyrae stars and the Cepheids, no stars of early spectral type should pulsate. We have seen that the term "pulsation" appears to be suitable for describing the cyclic variations which occur in the β Canis Majoris stars, but that the general tendency is that several periods are dominant at one time. Possibly the pulsations are being driven by the storage and release of energy in the ionisation zones due to the ultimate ionisation of the rather abundant light elements C, N, and O.

It has been postulated that the observed spectral variations and light variations are the result of recurring shock waves. ODGERS and KUSHWAHA (1958) showed that this was an adequate explanation for the observed changes in BW Vulpeculae but could give no cause for the generation of the shocks. LEDOUX (1958) has criticised the shock theory on the grounds that it is difficult to see how the generally high symmetry of the light and velocity curves can be reconciled with this hypothesis which must imply very strong shocks in the external layers. He also considers that there will be difficulties with preserving periods precisely. These criticisms are valid for stars such as γ Pegasi and δ Ceti which show very regular changes, but they loose force when one considers the more irregular members of the group such as ν Eridani, σ Scorpii and BW Vulpeculae.

LEDOUX (1951) has proposed a theory of non-radial oscillations into which rotation enters as a fundamental factor. This theory has been applied to β Canis Majoris but the results are not entirely satisfactory. CHANDRASEKHAR and LEBOVITZ (1962a, 1962b) have criticised this theory pointing out that it is unlikely that non-radial oscillations would be excited in preference to radial ones. They have presented an explanation of the beat phenomenon in β Canis Majoris stars in terms of non-radial oscillations which have slightly different frequencies as a result of the rotation of the star. CLEMENT (1965a, 1965b) has examined this theory and has shown that in order to obtain differences in period of the order of what is observed the stars would have to be rotating three to four times more rapidly than observations suggest. The arguments of McNamara and Hansen that the β Canis Majoris stars are truly slowly rotating stars appear rather strong. VAN HOOF (1964) has also considered the possibility that high rotational velocities may occur for β Canis Majoris stars and has rejected this possibility. He bases his arguments on the known characteristics of the spectroscopic orbits of σ Scorpii and 16 Lacertae and on the fact that with spectroscopic binaries it may generally be inferred that the axis of rotation of the star is perpendicular to the plane of the orbit.

SCHMALBERGER (1960) has pointed out that the β Canis Majoris stars seem to lie in the HR-diagram at about the place where the normal energy generation by the burning of hydrogen becomes exhausted and secondary contraction begins. He suggests that the ensuing structural change in the star might generate pulsations of the type which are observed. STOTHERS (1965) has developed this idea further by showing for a series of models that the relative periods of the various radial pulsational modes which could occur when the models are entering the stage of secondary contraction are those which are observed in some β Canis Majoris stars. Stothers does not investigate how the proposed radial pulsations could be started, the rate at which they would build up to the desired amplitude, nor whether they could be maintained for a sufficiently long period.

5. Stars related to the β Canis Majoris stars

The stars which are recognised to be β Canis Majoris stars are confined to the narrow spectral range B0.5 to B2 and they have sharp lines. At least three of them (σ Scorpii, θ Ophiuchi and 16 Lacertae) appear to be the brighter members of a spectroscopic binary. The question arises why the spectral range in which they occur is so narrow and why so few of them appear to be members of a binary system. Among the B stars in general at least fifty per cent are members of binary or multiple systems.

Another point about the β Canis Majoris stars is that they are all apparently bright. This may be a selection effect, because it is increasingly difficult to recognise short-period, small-range radial-velocity variations of faint stars, for the exposure times become too long. It is notable that for many β Canis Majoris stars the radial-velocity and light variations do not repeat exactly in fixed cycles. More than one harmonic motion must be adopted in order to represent the observed variations. Some stars behave more regularly than others.

One may ask: Why are cyclical variable stars not found at spectral types B3 and later and at spectral types B0 and earlier? It is interesting that two sharp-lined apparently bright B3IV stars, ι Herculis and 53 Piscium, are known to have irregular radial velocities. PETRIE and PETRIE (1939) have studied the radial velocity and spectrum of ι Herculis in detail with one run of 3^h40^m and another of 4^h40^m as well as observations on many other nights. They were able to show that the radial velocity of ι Herculis is variable with a range of about 15 km/sec, but they were not able to establish a period. MATTHEWS (1956) found very much the same thing for 53 Piscium. It is also suspected that 23 Sextantis shows similar variations. The radial velocity of the star 10 Lacertae, O9V, at the other end of the β Canis Majoris class, also appears to vary irregularly with a range of about 10 km/sec (GRYGAR, 1964; UNDERHILL and VAN HELDEN, 1965). At this time it is not clear if there is any relationship between these stars and the β Canis Majoris stars. However, the fact that the radial velocity of many sharp-lined stars varies within short periods of time indicates that the atmospheres of these stars are not in strict hydrostatic equilibrium and that some phenomenon which results in a moving atmosphere occurs quite commonly among the sharp-lined OB stars.

THE BASIS OF THE THEORY OF NORMAL STELLAR SPECTRA

It is assumed that a normal stellar atmosphere may be represented by plane parallel layers of gas of constant composition which are in mechanical equilibrium under the acceleration of gravity. The angular dependence of the radiation field is a function only of the angle, ϑ, between the direction of the beam and the normal to the layers and not of the azimuthal angle, φ. The geometric depth in the atmosphere is measured by the coordinate, z, which is measured outwards from the centre of the star.

One may assume a temperature law, $T(z)$, and then find the pressure at each depth in the atmosphere by solving the equation of hydrostatic equilibrium, taking account, by means of Saha's law, of the degree of ionisation of the gas at each level in the atmosphere and, if necessary, of the radiation pressure gradient. A model atmosphere consists of a set of values of T, P_g, and P_e at each level in the atmosphere. The depth of the level in the atmosphere is usually measured in terms of a characteristic optical depth τ. Here τ is the optical depth measured by a suitably chosen representative absorption coefficient. Optical depth is measured inwards from the boundary of the star, that is from the point $z=\infty$. Thus

$$\tau = -\int_{z}^{\infty} \kappa(z)\rho(z)\,\mathrm{d}z\,, \tag{A-1}$$

where $\kappa(z)$ is a representative mass absorption coefficient and $\rho(z)$ is the density at level z. Both $\kappa(z)$ and $\rho(z)$ can be found if one has a model. The choice of $\kappa(z)$ is arbitrary and it is usually made in order to facilitate the computation of some part of the radiation field of the star in a particular way. The appropriate temperature law for model atmospheres representing O and B stars is usually found by enforcing the condition that radiative equilibrium exist.

At any particular frequency ν in the spectrum, the monochromatic opacity may be due to absorption in the continuous spectrum, to absorption in one or more lines, or to scattering. Thus one has the following coefficients: $\kappa_\nu(z)$ which is the monochromatic mass absorption coefficient in the continuous spectrum; $l_\nu(z)$ which is the monochromatic mass absorption coefficient in the line spectrum, and $\sigma_\nu(z)$ which is the monochromatic mass scattering coefficient. The scattering is usually assumed to be coherent and isotropic, but more complex and physically real situations can be handled.

The monochromatic optical depth in a frequency ν is

$$t_\nu = -\int_z^\infty \left[\kappa_\nu(z) + l_\nu(z) + \sigma_\nu(z)\right] \rho \, \mathrm{d}z\,, \tag{A-2}$$

or

$$t_\nu = \int_0^\tau \frac{\kappa_\nu(z) + l_\nu(z) + \sigma_\nu(z)}{\kappa(z)} \, d\tau\,. \tag{A-3}$$

Thus given a model and expressions for κ_ν, l_ν, and σ_ν in terms of the composition of the atmosphere and the temperature and electron pressure at every point in the atmosphere, one may calculate the monochromatic optical depth at every depth in the atmosphere for all desired frequencies.

The emergent spectrum from the model atmosphere is found by solving that equation of radiative transfer which is appropriate to the physical interactions which occur between photons and atoms and ions as the radiation field passes through the atmosphere. A precise mathematical description of what occurs leads to a very complex equation that cannot be solved in all generality. A useful survey of the theoretical spectra from models representing O and B stars can be obtained by assuming that the radiation interacts with matter as though isotropic, coherent scattering occurs (due to the electrons) and as though the matter were in thermodynamic equilibrium at the local temperature T. Then one obtains the Milne-Eddington equation of transfer:

$$\cos\vartheta \frac{\mathrm{d}I_\nu(t_\nu, \vartheta)}{\mathrm{d}t_\nu} = I_\nu(t_\nu, \vartheta) - S_\nu(t_\nu)\,, \tag{A-4}$$

where $S_\nu(t_\nu)$ is the source function and

$$S_\nu(t_\nu) = \frac{\sigma_e}{2(\kappa_\nu + l_\nu + \sigma_e)} \int_0^\pi I_\nu(t_\nu, \vartheta) \sin\vartheta \, \mathrm{d}\vartheta + \frac{\kappa_\nu + l_\nu}{\kappa_\nu + l_\nu + \sigma_e} B_\nu(t_\nu)\,. \tag{A-5}$$

Here $I_\nu(t_\nu, \vartheta)$ is the specific intensity at depth t_ν in the model in a beam directed in the direction ϑ, and $B_\nu(t_\nu)$ is the Planck function at depth t_ν in the model.

One finds that the emergent spectrum from a star is given by $\pi F_\nu(0)$ where

$$F_\nu(0) = 2 \int_0^\infty S_\nu(t) E_2(t) \, \mathrm{d}t\,. \tag{A-6}$$

Here $E_2(t)$ is the second exponential integral. The derivation of these equations can be found in any standard text on astrophysics. Equation (A-6) represents integration over the face of the star as well as integration in depth.

A line profile is given by the variation of the function $F_\nu(0)/F_c(0)$ across a line. Here $F_c(0)$ is the emergent flux in the continuous spectrum in the neighbourhood of the line. The equivalent width of a line is obtained by integrating across a line. Thus

$$W_\nu = \int_a^b \frac{F_c(0) - F_\nu(0)}{F_c(0)} \, \mathrm{d}\nu \tag{A-7}$$

where a and b give the limits of the line.

The above set of equations (A-4) to (A-7) can be solved numerically to give the theoretical spectrum from a model. One requires explicit expressions for the line absorption coefficient and for the continuous absorption coefficient in terms of T, P_e, and the abundances of the elements. These are found from physics. The method of model atmospheres makes use of this direct method of attack and it can be applied whenever a large, fast computing machine is available to do the large amount of numerical work. Particular attention is paid to obtaining accurate forms for the various absorption coefficients. Various forms have been proposed depending upon the line broadening mechanisms that may be active in the stellar atmosphere.

The above theory assumes

1. that the populations of the various atoms and ions in their possible excited states are those appropriate to thermodynamic equilibrium at the local temperature and pressure, and

2. that the transfer of radiation is adequately described by the Milne-Eddington transfer equation.

Neither assumption gives a good representation of the true physical situation over the whole atmosphere, although both assumptions appear to be reasonably good for many spectral lines. The anomalous line intensities which are observed in some spectra such as He I in extended atmospheres and which are called dilution effects are an illustration of the breakdown of assumption 1. The theory sketched above is not valid for interpreting stellar spectra in which dilution effects occur. In such a case, one must determine the level populations from the equation of statistical equilibrium. The question of the level populations cannot truly be divorced from the question of the monochromatic radiation field, see Chapter VII.

It is often desired to obtain a theoretical estimate of the relationship between the equivalent widths of many lines, the abundances of the atoms and the temperature and pressure structure of the model. Several simplified methods exist for doing this. They are described in detail in the standard texts on astrophysics (see, for instance, ALLER, 1960 for a summary of the methods).

The method of weighting functions and saturation functions was introduced to speed the calculation of equivalent widths in the case that the line was not strong, and the line absorption coefficient had the form

$$l_\nu = \frac{\sqrt{\pi} e^2}{mc} \cdot \frac{f N^*}{\Delta\nu_D} \cdot H(a, v) . \tag{A-8}$$

Here $H(a, v)$ is the Voigt function and it gives the combined broadening effects of the Doppler broadening of the lines due to the thermal motion of the atoms and of collisional or radiative damping which can be described by a dispersion profile. In equation (A-8), N^* is the number of atoms per gram of star material excited to the lower level of the line, $\Delta\nu_D$ is the thermal Doppler half-width of the line, thus

$$\Delta\nu_D = \frac{\nu}{c}\sqrt{\frac{2kT}{m_A}}, \tag{A-9}$$

where m_A is the mass of the atoms, and a and v are

$$a = \gamma/4\pi\Delta\nu_D, \tag{A-10}$$

$$v = (\nu - \nu_0)/\Delta\nu_D. \tag{A-11}$$

Here γ is the damping constant for the line and ν_0 is the frequency at the centre of the line.

It is to be noted that an equivalent width is a number which is the result of three integrations, namely over depth in the atmosphere, over the face of the star, thus over the angle ϑ, and over frequency ν. If the line is not strong, that is if $l_\nu \leqslant \kappa_\nu$, one can separate the various integrals entering the expression for equivalent width, equation (A-7), and prepare tables of those integrals which are the same for all lines in the spectrum in one region of frequency. Then the final integration to find W_ν may be performed rapidly for different lines and one may readily explore the variation of W_ν, the equivalent width, with the abundance N of an element. Thus the method of weighting functions and of saturation functions is a means of computing rather rapidly the curve of growth for spectral lines in a model atmosphere. The temperature-pressure structure of the model is an important parameter in determining the weighting functions and saturation functions.

The question of predicting equivalent widths may be simplified even further by adopting a simple analytical relationship between depth in the line at frequency ν and the value of the line (sometimes also the continuous) absorption coefficient. Such a simple relationship may come from an analytical solution of the equation of transfer under simplifying conditions or it may be an empirical formula. Unsöld has worked with the Minnaert interpolation formula

$$\frac{1}{R_\nu} = \frac{1}{l_\nu} + \frac{1}{R_c}. \tag{A-12}$$

Here R_ν is the depth in the line at frequency ν and R_c is the central depth of the line. Others, for instance Menzel (1936) and Wrubel (1949, 1954) have used other expressions of a similar type.

This type of analysis is particularly useful when the atmosphere is represented by a single layer of gas at a characteristic temperature and electron pressure. It is, however, not well suited to revealing information about the physics of line formation because the details of the process are eliminated by the analytical formula which is used.

The theory of the curve of growth developed by Unsöld and used in his methods of coarse analysis performs the integration

$$W_\nu = \int_0^\infty R_\nu \, \mathrm{d}\nu = \int_0^\infty \frac{R_c l_\nu}{R_c + l_\nu}. \qquad \text{(A-13)}$$

Here the line absorption coefficient is given by equation (A-8). One can readily derive the dependence of W_ν on N, the abundance of the element. In all this work N^*, the number of atoms excited to the lower level of the line, is calculated using Boltzmann's law. Saha's law is used to find the degree of ionisation. Both these laws are valid for thermal equilibrium.

REFERENCES

ABHYANKAR, K. D.: 1964, *Astrophys. J.* **140,** 1368.
ABT, H. A.: 1957, *Astrophys. J.* **126,** 138.
ABT, H. A. and HUNTER, J. H.: 1962, *Astrophys. J.* **136,** 381.
ADAMS, W. S.: 1900, *Astrophys. J.* **11,** 309.
ADAMS, W. S.: 1949, *Astrophys. J.* **109,** 354.
ADAMS, W. S. and DUNHAM, T.: 1938, *Astrophys. J.* **87,** 102.
ADAMS, W. S. and MERRILL, P. W.: 1957, *Astrophys. J.* **125,** 102.
AITKEN, R. G.: 1935, *Binary Stars*, 2nd edition, McGraw Hill, New York, Chapter 6.
ALEXANDER, J. D. H., BOWEN, P. J., and HEDDLE, D. W. O.: 1963, *Space Research* **3** (ed. by H. PRIESTER), North-Holland Publ. Co., Amsterdam, p. 1068.
ALLER, L. H.: 1943, *Astrophys. J.* **97,** 135.
ALLER, L. H.: 1946, *Astrophys. J.* **104,** 347.
ALLER, L. H.: 1949, *Astrophys. J.* **109,** 244.
ALLER, L. H.: 1954, 'Les processes nucléaires dans les astres', *Mém. Soc. Roy. Sci. Liège* **14,** 337.
ALLER, L. H.: 1956a, *Astrophys. J.* **123,** 117.
ALLER, L. H.: 1956b, *Astrophys. J.* **123,** 133.
ALLER, L. H.: 1960, *Stars and Stellar Systems* **6** (ed. by J. L. GREENSTEIN), University of Chicago Press, Chicago, p. 156.
ALLER, L. H., ELSTE, G., and JUGAKU, J.: 1957, *Astrophys. J. Suppl.* **3,** 1.
ALLER, L. H. and FAULKNER, D. J.: 1963, *Astron. J.* **68,** 533.
ALLER, L. H., FAULKNER, D. J. and ROOD, H.: 1962, *Final Report, Abundances of Elements in Stars and Nebulae*, University of Michigan, p. 39.
ALLER, L. H. and JUGAKU, J.: 1959, *Astrophys. J. Suppl.* **4,** 109.
ALTER, G., RUPRECHT, J., and VANÝSEK, V.: 1958, *Catalogue of Star Clusters and Associations*, Czech. Acad. Sci., Prague.
AMBARTZUMIAN, V. A.: 1949, *Astron. Ž.* **26,** 3.
ANDRILLAT, Y.: 1955, Thesis, Université de Lyon, *Ann. Astrophys. Suppl.* No. **2.**
ANDRILLAT, Y.: 1958, *Etoiles à raies d'émission*, Université de Liège, p. 61.
ANDRILLAT, Y.: 1964, *Ann. Astrophys.* **27,** 486.
ANGER, C. J.: 1933, *Harvard Bull.* No. **891.**
ARKHIPOVA, V. P.: 1963, *Astron. Ž.* **40,** 71; *Sov. Astron. – A.J.* **7,** 51.
ARP, H. C.: 1958, *Handbuch der Physik* **51,** 75.
ARP, H. C. and VAN SANT, C. T.: 1958, *Astron. J.* **63,** 341.
ATHAY, R. G. and ZIRKER, J. B.: 1962, *Astrophys. J.* **136,** 242.
BAKER, E. A.: 1949, *Publ. Roy. Obs. Edinburgh* **1,** No. 2.
BALDWIN, R. B.: 1941, *Astrophys. J.* **93,** 420.
BAPPU, M. K. V.: 1951, *Astron. J.* **56,** 120.
BAPPU, M. K. V.: 1958, *Etoiles à raies d'émission*, Université de Liège, p. 40.
BAPPU, M. K. V., CHANDRA, S., SANWAL, N. B., and SINVHAL, S. D.: 1961, *Monthly Notices Roy. Astron. Soc.* **123,** 521.
BAPPU, M. K. V. and MENZEL, D. H.: 1954, *Astrophys. J.* **119,** 508.
BAPPU, M. K. V. and SINVHAL, S. D.: 1955, *Astron. J.* **60,** 152.
BARBIER, D. and CHALONGE, D.: 1939, *Ann. Astrophys.* **2,** 254.
BARBIER, D. and CHALONGE, D.: 1941, *Ann. Astrophys.* **4,** 30.
BARBIER, D., CHALONGE, D., and CANAVAGGIA, R.: 1947, *Ann. Astrophys.* **10,** 195.
BARNING, F. J. M.: *Bull. Astron. Inst. Neth.* **17, 22.**

BARRETT, A. H., MEEKS, M. L., and WEINREB, S.: 1964, *Astron. J.* **69**, 134.
BATTEN, A. H.: 1961, *J. Roy. Astron. Soc. Can.* **55**, 120.
BEALS, C. S.: 1929, *Monthly Notices Roy. Astron. Soc.* **90**, 202.
BEALS, C. S.: 1930, *Publ. Dominion Astrophys. Obs.* **4**, 228.
BEALS, C. S.: 1931, *Monthly Notices Roy. Astron. Soc.* **91**, 966.
BEALS, C. S.: 1934, *Publ. Dominion Astrophys. Obs.* **6**, 95.
BEALS, C. S.: 1938, *Trans. I.A.U.* **6**, 248.
BEALS, C. S.: 1940, *J. Roy. Astron. Soc. Can.* **34**, 169.
BEALS, C. S.: 1942, *Monthly Notices Roy. Astron. Soc.* **102**, 96.
BEALS, C. S.: 1950, *Publ. Dominion Astrophys. Obs.* **9**, 1.
BECKER, W.: 1938, *Z. Astrophys.* **15**, 225.
BECKER, W.: 1948, *Astrophys. J.* **107**, 278.
BEČVÁŘ, A.: 1958, *Atlas Eclipticalis*, Czech. Acad. Sci., Prague.
BEČVÁŘ, A.: 1962, *Atlas Borealis*, Czech. Acad. Sci., Prague.
BEER, A.: 1961, *Monthly Notices Roy. Astron. Soc.* **123**, 191.
BEER, A.: 1964, *Monthly Notices Roy. Astron. Soc.* **128**, 261.
BELTON, M. J. S. and WOOLF, N. J.: 1965, *Astrophys. J.* **141**, 145.
BERGER, J.: 1956, *Compt. Rend.* **242**, 2300; *Contrib. Inst. Astrophys. Paris*, A, No. **217**.
BERGER, J. and GREENSTEIN, J. L.: 1963, *Publ. Astron. Soc. Pacific* **75**, 336.
BERTIAU, F. C.: 1958, *Astrophys. J.* **128**, 533.
BERTOLA, F.: 1963, *Contrib. Obs. Astrophys. Asiago* No. **137**.
BIDELMAN, W. P.: 1943, *Astrophys. J.* **98**, 61.
BIDELMAN, W. P.: 1947, *Astrophys. J.* **105**, 492.
BIDELMAN, W. P.: 1952, *Astrophys. J.* **116**, 227.
BIDELMAN, W. P.: 1960a, *Publ. Astron. Soc. Pacific* **72**, 24.
BIDELMAN, W. P.: 1960b, *Publ. Astron. Soc. Pacific* **72**, 471.
BIDELMAN, W. P. and CORLISS, C. H.: 1962, *Astrophys. J.* **135**, 968.
BINNENDIJK, L.: 1946, *Ann. Sterrewacht Leiden.* **19**, 5.
BLAAUW, A.: 1946, *Publ. Astron. Lab. Groningen* No. **52**.
BLAAUW, A.: 1952, *Bull. Astron. Inst. Neth.* **11**, 405.
BLAAUW, A.: 1956, *Astrophys. J.* **123**, 408.
BLAAUW, A.: 1961, *Bull. Astron. Inst. Neth.* **15**, 265.
BLAAUW, A.: 1963, *Stars and Stellar Systems* **3** (ed. by K. A. STRAND), University of Chicago Press, Chicago, p. 383.
BLAAUW, A. and MORGAN, W. W.: 1953, *Astrophys. J.* **117**, 256.
BLAAUW, A. and SAVEDOFF, M.: 1953, *Bull. Astron. Inst. Neth.* **12**, 69.
BLANCO, V. M.: 1956, *Astrophys. J.* **123**, 64.
BOCKASTEN, K.: 1955, *Arkiv Fysik* **9**, 457.
BOK, B. J.: 1964, *Observatory* **84**, 235.
BORGMAN, J.: 1960, *Bull. Astron. Inst. Neth.* **15**, 255.
BORGMAN, J.: 1961, *Bull. Astron. Inst. Neth.* **16**, 99.
BORGMAN, J.: 1963, *Bull. Astron. Inst. Neth.* **17**, 111.
BOTTO, P., and HACK, M.: 1962, *Mem. Soc. Astron. Ital.* **33**, 159; *Contrib. Milano-Merate Obs.* No. **194**.
BOWEN, I. S.: 1935, *Astrophys. J.* **81**, 1.
BOWEN, I. S.: 1962, *Stars and Stellar Systems* **2** (ed. by W. A. HILTNER), University of Chicago Press, Chicago, p. 34.
BOYARCHUK, A. A. and KOPYLOV, I. M.: 1964, *Crimea Obs. Publ.* **31**, 44.
BURBIDGE, E. M. and BURBIDGE, G. R.: 1955, *Astrophys. J.* **122**, 89.
BURGER, H. C. and VAN CITTERT, P. H.: 1932, *Z. Physik* **79**, 722.
BURGER, H. C. and VAN CITTERT, P. H.: 1933, *Z. Physik*, **81**, 428.
BUSCOMBE, W.: 1962, *Monthly Notices Roy. Astron. Soc.* **124**, 189.
BUSCOMBE, W.: 1965, *Monthly Notices Roy. Astron. Soc.* **129**, 1.
BUTLER, H. E. and SEDDON, H.: 1957, *Publ. Roy. Obs. Edinburgh* **2**, No. 4.
BUTLER, H. E. and SEDDON, H.: 1959, *Publ. Roy. Obs. Edinburgh* **2**, No. 5.
BUTLER, H. E. and THOMPSON, G. I.: 1960, *Publ. Roy. Obs. Edinburgh* **2**, No. 6.
BYRAM, E. T., CHUBB, T. A., FRIEDMAN, H., and KUPPERIAN, J. E.: 1957, *Astron. J.* **62**, 9.

CAMPBELL, W. W.: 1928, *Lick Obs. Publ.* **16**.
CAYREL, R.: 1958, *Ann. Astrophys. Suppl.* No. **6**.
CHADEAU, C.: 1960, *Ann. Astrophys.* **23,** 935.
CHALONGE, D. and DIVAN, L.: 1952, *Ann. Astrophys.* **15,** 201.
CHANDRASEKHAR, S.: 1934, *Monthly Notices Roy. Astron. Soc.* **94,** 522.
CHANDRASEKHAR, S.: 1939, *An Introduction to the Study of Stellar Structure*, University of Chicago Press, Chicago, Chapter 6.
CHANDRASEKHAR, S. and LEBOVITZ, N. R.: 1962a, *Astrophys. J.* **136,** 1082.
CHANDRASEKHAR, S. and LEBOVITZ, N. R.: 1962b, *Astrophys. J.* **136,** 1105.
CHUBB, T. A. and BYRAM, E. T.: 1963, *Astrophys. J.* **138,** 617.
CLEMENT, M. J.: 1965a, *Astrophys. J.* **141,** 210.
CLEMENT, M. J.: 1956b, *Astrophys. J.* **141,** 1443.
CODE, A D.: 1960, *Stars and Stellar Systems* **6** (ed. by J. L. GREENSTEIN), University of Chicago Press, Chicago, p. 50.
CODE, A. D. and BLESS, R. C.: 1964, *Astrophys. J.* **139,** 787.
CODE, A. D. and LILLER, W. C.: 1962, *Stars and Stellar Systems* **2** (ed. by W. A. HILTNER), University of Chicago Press, Chicago, p. 281.
COLEMAN, C. D., BOZMAN, W. R., and MEGGERS, W. F.: 1960, *Nat. Bur. Std., Monograph* No. **3**.
COUSINS, A. W. J., EGGEN, O. J. and STOY, R. H.: 1961, *Roy. Obs. Bull.* No. **25**.
COWLEY, A. and ALLER, L. H.: 1962, *Final Report Abundances of Elements in Stars and Nebulae,* The University of Michigan, p. 57.
COWLING, T. G.: 1938, *Monthly Notices Roy. Astron. Soc.* **98,** 734.
CRAWFORD, D. L.: 1958, *Astrophys. J.* **128,** 185.
CRAWFORD, D. L.: 1961, *Astrophys. J.* **133,** 860.
CRAWFORD, D. L.: 1963a, *Astrophys. J.* **137,** 523.
CRAWFORD, D. L.: 1963b, *Astrophys. J.* **137,** 530.
CURTIS, R. H.: 1932, *Handbuch der Astrophysik* **5,** 1.
DAVIS, L. and GREENSTEIN, J. L.: 1951, *Astrophys. J.* **114,** 206.
DE GROOT, M. and UNDERHILL, A. B.: 1963, *Bull. Astron. Inst. Neth.* **17,** 280.
DE JAGER, C.: 1963, *Bull. Astron. Inst. Neth.* **17,** 1.
DE JAGER, C. and NEVEN, L.: 1964, *Les spectres infrarouges des astres*, Université de Liège, p. 213.
DE JAGER, C. and NEVEN, L.: 1966, *Bull. Astron. Inst. Neth.* **18,** 295.
DE VAUCOULEURS, A.: 1957, *Monthly Notices Roy. Astron. Soc.* **117,** 449.
DIVAN, L.: 1954, *Ann. Astrophys.* **17,** 456.
DIVAN, L.: 1956, *Ann. Astrophys.* **19,** 255.
DOAZON, V.: 1965, *Ann. Astrophys.* **28,** 1.
DOUGLAS, A. E. and MORTON, J. R.: 1960, *Astrophys. J.* **131,** 1.
DUKE, D.: 1951, *Astrophys. J.* **113,** 100.
DUNHAM, T. H.: 1956, *Vistas in Astronomy* **2** (ed. by A. BEER), Pergamon Press, London, p. 1223.
EBBIGHAUSEN, E. G.: 1958, *Astrophys. J.* **128,** 598.
EBBIGHAUSEN, E. G. and STRUVE, O.: 1956, *Astrophys. J.* **124,** 507.
EDDINGTON, A. S.: 1926, *The Internal Constitution of the Stars*, Cambridge University Press, p. 135.
EDLÉN, B.: 1953, *J. Opt. Soc. Amer.* **43,** 339.
EDLÉN, B.: 1956: *Vistas in Astronomy* **2** (ed. by A. BEER), Pergamon Press, London, p. 1456.
EGGEN, O. J.: 1961, *Roy. Obs. Bull.* No. **41**.
ELSTE, G.: 1953, *Z. Astrophys.* **33,** 39.
EVANS, D. S., MENZIES, A., and STOY, R. H.: 1957, *Monthly Notices Roy. Astron. Soc.* **117,** 534.
EVANS, D. S., MENZIES, A., and STOY, R. H.: 1959, *Monthly Notices Roy. Astron. Soc.* **119,** 638.
EWEN, H. I. and PURCELL, E. M.: 1951, *Nature* **168,** 356.
FATH, E. A.: 1935, *Lick. Obs. Bull.* **17,** 116.
FEAST, M. W.: 1958, *Monthly Notices Roy. Astron. Soc.* **118,** 618.
FEAST, M. W.: 1963, *Monthly Notices Roy. Astron. Soc.* **126,** 11.
FEAST, M. W. and SHUTTLEWORTH, M.: 1965, *Monthly Notices Roy. Astron. Soc.* **130,** 243.
FEAST, M. W. and THACKERAY, A. D.: 1958, *Monthly Notices Roy. Astron. Soc.* **118,** 125.
FEAST, M. W. and THACKERAY, A. D.: 1963, *Mem. Roy. Astron. Soc.* **68,** 173.
FEAST, M. W., THACKERAY, A. D., and WESSELINK, A. J.: 1955, *Mem. Roy. Astron. Soc.* **67,** 51.
FEAST, M. W., THACKERAY, A. D., and WESSELINK, A. J.: 1957, *Mem. Roy. Astron. Soc.* **68,** 1.

FEHRENBACH, C.: 1958, *Handbuch der Physik* **50,** 1.
FEINSTEIN, A.: 1964, *Publ. Astron. Soc. Pacific* **76,** 399.
FOSTER, J. S. and DOUGLAS, A. V.: 1939, *Monthly Notices Roy. Astron. Soc.* **99,** 150.
FROST, E. B.: 1902, *Astrophys. J.* **15,** 340.
FROST, E. B.: 1906, *Astrophys. J.* **24,** 259.
GAPOSCHKIN, S.: 1940, *Publ. Am. Astron. Soc.* **10,** 52.
GAPOSCHKIN, S.: 1958, *Handbuch der Physik* **50,** 225.
GASCOIGNE, S. C. B. and EGGEN, O. J.: 1957, *Monthly Notices Roy. Astron. Soc.* **117,** 430.
GEHRELS, T.: 1960, *Astron. J.* **65,** 470.
GEHRELS, T. and TESKA, T. M.: 1963, *Appl. Opt.* **2,** 67.
GERASIMOVIČ, B. P.: 1929, *Harvard Circ.* No. **339.**
GHOBROS, R. A.: 1962, *Z. Astrophys.* **56,** 113.
GLAD, S.: 1952, *Arkiv Fysik* **7,** 7.
GLENNON, B. M. and WIESE, W. L.: 1962, 'Bibliography on Atomic Transition Probabilities', *Nat. Bur. Std., Monograph* No. **50.**
GÖKGÖZ, A., HACK, M., and KENDIR, I.: 1963, *Mem. Soc. Astron. Ital.* **34,** 87.
GREAVES, W. M. H.: 1948, *Monthly Notices Roy. Astron. Soc.* **108,** 131.
GREAVES, W. M. H., BAKER, E. A., and WILSON, R.: 1955, *Publ. Roy. Obs. Edinburgh* **1,** No. 6.
GREENSTEIN, J. L. and ALLER, L. H.: 1950, *Astrophys. J.* **111,** 328.
GREENSTEIN, J. L. and OKE, J. B.: 1961, *Astrophys. J.* **133,** 349.
GREENSTEIN, J. L. and WALLERSTEIN, G.: 1958, *Astrophys. J.* **127,** 237.
GRIEM, H. R., BARANGER, M., KOLB, A. C., and OERTEL, G.: 1962, *Phys. Rev.* **125,** 177.
GROTH, H. G.: 1960, *Z. Astrophys.* **51,** 206 and 231.
GRYGAR, J.: 1964, *Bull. Astron. Inst. Neth.* **17,** 305.
GUILLAUME, C.: 1965, *Bull. Astron. Inst. Neth.* **18,** 175.
GUILLAUME, C., VAN RENSBERGEN, W., and UNDERHILL, A. B.: 1965, *Bull. Astron. Inst. Neth.* **18,** 106.
GULLEDGE, I. S. and PACKER, D. M.: 1963, *Astron. J.* **68,** 537.
GÜNTHER, S.: 1949, *Z. Astrophys.* **26,** 31.
GÜNTHER, S.: 1950, *Z. Astrophys.* **27,** 107.
GURM, H. S.: 1963, *Monthly Notices Roy. Astron. Soc.* **126,** 419.
GUTHNICK, P.: 1913, *Astron. Nachr.* **196,** 357.
GUTHRIE, B. N. G.: 1963, *Publ. Roy. Obs. Edinburgh* **3,** No. 4.
GUTHRIE, B. N. G.: 1965, *Publ. Roy. Obs. Edinburgh* **3,** No. 8.
HACK, M.: 1957, *Mem. Soc. Astron. Ital.* **28,** 71.
HACK, M.: 1963, *Mem. Soc. Astron. Ital.* **34,** 1; *Contrib. Milano-Merate Obs.* No. **202.**
HACK, M., AYDIN, C. and ISLIK, S.: 1964, *Mem. Soc. Astron. Ital.* **35,** 331.
HALL, J. S.: 1960, *Lowell Obs. Bull.* **4,** 264.
HALL, J. S. and MIKESELL, A. H.: 1950, *Publ. U.S. Naval Obs.* **17,** Part 1.
HANBURY BROWN, R., HAZARD, C., DAVIS, J., and ALLEN, L. R.: 1964, *Nature* **201,** 1111.
HANBURY BROWN, R. and TWISS, R. Q.: 1958, *Proc. Roy. Soc. A.* **248,** 199 and 222.
HARDIE, R. H.: 1950, *Astrophys. J.* **112,** 542.
HARDIE, R. H. and CRAWFORD, D. L.: 1961, *Astrophys. J.* **133,** 843.
HARDIE, R. H. and SEYFFERT, C. K.: 1959, *Astrophys. J.* **129,** 601.
HARDOP, J.: 1965, *Z. Astrophys.* in press.
HARPER, W. E.: 1937, *Publ. Dominion Astrophys. Obs.* **7,** 1.
HARRIS, D. L.: 1955, *Astrophys. J.* **121,** 554.
HARRIS, D. L., STRAND, K. A., and WORLEY, C. E.: 1963, *Stars and Stellar Systems* **3** (ed. by K. A. STRAND), University of Chicago Press, Chicago, p. 273.
HARTMANN, J.: 1904, *Astrophys. J.* **19,** 268.
HEARD, J. F. and MCRAE, D. A.: 1957, *J. Roy. Astron. Soc. Can.* **51,** 29; *Contrib. Dominion Astrophys. Obs.* No. **53.**
HECKMANN, O., DIECKVOSS, W. and KOX, H.: 1956, *Astron. Nachr.* **283,** 109.
HEGER, M. L.: 1919, *Lick Obs. Bull.* **10,** 141.
HENROTEAU, F.: 1918, *Lick Obs. Bull.* **9,** 155.
HERBIG, G. H.: 1954, *Astrophys. J.* **119,** 483.
HERBIG, G. H.: 1962a, *Astrophys. J.* **135,** 736.
HERBIG, G. H.: 1962b, *Advan. Astron. Astrophys.* **1,** 47.

HERBIG, G. H.: 1962c, *Astrophys. J.* **135,** 965.
HERBIG, G. H.: 1963, *Astrophys. J.* **137,** 200.
HERBIG, G. H.: 1964, *Trans. I.A.U.* **XIIA,** 562.
HERMAN, R. and DUVAL, M.: 1963, *Publ. Obs. Haute Provence* **6,** No. 28.
HERRICK, S.: 1935, *Lick Obs. Bull.* **17,** 85.
HERZBERG, G.: 1954, *Les particles solides dans les astres,* Université de Liège, p. 291.
HERZBERG, G.: 1957, *Mem. Roy. Sci. Liège,* 4ième série, t. XV, fasc. unique, 293.
HILL, P. W.: 1964, *Monthly Notices Roy. Astron. Soc.* **129,** 113.
HILL, P. W.: 1965, *Monthly Notices Roy. Astron. Soc.* **129,** 137.
HILTNER, W. A.: 1944a, *Astrophys. J.* **99,** 103.
HILTNER, W. A.: 1944b, *Astrophys. J.* **99,** 273.
HILTNER, W. A.: 1945a, *Astrophys. J.* **101,** 356.
HILTNER, W. A.: 1945b, *Astrophys. J.* **102,** 492.
HILTNER, W. A.: 1949a, *Astrophys. J.* **110,** 95.
HILTNER, W. A.: 1949b, *Astrophys. J.* **110,** 443.
HILTNER, W. A.: 1950, *Astrophys. J.* **112,** 477.
HILTNER, W. A.: 1951, *Astrophys. J.* **113,** 317.
HILTNER, W. A.: 1956, *Astrophys. J. Suppl.* **2,** 389.
HILTNER, W. A. and JOHNSON, H. L.: 1956, *Astrophys. J.* **124,** 367.
HOAG, A. A., JOHNSON, H. L., IRIARTE, B., MITCHELL, R. I., HALLAM, K. L., and SHARPLESS, S.: 1961, *Publ. U.S. Naval Obs.* **17,** 349.
HOGG, H. S.: 1959, *Handbuch der Physik* **53,** 129.
HOUCK, T. E.: 1956, Thesis, University of Wisconsin.
HUANG, S. S.: 1962a, *Astrophys J.* **136,** 903.
HUANG, S. S.: 1962b, *Astron. J.* **67,** 578.
HUANG, S. S.: 1963, *Astrophys. J.* **138,** 342.
HUANG, S. S. and STRUVE, O.: 1960, *Stars and Stellar Systems* **6** (ed. by J. L. GREENSTEIN), University of Chicago Press, p. 321.
HUFFER, C. M. and COLLINS, G. W.: 1962, *Astrophys. J. Suppl.* **7,** 351.
HUNGER, K.: 1955, *Z. Astrophys.* **36,** 42.
HUTCHINGS, J. B.: 1964, *Nature,* **202,** 992.
IBEN, I.: 1965, *Astrophys. J.* **141,** 993.
INGLIS, D. R. and TELLER, E.: 1939, *Astrophys. J.* **90,** 439.
JASCHEK, C., JASCHEK, M. and KUCEWICZ, B.: 1964, *Z. Astrophys.* **59,** 108.
JASCHEK, M., JASCHEK, C., and LAVAGNINO, C. J.: 1963, *Publ. Astron. Soc. Pacific,* **75,** 15.
JOHNSON, H. L.: 1955, *Ann. Astrophys.* **18,** 292.
JOHNSON, H. L.: 1957, *Astrophys. J.* **126,** 121.
JOHNSON, H. L.: 1958, *Lowell Obs. Bull.* **4,** 37.
JOHNSON, H. L.: 1959, *Lowell Obs. Bull.* **4,** 87.
JOHNSON, H. L.: 1965, *Astrophys. J.* **141,** 923.
JOHNSON, H. L. and BORGMAN, J.: 1963, *Bull. Astron. Inst. Neth.* **17,** 115.
JOHNSON, H. L. and HILTNER, W. A.: 1956, *Astrophys. J.* **123,** 267.
JOHNSON, H. L. and IRIARTE, B.: 1958, *Lowell Obs. Bull.* **4,** 47.
JOHNSON, H. L. and MITCHELL, R. I.: 1962, *Comm. Lunar and Planetary Lab.* **1,** 73.
JOHNSON, H. L. and MORGAN, W. W. 1955, *Astrophys. J.* **122,** 429.
JOHNSON, M.: 1954, *Observatory* **74,** 124.
JUGAKU, J.: 1959, *Contrib. Inst. Astrophys. Kyoto,* No. **85.**
JUGAKU, J. and SARGENT, W. L. W.: 1961, *Publ. Astron. Soc Pacific,* **73,** 249.
JUGAKU, J., SARGENT, W. L. W., and GREENSTEIN, J. L.: 1961, *Astrophys. J.* **134,** 783.
KEEPING, E. S.: 1947, *Publ. Dominion Astrophys. Obs.* **7,** 349.
KENT, J. T., TAACK, G. B., and LARSON, D. C.: 1963, NASA Tech. Report R-158.
KLEMOLA, A. R.: 1961, *Astrophys. J.* **134,** 130.
KLOCK, B. L.: 1965, *Astron. J.* **70,** 476.
KOGURE, T.: 1959a, *Publ. Astron. Soc. Japan* **11,** 127.
KOGURE, T.: 1959b, *Publ. Astron. Soc. Japan* **11,** 278.
KOGURE, T.: 1961, *Publ. Astron. Soc. Japan* **13,** 335.
KOPYLOV, I. M.: 1958a, *Crimea Obs. Publ.* **20,** 123.

KOPYLOV, I. M.: 1958b, *Crimea Obs. Publ.* **20,** 156.
KOURGANOFF, V.: 1958, *Etoiles à raies d'émission*, Université de Liège, p. 117.
KRON, G. E. and GORDON, K. C.: 1950, *Astrophys. J.* **111,** 454.
KUHI, L. V.: 1964, *Astrophys. J.* **140,** 1409.
KUIPER, G. P.: 1938, *Astrophys. J.* **88,** 429.
KUMAJGORODSKAJA, R. N.: 1960, *Crimea Obs. Publ.* **24,** 91.
KUMAJGORODSKAJA, R. N.: 1962, *Crimea Obs. Publ.* **28,** 135.
LACOARRET, M.: 1965, *Ann. Astrophys.* **28,** 321.
LALLEMAND, A., DUCHESNE, M., and WALKER, M. F.: 1960, *Publ. Astron. Soc. Pacific* **72,** 268.
LEDOUX, P.: 1951, *Astrophys. J.* **114,** 373.
LEDOUX, P.: 1958, *Handbuch der Physik* **51,** 579.
LEDOUX, P. and WALRAVEN, T.: 1958, *Handbuch der Physik* **51,** 353.
LEHMANN-FILHÉS, R.: 1894, *Astron. Nachr.* **136,** 16.
LIMBER, D. N.: 1964, *Astrophys. J.* **139,** 1251.
LINDHOLM, E. H.: 1957, *Astrophys. J.* **126,** 588.
LORD, H. C.: 1897, *Astrophys. J.* **5,** 348.
LOW, F. J.: 1965, *Astrophys. J.* **141,** 326.
LUNDMARK, K. and LUYTEN, W. J.: 1927, *Ark. Mat. A. Stockholm* **20A,** No. 18.
LYNDS, C. R.: 1959, *Astrophys. J.* **129,** 674.
LYNDS, C. R., SAHADE, J., and STRUVE, O.: 1956, *Astrophys. J.* **124,** 321.
MCCLEAN, F.: 1898, *Phil. Trans. Roy. Soc.* **191,** 127.
MCCUSKEY, S. W. and HOUK, N.: 1964, *Astron. J.* **69,** 412.
MCDONALD, J. K.: 1947, *Publ. Dominion Astrophys. Obs.* **7,** 311.
MCDONALD, J. K.: 1948, *J. Roy. Astron. Soc. Can.* **42,** 220; *Contrib. Dominion Astrophys. Obs.* No. **12.**
MCDONALD, J. K.: 1953, *Publ. Dominion Astrophys. Obs.* **9,** 269.
MCNAMARA, D. H.: 1953, *Publ. Astron. Soc. Pacific* **65,** 286.
MCNAMARA, D. H.: 1955, *Astrophys. J.* **122,** 95.
MCNAMARA, D. H.: 1956, *Publ. Astron. Soc. Pacific* **68,** 158.
MCNAMARA, D. H.: 1963, *Astrophys. J.* **137,** 316.
MCNAMARA, D. H. and HANSEN, K.: 1961, *Astrophys. J.* **134,** 207.
MCNAMARA, D. H. and LARSSON, H. J.: 1962, *Astrophys. J.* **135,** 748.
MCNAMARA, D. H., STRUVE, O., and BERTIAU, F. C.: 1955, *Astrophys. J.* **121,** 326.
MALVILLE, J. M.: 1964, *Astrophys. J.* **139,** 198.
MANNINO, G. and HUMBLET, J.: 1955, *Ann. Astrophys.* **18,** 237.
MARTIN, W. C.: 1960, *J. Res. Nat. Bur. Std. A*, **64,** 19.
MATHIS, J. S.: 1957, *Astrophys. J.* **125,** 318.
MATTHEWS, R. T.: 1956, *Publ. Astron. Soc. Pacific* **68,** 455.
MENDOZA, E. E.: 1958, *Astrophys. J.* **128,** 207.
MENZEL, D. H.: 1936, *Astrophys. J.* **84,** 462.
MERRILL, P. W.: 1934, *Astrophys. J.* **79,** 183.
MERRILL, P. W.: 1936, *Astrophys. J.* **83,** 126.
MERRILL, P. W.: 1953, *Astrophys. J.* **117,** 7.
MERRILL, P. W. and BURWELL, C. G.: 1933, *Astrophys. J.* **78,** 87.
MERRILL, P. W. and BURWELL, C. G.: 1943, *Astrophys. J.* **98,** 153.
MERRILL, P. W. and BURWELL, C. G.: 1949, *Astrophys. J.* **110,** 387.
MERRILL, P. W. and BURWELL, C. G.: 1950, *Astrophys. J.* **112,** 72.
MERRILL, P. W. and SANFORD, R. F.: 1944, *Astrophys. J.* **100,** 14.
MERRILL, P. W. and WILSON, O. C.: 1934, *Astrophys. J.* **80,** 19.
MERRILL, P. W. and WILSON, O. C.: 1938, *Astrophys. J.* **87,** 9.
MEURERS, J. and SANDMANN, H. J.: 1963, *Ver. Univ. Sternw. Bonn*, No. **65.**
MEYER, W. F.: 1934, *Publ. Astron. Soc. Pacific* **46,** 202.
MICZAIKA, G.: 1948, *Z. Naturforsch.* **3a,** 241.
MICZAIKA, G.: 1950, *Astrophys. J.* **112,** 361.
MIHALAS, D.: 1964, *Astrophys. J.* **140,** 885.
MIHALAS, D.: 1965, *Astrophys. J. Suppl.* **9,** 321.
MILONE, L. A.: 1965, *Z. Astrophys.* **62,** 1.
MITCHELL, R. I.: 1957, *Publ. Astron. Soc. Pacific* **69,** 392.

MITCHELL, R. I. and JOHNSON, H. L.: 1957, *Astrophys. J.* **125**, 414.
MIYAMOTO, S.: 1952, *Contrib. Inst. Astrophys. Kyoto* No. **24**.
MOORE, C. E.: 1945, *Publ. Princeton Univ. Obs.* No. **20**.
MOORE, J. H. and NEUBAUER, L. J.: 1948, *Lick Obs. Bull.* **20**, 1.
MORGAN, W. W.: 1933, *Astrophys. J.* **77**, 291.
MORGAN, W. W., CODE, A. D., and WHITFORD, A. E.: 1955, *Astrophys. J. Suppl.* **2**, 41.
MORGAN, W. W. and HARRIS, D. L.: 1956, *Vistas in Astronomy* **2** (ed. by A. BEER), Pergamon Press, London, p. 1124.
MORGAN, W. W. and JOHNSON, H. L.: 1953, *Astrophys. J.* **117**, 313.
MORGAN, W. W. and JOHNSON, H. L.: 1955, *Astrophys. J.* **122**, 142.
MORGAN, W. W. and KEENAN, P. C.: 1951, *Astrophysics* (ed. by J. A. HYNEK), McGraw Hill, New York, p. 12.
MORGAN, W. W., KEENAN, P. C. and KELLMAN, E.: 1943, *An Atlas of Stellar Spectra*, University of Chicago Press, Chicago.
MORGAN, W. W. and LODEN, K.: 1966, *Vistas in Astronomy* **8** (ed. by A. BEER and K. A. STRAND), Pergamon Press, London, p. 83.
MORGAN, W. W., WHITFORD, A. E., and CODE, A. D.: 1953, *Astrophys. J.* **118**, 318.
MORRIS, P. M.: 1961, *Monthly Notices Roy. Astron. Soc.* **122**, 325.
MORTON, D. C.: 1965, *Astrophys. J.* **141**, 73.
MORTON, D. C. and MIHALAS, D.: 1965, *Astrophys. J.* **142**, 253.
MÜLLER, C. A. and OORT, J. H.: 1951, *Nature* **168**, 357.
MÜNCH, G.: 1950, *Astrophys. J.* **112**, 266.
MÜNCH, G.: 1957, *Astrophys. J.* **125**, 42.
MÜNCH, G.: 1964, *Astrophys. J.* **140**, 107.
MÜNCH, G. and FLATHER, E.: 1957, *Publ. Astron. Soc. Pacific* **69**, 142.
NANDY, K.: 1964, *Publ. Roy. Obs. Edinburgh* **3**, 141.
NEUBAUER, F. J.: 1943, *Astrophys. J.* **97**, 300.
NEUBAUER, F. J. and ALLER, L. H.: 1948, *Astrophys. J.* **107**, 281.
NEVEN, L. and DE JAGER, C.: 1954, *Bull. Astron. Inst. Neth.* **12**, 103.
NEVEN, L. and DE JAGER, C.: 1957, *Rech. Astron. Obs. Utrecht* **13**, Part 4.
NEY, E. P. and GOULD, R. J.: 1964, *Astrophys. J.* **140**, 388.
ODGERS, G. J.: 1955, *Publ. Dominion Astrophys. Obs.* **10**, 215.
ODGERS, G. J. and KUSHWAHA, R. S.: 1958, *Publ. Dominion Astrophys. Obs.* **11**, 185.
OKE, J. B.: 1954, *Astrophys. J.* **120**, 22.
OKE, J. B.: 1964, *Astrophys. J.* **140**, 689.
ONDERLIČKA, B.: 1958, *Etoiles à raies d'émission*, Université de Liège, p. 68.
OOSTERHOFF, P. TH.: 1937, *Leiden Ann.* **17**, No. 1.
OVENDEN, M. W.: 1963, *Monthly Notices Roy. Astron. Soc.* **126**, 77; *Contrib. Dominion Astrophys. Obs.* No. **81**.
PADDOCK, G. F.: 1935, *Lick Obs. Bull.* **17**, 99.
PAGEL, B. E. J.: 1956, *Monthly Notices Roy. Astron. Soc.* **116**, 10.
PAGEL, B. E. J.: 1960, *Vistas in Astronomy* **3** (ed. A. BEER), Pergamon Press, London, p. 203.
PAWSEY, J. L.: 1951, *Nature*, **168**, 358.
PAYNE, C. H.: 1933, *Z. Astrophys.* **7**, 1.
PAYNE-GAPOSCHKIN, C. H.: 1935, *Trans. I.A.U.* **5**, 187.
PEARCE, J. A.: 1926, *Publ. Dominion Astrophys. Obs.* **3**, 275.
PEARCE, J. A.: 1930, *Publ. Dominion Astrophys. Obs.* **5**, 110.
PEARCE, J. A.: 1932, *Monthly Notices Roy. Astron. Soc.* **92**, 877.
PEARCE, J. A.: 1957, *J. Roy. Astron. Soc. Can.* **51**, 59; Contrib. Dominion Astrophys. Obs. No. **53**.
PECKER, J. C.: 1950, *Ann. Astrophys.* **13**, 433.
PENZIAS, A. A.: 1964, *Astron. J.* **69**, 146.
PERRINE, C. D.: 1920, *Astrophys. J.* **52**, 39.
PETRIE, R. M.: 1946, *J. Roy. Astron. Soc. Can.* **40**, 325; *Contrib. Dominion Astrophys. Obs.* No. **4**.
PETRIE, R. M.: 1947a, *Publ. Dominion Astrophys. Obs.* **7**, 321.
PETRIE, R. M.: 1947b, *J. Roy. Astron. Soc. Can.* **41**, 311; *Contrib. Dominion Astrophys. Obs.* No. **10**.
PETRIE, R. M.: 1948, *J. Roy. Astron. Soc. Can.* **42**, 213; *Contrib. Dominion Astrophys. Obs.* No. **11**.
PETRIE, R. M.: 1950, *Publ. Dominion Astrophys. Obs.* **8**, 341.

PETRIE, R. M.: 1952, *Publ. Dominion Astrophys. Obs.* **9,** 251.
PETRIE, R. M.: 1953, *Publ. Dominion Astrophys. Obs.* **9,** 297.
PETRIE, R. M.: 1954a, *J. Roy. Astron. Soc. Can.* **48,** 185.
PETRIE, R. M.: 1954b, *Publ. Dominion Astrophys. Obs.* **10,** 39.
PETRIE, R. M.: 1955, *Trans. I.A.U.* **9,** 439.
PETRIE, R. M.: 1956, *Vistas in Astronomy* **2** (ed. by A. BEER), Pergamon Press, London, p. 1346.
PETRIE, R. M.: 1958a, *Astron. J.* **63,** 189.
PETRIE, R. M.: 1958b, *Monthly Notices Roy. Astron. Soc.* **118,** 342; *Contrib. Dominion Astrophys. Obs.* No. **61.**
PETRIE, R. M.: 1959, *Publ. Dominion Astrophys. Obs.* **10,** 459.
PETRIE, R. M.: 1960, *Ann. Astrophys.* **23,** 744; *Contrib. Dominion Astrophys. Obs.* No. **67.**
PETRIE, R. M.: 1961, *Monthly Notices Roy. Astron. Soc.* **123,** 501; *Contrib. Dominion Astrophys. Obs.* No. **72.**
PETRIE, R. M.: 1962, *Stars and Stellar Systems* **2** (ed. by W. A. HILTNER), University of Chicago Press, Chicago, p. 560.
PETRIE, R. M.: 1964, *Publ. Dominion Astrophys. Obs.* **12,** 317.
PETRIE, R. M.: 1965, *Publ. Astron. Soc. Pacific* **77,** 97.
PETRIE, R. M., CUTTLE, P. M., and ANDREWS, D. H.: 1956, *Astron. J.* **61,** 289; *Contrib. Dominion Astrophys. Obs.* No. **48.**
PETRIE, R. M. and LAIDLER, D. M.: 1952, *Publ. Dominion Astrophys. Obs.* **9,** 181.
PETRIE, R. M. and MAUNSELL, C. D.: 1949, *Publ. Dominion Astrophys. Obs.* **8,** 253.
PETRIE, R. M. and MOYLS, B. N.: 1956, *Publ. Dominion Astrophys. Obs.* **10,** 287.
PETRIE, R. M. and PEARCE, J. A.: 1962, *Publ. Dominion Astrophys. Obs.* **12,** 1.
PETRIE, R. M. and PETRIE, W. 1939, *Publ. Dominion Astrophys. Obs.* **7,** 189.
PLASKETT, H. H.: 1922, *Publ. Dominion Astrophys. Obs.* **1,** 325.
PLASKETT, J. S.: 1919, *Publ. Dominion Astrophys. Obs.* **1,** 81.
PLASKETT, J. S.: 1922, *Publ. Dominion Astrophys. Obs.* **2,** 147.
PLASKETT, J. S. and PEARCE, J. A.: 1930a, *Publ. Dominion Astrophys. Obs.* **5,** 1.
PLASKETT, J. S. and PEARCE, J. A.: 1930b, *Publ. Dominion Astrophys. Obs.* **5,** 99.
PLASKETT, J. S. and PEARCE, J. A.: 1934, *Publ. Dominion Astrophys. Obs.* **5,** 241.
PLAUT, L.: 1950, *Publ. Kapteyn Astron. Lab.* No. **54.**
PLAUT, L.: 1953, *Publ. Kapteyn Astron. Lab.* No. **55.**
POPPER, D. M.: 1942, *Publ. Astron. Soc. Pacific* **54,** 160.
POPPER, D. M.: 1944, *Astrophys. J.* **100,** 94.
POPPER, D. M.: 1946, *Publ. Astron. Soc. Pacific* **58,** 370.
POPPER, D. M.: 1947, *Publ. Astron. Soc. Pacific* **59,** 320.
POPPER, D. M.: 1957, *J. Roy. Astron. Soc. Can.* **51,** 51; *Contrib. Dominion Astrophys. Obs.* No. **53**
POSENER, D. W.: 1959, *Australian J. Phys.* **12,** 184.
POTTASCH, S. R.: 1961, *Ann. Astrophys.* **24,** 159.
REDMAN, R. O.: 1930, *Publ. Dominion Astrophys. Obs.* **4,** 341.
RINGUELET-KASWALDER, A. E.: 1963, *Publ. Astron. Soc. Pacific* **75,** 323.
ROBERTS, M. S.: 1957, *Publ. Astron. Soc. Pacific* **69,** 59.
ROBERTS, M. S.: 1958, *Etoiles à raies d'émission*, Université de Liège, p. 76.
ROBERTS, M. S.: 1962, *Astron. J.* **67,** 79.
RODGERS, A. W.: 1961, *Monthly Notices Roy. Astron. Soc.* **122,** 413.
RODGERS, A. W. and BELL, R. A.: 1962, *Observatory*, **82,** 26.
ROMAN, N. G.: 1947, *Astrophys. J.* **106,** 311.
ROMAN, N. G.: 1951, *Astrophys. J.* **114,** 492.
ROSS, L. W.: 1961, *Publ. Astron. Soc. Pacific* **73,** 354.
ROSSATI, F.: 1963, *Mem. Astron. Soc. Ital.* **34,** 349.
ROTTENBERG, J. A.: 1952, *Monthly Notices Roy. Astron. Soc.* **112,** 125.
RUBLEV, S. V.: 1961, *Astron. Ž.* **37,** 828; *Sov. Astron. – A.J.* **4,** 780.
RUBLEV, S. V.: 1962, *Astron. Ž.* **39,** 879; *Sov. Astron. – A.J.* **6,** 686.
RUBLEV, S. V.: 1963, *Astron. Ž.* **40,** 100; *Sov. Astron. – A.J.* **7,** 492.
RUBLEV, S. V.: 1964, *Astron. Ž.* **41,** 63; *Sov. Astron. – A.J.* **8,** 45.
RUBIN, V. C. and BURLEY, J.: 1964, *Astron. J.* **69,** 80.

RUBIN, V. C., BURLEY, J., KIASATPOOR, A., KLOCK, B., PEASE, G., RUTSCHEIDT, E., and SMITH, C.: 1962, *Astron. J.* **67**, 491.
RUDKJØBING, M.: 1947, *Publ. Copenhagen Obs.* No. **145**.
RUDNICK, P.: 1936, *Astrophys. J.* **83**, 439.
SAHADE, J.: 1958, *Etoiles à raies d'émission*, Université de Liège, p. 46.
SAHADE, J. and STRUVE, O.: 1956, *Publ. Astron. Soc. Pacific* **68**, 451.
SAITO, S.: 1954, *Contrib. Inst. Astrophys. Kyoto*, No. **48**.
SAITO, S.: 1956, *Contrib. Inst. Astrophys. Kyoto*, No. **69**.
SANDBERG, H. E. and MCNAMARA, D. H.: 1960, *Publ. Astron. Soc. Pacific* **72**, 508.
SANFORD, R. F. and WILSON, O. C.: 1939, *Astrophys. J.* **90**, 235.
SARGENT, W. L. W. and JUGAKU, J.: 1961, *Astrophys. J.* **134**, 777.
SCHLESINGER, F.: 1899, *Astrophys. J.* **10**, 1.
SCHLESINGER, F.: 1909, *Publ. Allegheny Obs.* **1**, 123.
SCHLESINGER, F.: 1913, *Publ. Allegheny Obs.* **3**, 23.
SCHLESINGER, F. and UDICK, S.: 1912, *Publ. Allegheny Obs.* **2**, 155.
SCHMAHL, G.: 1963, *Z. Astrophys.* **58**, 165.
SCHMALBERGER, D. C.: 1960, *Astrophys. J.* **132**, 591.
SCHNELLER, H.: 1960, *Geschichte und Literatur des Lichtwechsels der veränderlichen Sternen* **5**, Akademie-Verlag, Berlin.
SERKOWSKI, K.: 1963, *Astrophys. J.* **138**, 1035.
SEYFERT, C. K. and POPPER, D. M.: 1941, *Astrophys. J.* **93**, 461.
SHAJN, G. and STRUVE, O.: 1929, *Monthly Notices Roy. Astron. Soc.* **89**, 222.
SHANE, C. D.: 1932, *Lick Obs. Bull.* No. **449**.
SHARPLESS, S.: 1952, *Astrophys. J.* **116**, 251.
SHARPLESS, S.: 1954, *Astrophys. J.* **119**, 200.
SHARPLESS, S.: 1963, *Stars and Stellar Systems* **3** (ed. by K. A. STRAND), University of Chicago Press, Chicago, p. 225.
SINNERSTAD, U.: 1961a, *Ann. Stockholm Obs.* **21**, No. 6.
SINNERSTAD, U.: 1961b, *Ann. Stockholm Obs.* **22**, No. 2.
SLETTEBAK, A.: 1949, *Astrophys. J.* **110**, 498.
SLETTEBAK, A.: 1954, *Astrophys. J.* **119**, 146.
SLETTEBAK, A.: 1956, *Astrophys. J.* **124**, 173.
SLETTEBAK, A. and HOWARD, R. F.: 1955, *Astrophys. J.* **121**, 102.
SMITH, B.: 1943, *Astrophys. J.* **98**, 82.
SMITH, H. J.: 1955, Thesis, Harvard University.
SOBOLEV, V. V.: 1947, *Moving Stellar Envelopes*, Leningrad State University, translated from the Russian by S. GAPOSCHKIN, 1960, Harvard University Press.
STEBBINS, J. and KRON, G. E.: 1954, *Astrophys. J.* **120**, 189.
STEBBINS, J., and WHITFORD, A. E.: 1943, *Astrophys. J.* **98**, 20.
STEBBINS, J., and WHITFORD, A. E.: 1945, *Astrophys. J.* **102**, 318.
STEBBINS, J., WHITFORD, A. E. and HUFFER, C. M.: 1940, *Astrophys. J.* **91**, 20.
STECHER, T. P. and MILLIGAN, J. E.: 1962, *Astrophys. J.* **136**, 1.
STERNE, T. E.: 1939, *Monthly Notices Roy. Astron. Soc.* **99**, 451.
STERNE, T. E.: 1941, *Proc. Nat. Acad. Sci.* **27**, 175.
STIBBS, D. W. N.: 1956, *Monthly Notices Roy. Astron. Soc.* **116**, 453.
STOECKLY, R. and DRESSLER, K.: 1964, *Astrophys. J.* **139**, 240.
STOTHERS, R.: 1965, *Astrophys. J.* **141**, 671.
STRAND, K. A.: 1958, *Astrophys. J.* **128**, 14.
STROM, E. S.: 1964, Thesis, Harvard, University.
STRÖMGREN, B.: 1958, *Stellar Populations* (ed. by D. J. K. O'Connell, S. J.), North-Holland Publ. Co., Amsterdam, p. 385.
STRÖMGREN, B.: 1963, *Quart. J. Roy. Astron. Soc.* **4**, 8.
STRUVE, O.: 1931, *Astrophys. J.* **74**, 225.
STRUVE, O.: 1939, *Astrophys. J.* **90**, 699.
STRUVE, O.: 1942, *Astrophys. J.* **95**, 134.
STRUVE, O.: 1943, *Astrophys. J.* **98**, 98.
STRUVE, O.: 1944a, *Astrophys. J.* **100**, 189.

Struve, O.: 1944b, *Astrophys. J.* **100,** 384.
Struve, O.: 1950, *Astrophys. J.* **112,** 520.
Struve, O.: 1955, *Publ. Astron. Soc. Pacific*, **67,** 135.
Struve, O. and Horak, H. G.: 1949, *Astrophys. J.* **110,** 447.
Struve, O. and Huang, S. S.: 1958, *Handbuch der Physik* **50,** 243.
Struve, O., Huang, S. S., and Zebergs, V.: 1959, *Astrophys. J.* **129,** 314.
Struve, O., McNamara, D. H., Kraft, R. P., Kung, S. M., and Williams, A. D.: 1952, *Astrophys. J.* **116,** 81.
Struve, O., McNamara, D. H., Kung, S. M. and Beymer, C.: 1953, *Astrophys. J.* **118,** 39.
Struve, O., McNamara, D. H., Kung, S. M., van Hoof, A., and Deurinck, R.: 1954, *Astrophys. J.* **119,** 198.
Struve, O. and Roach, F. E.: 1939, *Astrophys. J.* **90,** 727.
Struve, O. and Rudkjøbing, M.: 1948, *Astrophys. J.* **108,** 537.
Struve, O. and Sahade, J.: 1957, *Astrophys. J.* **125,** 689.
Struve, O., Sahade, J., and Huang, S. S.: 1958, *Astrophys. J.* **127,** 148.
Struve, O., Sahade, J., and Zebergs, V.: 1959, *Astrophys. J.* **129,** 59.
Struve, O. and Swings, P.: 1940, *Astrophys. J.* **91,** 546.
Struve, O. and Swings, P.: 1941, *Astrophys. J.* **94,** 99.
Struve, O. and Wurm, K.: 1938, *Astrophys. J.* **88,** 84.
Struve, O. and Zebergs, V.: 1955, *Astrophys. J.* **122,** 134.
Struve, O. and Zebergs, V.: 1959, *Astrophys. J.* **129,** 668.
Struve, O. and Zebergs, V.: 1960, *Astrophys. J.* **132,** 87.
Swings, P.: 1942, *Astrophys. J.* **95,** 112.
Swings, P.: 1948, *Ann. Astrophys.* **11,** 228.
Swings, P. and Jose, P. D.: 1950, *Astrophys. J.* **111,** 513.
Tanner, R. W.: 1948, *J. Roy. Astron. Soc. Can.* **42,** 177.
Thackeray, A. D.: 1954, *Monthly Notices Roy. Astron. Soc.* **114,** 93.
Thackeray, A. D.: 1964, *Quart. J. Roy. Astron. Soc.* **5,** 275.
Thackeray, A. D. and Wesselink, A. J.: 1952, *Observatory* **72,** 248.
Thomas, R. N.: 1949, *Astrophys. J.* **109,** 500.
Torreson, Y. G.: 1960, *Arkiv Fysik* **18,** 389.
Traving, G.: 1955, *Z. Astrophys.* **36,** 1.
Traving, G.: 1957, *Z. Astrophys.* **41,** 215.
Traving, G.: 1962, *Astrophys. J.* **135,** 439.
Treanor, P. J.: 1960, *Monthly Notices Roy. Astron Soc.* **121,** 503.
Trumpler, R. J.: 1930, *Lick Obs. Bull.* **14,** 154.
Underhill, A. B.: 1946, *Astrophys. J.* **104,** 388.
Underhill, A. B.: 1948a, *Astrophys. J.* **107,** 337.
Underhill, A. B.: 1948b, *Astrophys. J.* **107,** 349.
Underhill, A. B.: 1948c, *Astrophys. J.* **108,** 476.
Underhill, A. B.: 1950, *Publ. Copenhagen Obs.* No. **151.**
Underhill, A. B.: 1951a, *Publ. Dominion Astrophys. Obs.* **8,** 357.
Underhill, A. B.: 1951b, *Can. J. Phys.* **29,** 447; *Contrib. Dominion Astrophys. Obs.* No. **23.**
Underhill, A. B.: 1952a, *Publ. Dominion Astrophys. Obs.* **9,** 139.
Underhill, A. B.: 1952b, *Publ. Dominion Astrophys. Obs.* **9,** 219.
Underhill, A. B.: 1953a, *J. Roy. Astron. Soc. Can.* **47,** 153; *Contrib. Dominion Astrophys. Obs.* No. **30.**
Underhill, A. B.: 1953b, *Publ. Dominion Astrophys. Obs.* **9,** 363.
Underhill, A. B.: 1955, *Publ. Dominion Astrophys. Obs.* **10,** 169.
Underhill, A. B.: 1956a, *Publ. Dominion Astrophys. Obs.* **10,** 201.
Underhill, A. B.: 1956b, *Publ. Dominion Astrophys. Obs.* **10,** 357.
Underhill, A. B.: 1957, *Astrophys. J.* **126,** 28; *Contrib. Dominion Astrophys. Obs.* No. **55.**
Underhill, A. B. 1958a, *Publ. Dominion Astrophys. Obs.* **11,** 143.
Underhill, A. B.: 1958b, *Etoiles à raies d'émission*, Université de Liège, p. 17.
Underhill, A. B.: 1959a, *Publ. Dominion Astrophys. Obs.* **11,** 209.
Underhill, A. B.: 1959b, *Publ. Dominion Astrophys. Obs.* **11,** 283.
Underhill, A. B.: 1960, *Publ. Dominion Astrophys. Obs.* **11,** 353.

UNDERHILL, A. B.: 1961, *Publ. Dominion Astrophys. Obs.* **11**, 405.
UNDERHILL, A. B.: 1962a, *Publ. Dominion Astrophys. Obs.* **11**, 433.
UNDERHILL, A. B.: 1962b, *Publ. Dominion Astrophys. Obs.* **11**, 467.
UNDERHILL, A. B.: 1962c, *Astrophys. J.* **136**, 14; *Contrib. Dominion Astrophys. Obs.* No. **76**.
UNDERHILL, A. B.: 1963a, *Bull. Astron. Inst. Neth.* **17**, 161.
UNDERHILL, A. B.: 1963b, *Space Sci. Rev.* **1**, 749.
UNDERHILL, A. B.: 1965, *Vistas in Astronomy* **8** (ed. by A. BEER and K. A. STRAND), Pergamon Press, London, p. 41.
UNDERHILL, A. B. and DE GROOT, M.: 1964, *Bull. Astron. Inst. Neth.* **17**, 453.
UNDERHILL, A. B. and DE GROOT, M.: 1965, *Rech. Astron. Obs. Utrecht*, **XVII**, No. 3.
UNDERHILL, A. B. and VAN HELDEN, R.: 1966, *Astron. J.* **71**, 183.
UNDERHILL, A. B. and PETRIE, W.: 1944, *J. Roy. Astron. Soc. Can.* **38**, 385.
UNDERHILL, A. B. and WALKER, G. A. H.: 1966, *Monthly Notices Roy. Astron. Soc.* **131**, 476.
UNSÖLD, A.: 1942a, *Z. Astrophys.* **21**, 22.
UNSÖLD, A.: 1942b, *Z. Astrophys.* **21**, 229.
UNSÖLD, A.: 1944a, *Z. Astrophys.* **23**, 75.
UNSÖLD, A.: 1944b, *Z. Astrophys.* **23**, 100.
UNSÖLD, A.: 1963, *Z. Astrophys.* **56**, 221.
UNSÖLD, A. and STRUVE, O.: 1940, *Astrophys. J.* **91**, 365.
VAN DE HULST, H. C.: 1946, *Bull. Astron. Inst. Neth.* **10**, 75.
VAN DE HULST, H. C. and REESINCK, J. J. M.: 1947, *Astrophys. J.* **106**, 121.
VAN DE KAMP, P.: 1960, *J. Roy. Astron. Soc. Can.* **54**, 167.
VAN DEN HEUVEL, E. H. P.: 1965, *Observatory*, **85**, 241.
VAN HOOF, A.: 1961a, *Z. Astrophys.* **53**, 106.
VAN HOOF, A.: 1961b, *Z. Astrophys.* **53**, 124.
VAN HOOF, A.: 1962a, *Kleine Veröffentl. Remeis-Sternw. Bamberg* No. **34**, 68.
VAN HOOF, A.: 1962b, *Z. Astrophys.* **56**, 15.
VAN HOOF, A.: 1962c, *Z. Astrophys.* **56**, 141.
VAN HOOF, A.: 1962d, *Z. Astrophys.* **56**, 244.
VAN HOOF, A.: 1962e, *Z. Astrophys.* **54**, 255.
VAN HOOF, A.: 1963, *Z. Astrophys.* **57**, 278.
VAN HOOF, A.: 1964, *Z. Astrophys.* **60**, 184.
VAN HOOF, A.: 1965, *Z. Astrophys.* **62**, 174.
VAN HOOF, A. and BLAAUW, A.: 1958, *Astrophys. J.* **128**, 273.
VAN HOOF, A. and BLAAUW, A.: 1964, *Bull. Astron. Inst. Neth.* **17**, 451.
VAN HOOF, A., PRETORIUS, W. S., and PIKOOS, C.: 1964, Information Bulletin No. 63, Commission 27, Intern. Astron. Union.
VAN HOOF, A. and STRUVE, O.: 1953, *Publ. Astron. Soc. Pacific*, **65**, 158.
VOIGT, H. H.: 1952, *Z. Astrophys.* **31**, 48.
VORONTSOV-VELYAMINOV, B. A.: 1945, *Astron. Ž.* **22**, 93.
VORONTSOV-VELYAMINOV, B. A.: 1958, *Etoiles à raies d'émission*, Université de Liège, p. 55.
WALKER, G. A. H.: 1963, *Monthly Notices Roy. Astron. Soc.* **125**, 141.
WALKER, M. F.: 1950, *Publ. Astron. Soc. Pacific* **63**, 35.
WALKER, M. F.: 1954, *Astrophys. J.* **120**, 58.
WALKER, M. F.: 1956a, *Astrophys. J. Suppl.* **2**, 365.
WALKER, M. F.: 1956b, *Publ. Astron. Soc. Pacific*, **68**, 154.
WALLERSTEIN, G.: 1959, *Astrophys. J.* **130**, 338.
WALRAVEN, T. and WALRAVEN, J. H.: 1960, *Bull. Astron. Inst. Neth.* **15**, 67.
WAMPLER, E. J.: 1962, *Astrophys. J.* **136**, 100.
WAMPLER, E. J.: 1963, *Astrophys. J.* **137**, 1071.
WATSON, J. C.: 1867, *Theoretical Astronomy*, Lippincott and Co., Philadelphia.
WEAVER, H. F.: 1955, *Astron. J.* **60**, 181.
WEAVER, H. F. and EBERT, A.: 1964, *Publ. Astron. Soc. Pacific* **76**, 6.
WEENEN, J.: 1950, *Bull. Astron. Inst. Neth.* **11**, 176.
WELLMANN, P.: 1951, *Z. Astrophys. J.* **30**, 71.
WESSELINK, A. J.: 1962, *Monthly Notices Roy. Astron. Soc.* **124**, 359.
WESTERLUND, B. E. and SMITH, L. F.: 1964, *Monthly Notices Roy. Astron. Soc.* **128**, 311.

WHITFORD, A. E.: 1948, *Astrophys. J.* **107,** 102.
WHITFORD, A. E.: 1958, *Astron. J.* **63,** 201.
WHITFORD, A. E.: 1962, *Handbuch der Physik* **54,** 240.
WHITNEY, C. A.: 1964, *Astron. J.* **69,** 564.
WILDEY, R. L.: 1964, *Astrophys. J. Suppl.* **8,** 439.
WILDEY, R. L. and MURRAY, B. C.: 1964, *Astrophys. J.* **139,** 435.
WILLIAMS, E. G.: 1936a, *Astrophys. J.* **83,** 279.
WILLIAMS, E. G.: 1936b, *Astrophys. J.* **83,** 305.
WILLSTROP, R. V.: 1965, *Mem. Roy. Astron. Soc.* **69,** 83.
WILSON, O. C.: 1939, *Publ. Astron. Soc. Pacific* **51,** 55.
WILSON, O. C.: 1942, *Astrophys. J.* **95,** 402.
WILSON, O. C.: 1948a, *Publ. Astron. Soc. Pacific* **60,** 383.
WILSON, O. C.: 1948b, *Publ. Astron. Soc. Pacific* **60,** 385.
WILSON, O. C.: 1949, *Astrophys. J.* **109,** 76.
WILSON, R.: 1956, *Publ. Roy. Obs. Edinburgh* **2,** No. 1.
WILSON, R.: 1957, *Publ. Roy. Obs. Edinburgh,* **2,** No. 3.
WILSON, R.: 1958, *Astrophys. J.* **128,** 57; *Contr. Dominion Astrophys. Obs.* No. **63.**
WILSON, R.: 1960, *Monthly Notices Roy. Astron. Soc.* **120,** 51; *Contrib. Dominion Astrophys. Obs.* No. **64.**
WOLF, C. and RAYET, 1867, *Compt. Rend.* **65,** 292.
WOLFF, R. J.: 1963, *Publ. Astron. Soc. Pacific* **75,** 485.
WOODS, M. L.: 1955, *Mem. Commonwealth Obs.* **3,** No. **12.**
WOOLF, N. J.: 1965, *Astrophys. J.* **141,** 155.
WOOLLEY, R. v. D. R. and EGGEN, O. J.: 1958, *Observatory* **78,** 149.
WRIGHT, K. O.: 1954, *Publ. Dominion Astrophys. Obs.* **10,** 1.
WRIGHT, K. O.: 1962, *Stars and Stellar Systems* **2** (ed. W. A. HILTNER), University of Chicago Press, Chicago, p. 83.
WRIGHT, K. O., LEE, E. K., JACOBSEN, T. V., and GREENSTEIN, J. L.: 1963, *Publ. Dominion Astrophys. Obs.* **12,** No. 7.
WRUBEL, M. H.: 1949, *Astrophys. J.* **109,** 66.
WRUBEL, M. H.: 1954, *Astrophys. J.* **119,** 51.
YESSIPOV, V. F.: 1962, *Astron. Ž.* **39,** 365.
YOUNG, R. K.: 1920, *Publ. Dominion Astrophys. Obs.* **1,** 239.
ZANSTRA, H. and WEENEN, J.: 1950, *Bull. Astron. Inst. Neth.* **11,** 165.
ZUG, R. S.: 1933, *Lick Obs. Bull.* **16,** 119.

INDEX OF STARS MENTIONED IN THE TEXT

STARS WITH CONSTELLATION NAMES

STARS WITH HD NUMBERS

STARS WITH OTHER DESIGNATIONS

INDEX OF SUBJECTS

GPSR Compliance

The European Union's (EU) General Product Safety Regulation (GPSR) is a set of rules that requires consumer products to be safe and our obligations to ensure this.

If you have any concerns about our products, you can contact us on ProductSafety@springernature.com

In case Publisher is established outside the EU, the EU authorized representative is:

Springer Nature Customer Service Center GmbH
Europaplatz 3
69115 Heidelberg, Germany

Batch number: 10584556

Printed by Printforce, the Netherlands